Economically Enabled Energy Management

Takeshi Hatanaka · Yasuaki Wasa ·
Kenko Uchida

Editors

Economically Enabled Energy Management

Interplay Between Control Engineering and Economics

 Springer

Editors
Takeshi Hatanaka
School of Engineering
Tokyo Institute of Technology
Meguro-ku, Tokyo, Japan

Yasuaki Wasa
School of Advanced Science
and Engineering
Waseda University
Shinjuku, Tokyo, Japan

Kenko Uchida
Research Institute of Science
and Engineering
Waseda University
Shinjuku, Tokyo, Japan

ISBN 978-981-15-3575-8   ISBN 978-981-15-3576-5   (eBook)
https://doi.org/10.1007/978-981-15-3576-5

This Springer imprint is published by the registered company Springer Nature Singapore Pte Ltd.
The registered company address is: 152 Beach Road, #21-01/04 Gateway East, Singapore 189721,
Singapore

Preface

The fundamental national strategy for Japanese energy management has undergone dramatic changes since the Great East Japan Earthquake and Tsunami disaster of 2011. The core challenge in realizing the reform of Japan's electricity system is electricity deregulation via energy market transactions that enhances the economic efficiency, security, and safety of decarbonized energy systems. The energy supply–demand networks of a next-generation energy management system (EMS) are composed of distributed and centralized energy generators, consumers, prosumers, aggregators, and distribution energy networks (e.g., traditional power plants, virtual power plants, and automated demand response (DR) systems). EMSs will use online sensing and will gather information via smart meters at each subsystem, interactively connecting both suppliers and consumers to allow them to engage in energy transactions. Hence, human end users will play a central role in the system, and EMSs will definitely function as economic systems based on human economic activity. In an EMS, human and local controllers can be regarded as agents and decision makers as or on behalf of generators, consumers, prosumers, and aggregators. These rational subsystems behave economically while optimizing physical performance, and thus, any analysis and synthesis of EMSs that allow matching the supply and demand of energy need to consider economic and physical perspectives as well as their intersection. In this book, we call the above concept an *economically enabled energy management system* as a whole.

Our goal is to describe an economic mechanism for integrating agents' strategic decisions and controls into pluralistic social welfare under conditions of energy deregulation, while maintaining the stability of the physical network. To analyze and synthesize economically enabled EMSs, which embody the ideal energy supply–demand transactions, it is essential to consider many central social issues, which also have both economic and physical aspects and intersect in important ways. Hence, the two main challenges of this book are to describe a fundamental theory and the technology needed to establish and realize energy supply–demand networks that integrate economic models with physical models, and to recommend policies based on experimental evidence regarding the economic and social effects of implementations.

This book includes contributions from a cross-disciplinary research team of control engineering and economics researchers. This team was formed to address economically enabled energy management. The research team is just one of the five super research teams working on the Japan Science and Technology Agency (JST) Core Research for Evolutional Science and Technology (CREST) EMS Project, which is being conducted from 2012 to 2020; the research mission of this team is the same as the objective of this book. This book is organized into an overview chapter followed by eleven technical chapters, each authored by leading researchers in power, economics, and control. The contributors to this book include eleven economists and twenty-two engineering scientists. The twelve chapters are divided into four broad topics: comprehensive views to allow interdisciplinary interpretation; economic viewpoints; integration mechanisms; and state-of-the-art control engineering mechanisms.

The first three chapters underscore the pressing need for economically efficient EMSs as well as academic work on this emerging research topic in the context of specific fields. A comprehensive overview and a discussion of general research opportunities in the field of economically enabled energy management are presented in Chap. 1. Chapter 2 focuses on balancing power markets to regulate electrical frequency and achieve power balance, from the viewpoint of power engineering. To avoid power imbalance and power flow congestion caused by full adoption of renewable sources, some novel mechanisms are proposed for realizing a balancing power market and achieving advanced frequency control. Through a detailed simulation of the Japanese power grid, the risk management needed for the use of renewable sources and the impact of the proposed balancing of power market operation are evaluated. Chapter 3 aims at identifying fundamental differences between model-based approaches in control engineering and those in economics from the viewpoint of the budget constraints of consumers. This analysis is crucial to promoting interdisciplinary research between economics and other engineering fields on a smart grid system.

The next three chapters (Chaps. 4–6) describe economics-oriented approaches to the subject. Chapter 4 analyzes the economic impact of various policy schemes for strategic pricing in network access and vertical integration and separation from the viewpoint of industrial organization, with the aim of achieving a high proportion of renewable energy sources. Chapter 5 investigates the impact of environmental taxation and subsidization on non-polluting renewable power plants in a wholesale electricity market that includes oligopolistic firms. Chapter 6 analyzes customers' electricity conservation behavior for DR requests through valuable web-based survey and field and laboratory experiments performed in collaboration with major construction companies from the viewpoint of behavioral economics.

The third part of this book (Chaps. 7–9) addresses the optimal energy market design, integrating both physical and economic models. Chapter 7 focuses on demand-side strategic aggregators as intermediaries between a wholesale electricity market and a massive number of consumers. The economic impacts of the aggregator's market power are solved by a hierarchical distributed optimization method and evaluated via numerical studies. Chapter 8 addresses a novel balancing market that includes incentives and a policy agreement between an independent system operator and strategic agents with individual dynamics, aiming to guarantee that physically ancillary services maximize social welfare economically. The presented incentive-based framework can overcome a dynamic moral hazard problem and a dynamic adverse selection problem by using mechanism design and contract theory. Chapter 9 proposes a novel dynamic pricing rules for balancing power markets by using distributed optimization techniques without exchanging private information among rational agents. Under the proposed rules, the social welfare of multiple regional markets can be theoretically optimized.

This book's last three chapters (Chaps. 10–12) focus mainly on the engineering aspects of next-generation energy management, though economic factors are also shown to play important roles. Chapter 10 highlights a real-time pricing mechanism based on nonlinear and stochastic model predictive control (MPC) used to achieve high-accuracy supply–demand balance and load–frequency regulation in physically coupled power systems. Chapter 11 proposes a distributed multi-agent optimization protocol suitable for large-scale power flow networks, toward the application of distributed EMSs. Chapter 12 presents a cyber-physical systems (CPS) design for enhancing the energy efficiency of heating, ventilation, and air-conditioning systems in multiple connected buildings while optimally balancing human comfort and energy efficiency. The present CPS framework enables passivity-based stability and convergence analysis.

The authors gratefully acknowledge financial support from JST CREST Grant No. JPMJCR15K2. We also sincerely appreciate the CREST research supervisor, Prof. Masayuki Fujita, and eight research area advisors for providing in-depth feedback and advice. Finally, we offer a big thanks to over one hundred team members, including domestic and foreign leading academic researchers, collaborators, colleagues, students, and members of industry, for promoting this science research project so strongly. It is our great pleasure to present this book.

Tokyo, Japan   Takeshi Hatanaka
                                                                                 Yasuaki Wasa
                                                                                 Kenko Uchida

Contents

Chapter 1
Economically Enabled Energy Management: Overview and Research Opportunities

Yasuaki Wasa, Kenta Tanaka, Takao Tsuji, Shunsuke Managi and Kenko Uchida

Abstract This chapter gives comprehensive overviews of the social context of energy in Japan and underscores the pressing need for economically efficient energy management systems. Next, because the aim is to overcome emerging worldwide issues, this chapter surveys past, present, and future academic research trends and their contributions to the study of power, economics, control, and system architecture. This chapter also provides a detailed outline of the book.

1.1 Introduction

Reform of Japan's electric system, which is based on deregulation, started in the 1990s and continues to this day. The reform is primarily intended to increase the economic efficiency of the system. The latest strategic energy plan of Japan [67] enacts the so-called 3E+S—economic efficiency, energy security, environment (work

Y. Wasa (✉)
Department of Electrical Engineering and Bioscience, Waseda University,
Tokyo 169-8555, Japan
e-mail: wasa@aoni.waseda.jp

K. Tanaka
Faculty of Economics, Musashi University, Tokyo 176-8534, Japan
e-mail: k-tanaka@cc.musashi.ac.jp

T. Tsuji
College of Engineering Science, Yokohama National University,
Kanagawa 240-8501, Japan
e-mail: t-tsuji@ynu.ac.jp

S. Managi
Urban Institute and Departments of Urban and Environmental Engineering, Kyushu University,
Fukuoka 819-0395, Japan
e-mail: managi@doc.kyushu-u.ac.jp

K. Uchida
Research Institute for Science and Engineering, Waseda University,
Tokyo 169-8555, Japan
e-mail: kuchida@waseda.jp

T. Hatanaka et al. (eds.), *Economically Enabled Energy Management*,
https://doi.org/10.1007/978-981-15-3576-5_1

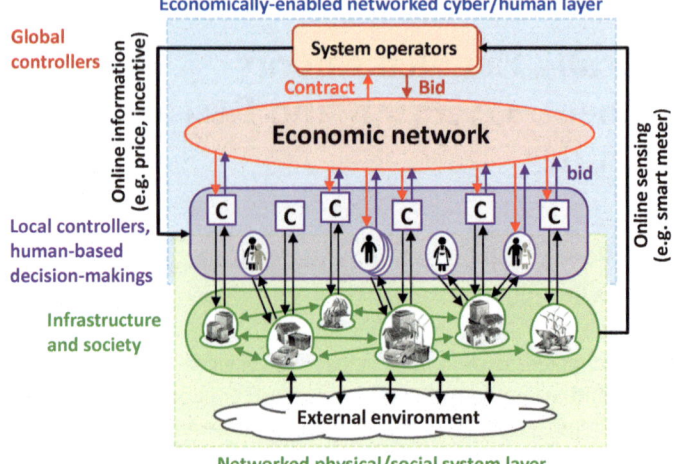

Fig. 1.1 Sketch of economically enabled energy management systems for the coming generation

toward decarbonization), and safety—as key targets of reconstruction and new construction for energy supply–demand systems. Safety, in particular, was taken up as a major goal of the system after the Great East Japan Earthquake and Tsunami disaster of 2011, in which damage to nuclear power plants caused the release of radioactive materials to the environment.

While this view of the strategic energy plan can serve as a common basis for the articles included in this book, another common view is that aspects of the economic system embodied by next-generation energy management systems (EMSs) will expand and humans (in the sense of either classical economics or behavioral economics) will play a central role in the system, as shown in Fig. 1.1. In this book, the EMS controls and integrates the different elements of energy supply–demand networks. These elements are distributed/centralized energy generators, consumers, prosumers, aggregators, and distribution energy networks. The aim of control and integration is to achieve social welfare as characterized by the 3E+S goals for energy supply–demand systems. Typical examples of sophisticated EMSs are home EMSs (HEMSs), building EMSs (BEMSs), factory EMSs (FEMSs), and community EMSs (CEMSs), which are sometimes called xEMSs.

Information and communication technology (ICT) has advanced rapidly during the last decade, and online sensing and information gathering via smart meters at each element are presently available, meaning that both suppliers and consumers will be interactively connected and transacting energy in next-generation EMSs. In any EMS, the human and local controllers can be regarded as agents, and they should thus be the decision-makers, acting as or on behalf of generators, consumers, prosumers, and aggregators. Therefore, within a deregulated energy market, we should have some economic mechanism (e.g., a market mechanism) to integrate the agents' strategic decisions and controls such that social welfare is considered. From the observations above, we can see that EMS is definitely an economic system based on human economic activity. Remember here that the goal of an EMS is to manage an appropriate energy supply–demand system that relies on a specific set of physical entities that generate, transmit, and consume energy. Thus, analysis and synthesis of EMSs that balance the energy supply–demand system will require consideration of both economic and physical perspectives and their combination, which is depicted in Fig. 1.1; this view of EMSs will be common to all articles in this book. In this book, we call an EMS based on the above concepts an *economically enabled energy management*. Note that economically enabled EMSs are a special class of a cyber-physical and human system (CPHS). See Sect. 1.3.4 for more details.

This chapter provides a comprehensive overview and describes research opportunities to realize economically enabled EMSs. Section 1.2 reviews the history of the Japanese power system and power market as well as global trends [7], with a particular focus on that of the USA [76]. In the same section, we also discuss the need for EMS development. In Sect. 1.3, we outline energy supply–demand systems to reveal the physical aspects and technological issues of EMSs from the viewpoint of power systems, economics, control and optimization methods, and interdisciplinary CPHSs. In Sect. 1.4, the organization of the book is given. In Sect. 1.5, we comment on some further research topics and challenges not discussed in this book.

1.2 Background of Japanese Power System and Power Markets

1.2.1 Structure of Power Industry in Japan

After World War II, the power system in Japan was divided into nine areas, each managed by an independent power company under a vertically integrated structure in which the generation, transmission and distribution, and retailing sectors were handled by the same company. In 1988, Okinawa Electric Power Co., Inc. was privatized, and the basic structure of the power industry as it now exists in Japan was formed. As shown in Fig. 1.2, ten control areas have been managed by power companies with some coordination through inter-area tie lines. Here, it should be noted that the eastern and western systems are operated at 50 Hz and 60 Hz, respectively, for historical reasons. Worldwide, deregulation of the power industry took hold in the 1990s. In Japan, institutional reforms were started in 1995, and the electricity rate of the ten companies has been gradually decreased by new players, such as "independent power producers" and "power producer and suppliers." The first power market in Japan, Japan Electric Power Exchange (JEPX), was established in 2003 as a result of the third set of institutional reforms. Since 2009, a day-ahead spot market, forward market, and intraday market have been available [48].

In 2011, all nuclear power plants in Japan were forced to stop after the Great East Japan Earthquake caused an accident at Fukushima Daiichi Nuclear Power Station,

Fig. 1.2 Overview of the
power system in Japan: ten
incumbent electric power
companies, with generation
capacities in 2017 July [4].
Japan is currently divided
into ten areas with utilities
supplied regionally and only
1.2 GW of transmission
capacity between the east
and west areas. The main
island is Honshu, and
Hokkaido, Shikoku, Kyushu,
and Okinawa are separate
islands

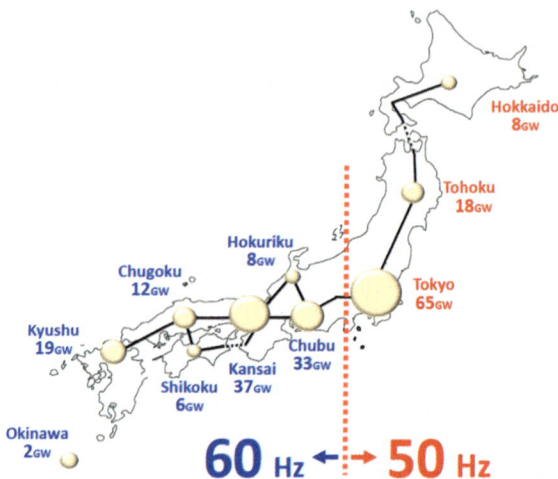

and a new national policy that further restructured the power industry was adopted
in order to secure a more stable power supply while reducing the electricity rate by
encouraging greater market participation and decentralizing resources. The content
of this restructuring consists of the following three steps.

- The establishment of the Organization for Cross-regional Coordination of Transmission System Operators (OCCTO), for facilitating operations among multiple
  control areas, in 2015 [77].
- Full liberalization in the retailing sector, in 2016.
- Separation of power generation and transmission to achieve neutrality of the transmission and distribution sector, in 2020.

The restructuring has proceeded on schedule so far, and in accordance with the
integration of variable renewable energy sources (VRES), different legal systems
have been discussed, such as the settlement of new markets to achieve balancing
power reserve requirements or ensure sufficient generation capacity.

In September 2018, the first blackout in the Hokkaido area due to an earthquake
occurred, the details of which will be described in Sect. 1.3.1. The importance of
power supply stability was reaffirmed by this event, and further discussion is ongoing
to promote liberalization while maintaining a stable power supply.

1.2.2 Renewable Energy Integration in Japan

In accordance with energy liberalization, the integration of random, intermittent, and
uncontrollable VRES, such as photovoltaic (PV) and wind power (WP) generation,
is spreading worldwide. Figure 1.3 shows the recent expansion of VRES in the world.
The cumulative capacities for PV and WP are over 400 and 500 GW, respectively, as

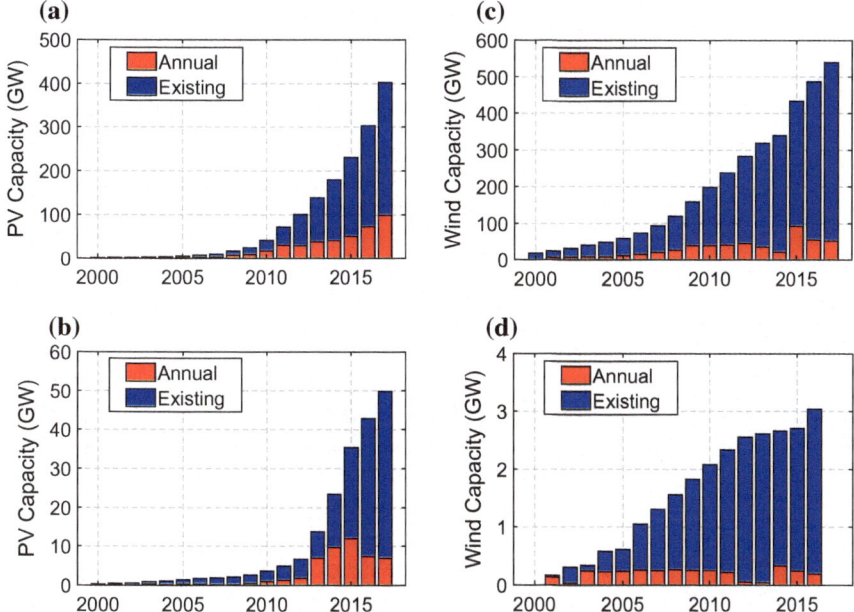

Fig. 1.3 Expansion of VRES integration worldwide and in Japan [46]. **a** Global PV expansion. **b** Japan PV expansion. **c** Global wind capacity expansion. **d** Japan wind capacity expansion

of the end of 2017. In Japan, in particular, PV integration has progressed rapidly since the 2011 earthquake due to a feed-in-tariff (FIT) mechanism introduced in 2012; in comparison, WP integration has not progressed much in Japan.

Due to the integration of VRES, the supply–demand balance is a challenging issue for many power systems. In Germany, for example, the total generation output from VRES is often high enough, particularly in off-peak seasons, that the market-clearing price becomes negative [55]. Because a variable premium method is applied in Germany for VRES generation and the premium is lowered to zero when a negative market price endures for at least six consecutive hours, VRES owners are motivated to curtail power generation themselves when the amount generated by VRES is too large, aiming to avoid the zero premium. In Japan, a large amount of PV capacity has been already installed in some control areas, such as Kyushu and Shikoku. The impact of PV integration has appeared over time. For example, a global flicker was observed in the Kyushu area in 2017. This phenomenon was caused by the islanding detection functionality equipped as part of a power-conditioning subsystem (PCS). It is mandated in Japan's grid code that power-conditioning subsystems have an islanding detection mechanism based on the frequency feedback method with step injection of reactive power. Synchronous reactive power injection is used by this method to automatically detect the islanding condition, which does not result in negative effects on power systems when the amount of PV integration is limited. However, in the 2017 event, the total amount of reactive power control became high enough that the voltage was oscillated globally and continuously by this method. The impact of VRES integration on supply–demand balance control can have more serious impacts in the present system. VRES generation is given high priority, and the power is still purchased at a high price (via the FIT mechanism). Under the present Japanese regulation, a transmission system operator (TSO) can ask VRES owners to curtail power generation in advance only when that generation would threaten the stability of the power supply. The first curtailment of VRES generation was implemented in 2018 in Kyushu due to an excess supply of power from PVs.

1.2.3 Power Markets in Japan

JEPX was established in 2003, and both day-ahead and intraday spot markets are presently available. The Japanese government has tried to promote power trade through market mechanisms, and the share of transactions conducted through these markets relative to total power demand has increased steadily (standing at 18.4% as of June 2018 [26]). An important role of the market is to provide price signals that properly represent the supply–demand balance. When the market price is very high, market participants know that the ongoing power supply is not sufficient and they have incentives to put forth best efforts to either decrease demand or increase generation through frameworks such as demand response (DR) and virtual power plant (VPP).

In addition to these spot markets, there are plans to establish balancing power markets and capacity markets in 2021 and 2024, respectively. Due to the integration of VRES, TSOs will require larger balancing reserves to maintain the supply–demand balance at a proper frequency although the inverse phenomenon has been reported in the case of the so-called German paradox [58]. Balancing power market is important for procuring the reserve from a variety of market participants in an economical fashion. Table 1.1 shows the product list of balancing power markets in Japan as of April 2019. Imbalance fluctuations can be classified into multiple timescales, and the requirements for each product are shown in the table. For each balancing reserve, the reserve capacity for balancing control, ΔkW, is cleared in advance. At the time of actual operation, a control signal is sent to the corresponding generators or loads, based on the merit order of the price per kilowatt-hour. Although the details of the market mechanism are still being discussed, the kilowatt-hour cost could be updated up to the gate closure time to reflect the real-time supply–demand balance condition as accurately as possible.

Along with the new balancing power market, the imbalance charge mechanism will be updated in 2021. The main aim of balancing control is to compensate for imbalances caused by deviation from planned values among all trades (in terms of the 30-min average value), so the cost to procure the balancing power should be covered mainly by imbalance charges. At present, the imbalance charge is decided by the bidding curves of the day-ahead and intraday spot markets without any consideration of the supply–demand balance at the time of operation. Under the planned mechanism, imbalance charges will be linked to the cost of balancing power,

Table 1.1   Products in balancing power markets of Japan (as of April 2019)

	FCR	S-FRR	FRR	RR	RR-FIT
Control	Offline (local control)	Online (LFC)	Online (EDC)	Online (EDC)	Online
Monitoring	Online (and offline)	Online	Online	Online	Online (and offline)
Response time	Within 10 s	Within 5 min	Within 5 min	Within 15 min	Within 45 min
Duration	Over 5 min	Over 30 min	Over 30 min	3 h	3 h
Minimum capacity	5 MW (1 MW for offline)	5 MW	5 MW	5 MW	5 MW (1 MW for offline)

FCR frequency containment reserve, *S-FRR* synchronized frequency restoration reserve, *FRR* frequency restoration reserve, *RR* replacement reserve, *RR-FIT* replacement reserve for feed-in tariff, *LFC* load-frequency control, *EDC* economic dispatching control

meaning that imbalance charges will increase sharply when the supply–demand balance is severely skewed; this is already in place in other countries. For example, in the UK, the imbalance charge is given as the marginal cost to procure the balancing power. Moreover, an additional amount is added to the imbalance charge when the reliability index, such as the loss of load probability (LOLP), crosses a certain threshold. To allow participants to pursue economic rationality in response to changes in imbalance charges, the relevant information has to be available at the right time, and a method of information disclosure has been developed to do so in the UK.

As mentioned above, it is necessary to realize flexible supply–demand balance operations based on market principles in order to cope with further VRES integration. In particular, as shown in the UK case, it is essential to apply flexible price signals for market pricing and imbalance charges in response to power system conditions. Although there are various systems in other countries, the best mechanism will depend on the characteristics of the power system together with the relevant legal systems of the country. Because the power system in Japan has its own inherent characteristics, further discussion will be needed to find a more suitable mechanism for Japan, and this research project should contribute to decision-making on this topic.

1.3   Perspectives Toward Economically Enabled Energy Management

1.3.1   Power Perspective

Imbalance and Frequency in Conventional Power Systems
It is essential for a stable power supply that power systems maintain a balance, in real time, between supply and demand. Synchronous generators have conventionally been used to supply power, with all synchronous generators in the same interconnected alternating current (AC) network synchronized and rotating at the same

Fig. 1.4 Sketch of imbalance and frequency in a power system

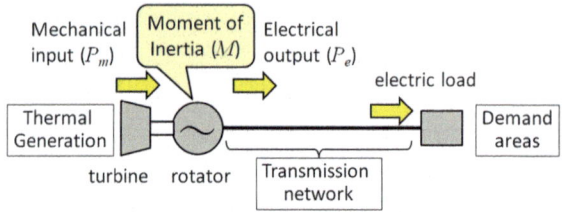

angular speed because of their inherent characteristics, which results in a large-scale equivalent generator, as shown in Fig. 1.4. Here, the generators can be regarded as a voltage source in an AC circuit in which the rotational speed decides the system frequency. The rotational speed of the equivalent generator (or synchronized system of generators) changes depending on the difference between the total mechanical input and the total electrical output according to the following equation [57]:

$$\frac{\mathrm{d}\Delta\omega}{\mathrm{d}t} = \frac{1}{M}(P_m - P_e - D\Delta\omega). \tag{1.1}$$

Here, $\Delta\omega$ is the deviation of the angular speed of generator from the nominal value (50 or 60 Hz), M is the moment of inertia, D is the damping coefficient, and P_m and P_e are the mechanical and electrical outputs, respectively.

When the electrical output increases and the mechanical input follows after a control delay, the system frequency decreases before the mechanical input catches up with the electrical output and vice versa. To keep the system frequency near the nominal value, automatic generation control (AGC) is applied. The control mechanisms can be classified into primary, secondary, and tertiary controls in many countries; these are known as governor-free (GF) control, load–frequency control (LFC), and economic dispatching control (EDC), respectively. GF and LFC work primarily to reduce frequency fluctuation, while economic and network aspects are considered in an EDC, which has a control time cycle that is longer than the cycle of other types of systems. Details about these controls are as follows [57].

GF By changing the valve opening to adjust the volume of steam provided by a boiler, the output can be controlled rapidly according to the droop characteristic, although the maximum control amount is limited (e.g., to 5% of rated capacity). GF can be realized for a system frequency that is the same across the entire AC network. Therefore, GF is capable of using only local information to operate and does not need dedicated communication lines.

LFC To change the generation output more sharply than possible with GF, the amount of fuel consumption can be changed, though this is slower. Thus, the possible control amount with LFC is normally much bigger than with GF, but the control response is inferior. Generators with high control speed are often specified as LFC generators, and control signals are sent by a TSO to all LFC generators. The control signal is often calculated by proportional–integral (PI) logic to eliminate the offset of frequency deviation.

EDC In the long term, generation dispatch should be decided according to cost minimization based on forecasted demand and output from VRES. Additionally, the power flow and other constraints can be taken into account such that necessary conditions (e.g., thermal limit of power flow, voltage stability, and synchronous stability) are met.

Here, note that generators must have control reserves in order to respond to the above control signals. In particular, the control margin in the upward direction has to be considered in the design because generators are normally operated at their rated power to maximize the efficiency of generation. Therefore, the fuel costs increase as a result of maintaining reserve capacity to allow balancing power. Moreover, some generating plants may have to be started to provide a sufficient amount of balancing power. TSOs carefully decide the amount of the reserve maintained for balancing power in unit commitment, aiming for cost minimization.

Imbalance and Frequency in Modern Power Systems
As VRES becomes further integrated, the supply–demand balance of power systems becomes more uncertain. It is well known that the uncertain output fluctuation of VRESs located at different places often cancels out each other because they are not synchronized. As a result, the magnitude of fluctuation of total generation output is proportional to the square root of the number of generators (denoted by n) if they fluctuate independently, but the magnitude of fluctuation is multiplied by n if the generators are fluctuated together. This effect is defined as the smoothing effect, and it works effectively when many VRESs with weakly correlated outputs are integrated into one AC network. Thanks to the smoothing effect and improvements in weather forecasting, it is possible to forecast VRES output with some certainty. However, the ramp phenomenon, which is defined as a change of at least 30% in aggregated wind power output over 6 h, is still difficult to predict, and the ramp phenomenon is a serious issue from the viewpoint of stable power system operation [93]. There are various projects to improve the accuracy of forecasting. In Japan, for example, a technique used to forecast the ramp phenomenon has been investigated in the national research and development (R&D) project "Grid Integration of Variable Renewable Energy: Mitigation Technologies on Output Fluctuations of Renewable Energy Generations in Power Grid" [93], and it was shown in [50] that the forecasting error was around 5% of rated capacity, in terms of mean average error, from actual measurement data spanning 11 months.

In power systems with VRES integration, the ratio of power produced by synchronous generators relative to all loads falls because the power from some synchronous generators is replaced with that from VRES. Consequently, the moment of inertia M in the AC networks is greatly reduced, and the system frequency changes more strongly in response to the same amount of imbalance, as shown in (1.1). This phenomenon has become more important recently as some countries achieve high levels of VRES integration [101]. For example, in Ireland, the total amount of wind power reached 3.4 GW in 2017 [41]. The maximum limit for the system non-synchronous penetration (SNSP) [2] level, defined as the rate of non-synchronous generators to the sum of loads and power exports, is set at 65% as of 2017, and power curtailment is often applied to avoid exceeding the SNSP limit [25]. The impact of reduced inertia is more important in contingency conditions after a fault. Details about the impact of the reduced inertia in contingency cases are described in the following.

Fast Frequency Response in Contingency Conditions

Serious frequency drop is caused by a significant loss of power supply, such as the loss of output from a generating plant if one or more of the generators is tripped. When the frequency drops below a critical threshold, many synchronous generators will automatically shut down to avoid mechanical and electrical damage. This additional loss of power supply will cause a further frequency drop, with blackout in the result of a cascading fault.

In September 2018, an earthquake triggered a massive blackout in the Hokkaido area of Japan at around 3 AM via this failure mechanism. The earthquake occurred during off-peak time, and the demand right before the earthquake was very low, at around 3 GW. Half of the demand was covered by three coal-fired generators at the Tomato-Atsuma thermal power station; two of the generators were shut down due to vibration immediately following the earthquake. Unfortunately, four transmission lines connecting the western and eastern areas tripped simultaneously due to the earthquake vibrations, and the eastern network became separated from many hydro-generating plants as a result. Because the total generation was bigger than demand in the islanding system, many hydro-generating plants tripped due to the resultant high frequency. Moreover, around 170 MW of wind power tripped; the total loss of generation was around 1.76 GW, resulting in a very sharp frequency drop in the system. The frequency was restored by load shedding by under-frequency relay (UFR) and power support from Tohoku Electric Power Company through a direct current (DC) connection. Using this power, the system operated for a while after the disturbance. However, another generator in the Tomato-Atsuma thermal power station tripped, resulting in a blackout. It took 45 h for power to be restored. Details of this blackout are given in the final report by the OCCTO Investigation Committee [78].

After the first restoration of frequency, the system frequency gradually decreased again and a low frequency of around 49.2 Hz continued for several minutes due to the load increase. There was almost no balancing power available from the conventional generators because many hydro-generators were stopped and some thermal generators could not respond to the control signal due to boiler troubles. Here, the supply–demand balance and reserve capacity might have been restored properly, before the additional generator tripped, if a suitable market principle had been in place for contingency situations. In one suitable system, the electricity rate could be increased in real time after a disturbance (reflecting the serious shortage of power supply), and then loads could be naturally reduced, possibly allowing the system operation to weather the additional loss of generation.

Other Control Resources

These days, various control resources are available on the demand side in addition to conventional generators, and the development of balancing power markets would be valuable to aggregate those resources economically. Some emerging technologies are described here.

VRES In general, wind turbine and PV systems can reduce their output below the power provided by wind or insolation. For wind turbines, the power input from wind speed can be reduced by applying pitch angle control to the blades. Moreover, wind turbines have characteristics that vary with rotational speed, and optimal rotational speed is maintained during normal operation. If the rotational speed is intentionally shifted from the optimal value, then the generation output can be reduced by a certain amount. In the case of PVs, optimal generation output is available with optimal voltage according to the current–voltage (I–V) characteristic curve, and power curtailment is available by shifting the voltage from the optimal value based on DC chopper control. This power curtailment is effective in the case of oversupply. In addition, if reserve capacity has been achieved by applying curtailment in advance, VRES can provide a balancing reserve in the upward direction. In fact, in some countries, it is mandated that large-scale VRES operators contribute to frequency control by applying the droop characteristic to generation control. Power curtailment is also useful for mitigating the ramping rate of steep changes in generation output.

Virtual Synchronous Generators (VSGs) Distributed generators (DGs) connected to the grid by converters are typically asynchronous generators that do not contribute to the moment of inertia. However, these DGs can behave as synchronous generators by changing the electrical output in response to a change in system frequency (as shown in the following equation) if they are rotational generators or are equipped with rechargeable batteries [101]:

$$\Delta P_e = \Delta P_m - M \frac{\mathrm{d}\Delta \omega}{\mathrm{d}t} - D\Delta \omega. \qquad (1.2)$$

The difference between the original generation and the converter output is caused by accelerating (or decelerating) the generator or charging (discharging) the battery. VSGs contribute to the inertia when using this control mechanism, allowing frequency fluctuation to be mitigated. The same approach can be applied to VRES if a control reserve has been achieved by active power curtailment. Here, even if curtailment has not been applied in advance, wind turbines can temporarily increase output by using kinetic energy stored in the rotating generator and blades (that is, by inertial response control) to mimic the dynamic behavior of synchronous generators. Using the above approaches, VSGs can contribute to a fast frequency response by rapidly increasing their generation output right after the loss of generation from other sources. As a result, it is expected that the rate of change of frequency (ROCOF) and the lowest frequency can be improved by using these control resources.

DR and VPP The concepts of DR and VPP have advanced together with the growth of aggregation businesses. DR and VPP provide flexibility to the grid, with controllable loads and generators. The penetration of electric vehicles and other batteries and energy storage systems is a great help in realizing flexible DR and VPP. Electrolyzers, which convert power into hydrogen, are also expected to contribute to matching supply and demand as a controllable load. Nowadays, balancing power is traded through balancing power markets in many countries, and DR and VPP are being used by market participants [32].

A mechanism to match supply and demand by using the above new resources is required to allow further integration of VRES. For this, the consideration of economic rationality is so important that any balancing control mechanism should be combined with an economic approach based on a well-coordinated legal framework.

1.3.2 Economic Perspective

Economic Analysis of Electricity and Energy Sector

Many economists have analyzed energy-related problems. Although energy economics has a long history, significant amounts of economic research into energy started in the 1970s. One of the big issues in energy economics is how to set up an efficient energy market. Drawing on developments in the industrial organization field, economists can analyze market structures under various technological constraints on electricity and other energy uses. Notably, the movement toward deregulation and privatization, which started in the USA and UK, has motivated policy discussions about energy markets.

Traditionally, economists have considered the effect of deregulation on public sectors, such as transportation, banking, and energy. General theoretical discussion led to a consensus that deregulation would improve efficiency for consumers and producers. Winston [104] summarizes the effect of efficiency improvement by deregulation. He mentions theories of regulation and suggests that deregulation enhances efficiency in one of two ways. First, he points out that one of the effects is decreased inefficienctly within regulated firms. Generally, regulated industries experience inefficient operation because the markets are less competitive. This inefficiency is called the "X-inefficiency" [60].

Second, the rents that accrue to well-organized groups benefiting from regulation will disappear with less regulated competition. That is, firms can reap excess profits within regulated markets. In contrast, consumer surpluses decrease. The reduction of consumer surpluses exceeds the increase in consumer surpluses, meaning that deregulation is expected to improve social surpluses.

This general theoretical research contributes to not only the discussion of energy market reform but also the construction of models that allow discussing the effect of constraints on the energy market in more detail. Drawing on these basic theories, energy economists have constructed models that consider several kinds of actual restrictions related to the energy industry. For example, some researchers have examined transmission constraints based on Cournot modeling (e.g., [17, 40]). These analyses contribute to the discussion about wholesale market reform and construction of electricity networks. As an example, Tanaka [98] simulates the Japanese wholesale electricity market as a transmission-constrained Cournot market. That study shows that upgrading the transmission line between the eastern and western regions of Japan, which acts as a bottleneck, could reduce the market power of large-scale electricity companies.

Empirical analysis is necessary due to the advancement of theoretical perspectives. As mentioned above, the theoretical perspective suggests that deregulation and privatization will increase the efficiency of the energy industry, and many empirical studies support this theoretical prediction. Substantial reforms that deregulated energy markets have occurred since the early 1980s in the USA and European countries. The Japanese government also initiated regulatory reforms in the mid-1990s. This deregulation was motivated by the consensus view that deregulation was a key factor in improving productivity in the industry. Numerous studies have analyzed the effects of deregulation on productivity, and some have analyzed the energy industry specifically (see [54, 69, 80, 97]).

Recent Focus of Energy Economists
These analyses could contribute to success in real policy-making. However, such studies have not provided a clear elucidation of demand-side behavior. Notably, policy reform of energy markets that depends on supply-side analysis can increase the efficiency of the industry and consumer surplus, but countries now face new problems related to energy, such as how to address climate change.

In 1995, most of the world's countries agreed to the Kyoto protocol, which requires the contracting parties to reduce their emission of greenhouse gases (GHGs). Developed countries need to adjust their mix of energy sources and promote more efficient use of energy in order to significantly reduce GHG emissions. Taking account of such issues, the recent study of energy economics seeks how to contribute to addressing new problems such as reduction targets for GHG emissions.

One of the problems to consider for future energy use is how to increase the share of the market comprising renewable energy sources. Now, many governments want to increase the use of power from VRESs to decrease CO_2 emissions. However, the increasing use of renewable energy leads to some significant problems. For example, the cost of shifting to renewable energies can be significant. Recent technological advancements have resulted in lower installation costs for VRES, but the primary types of VRES (notably PV and WP) introduce significant fluctuations to the supply amount available during each hour. Therefore, increasing the ratio of VRES incurs an adjustment cost for electricity companies. At worst, if there is an excessively rapid change in electric supply, electricity companies may experience outages, potentially causing sizable economic losses. To overcome this problem, economists have proposed that demand-side control be developed. In demand-side control, the amount of energy used by consumers is controlled. The most conventional form of such control is to change the fees according to the total change in demand. Much of the recent research by energy economists focuses on mechanisms for demand-side control, rather than for supply-side control.

For the past century, electricity pricing has not reflected the time variation in costs that characterize the electricity industry. The demand for electricity varies widely from hour to hour and from day to day. However, electricity supply and demand could not be adjusted by a price mechanism. Addressing this deficit, many researchers have discussed the importance of installing a price mechanism for electricity. A suitable price mechanism could contribute to electric load leveling. In particular, demand-side control is an important tool in circumstances where the electricity supply cannot be controlled because it is renewable energy that depends on changes in natural weather conditions. To demonstrate the importance of dynamic pricing, researchers have performed a field experiment with dynamic pricing. The results of that dynamic pricing experiment indicate that dynamic pricing can reduce peak demand [28, 39, 64].

Nevertheless, the literature notes that dynamic pricing cannot satisfactorily maximize the consumer surplus. One reason for this problem is information-theoretic. Electricity traditionally has low price elasticity on the demand side. Jessoe and Rapson [49] note that the reason for this phenomenon is the information problem: Real people cannot perfectly understand the costs of electricity use in real time. In addition, people do not know how much to pay for electricity use. To understand potential solutions, Jessoe and Rapson [49] tested the effect of providing high-frequency information about residential electricity usage while applying a dynamic pricing rule. They found that information feedback about electricity usage increased the price elasticity of demand.

Along other lines, many researchers have noted that electricity use behavior is affected by other factors. Pollitt and Shaprshadze [79] mentioned the importance of behavioral economics in understanding consumer energy use behavior. Indeed, some previous studies have already shown that psychological factors affect consumer behavior. For example, Allcott [6] found that a social comparison scheme encouraged electricity conservation behavior. The social comparison involved receiving information about the energy use of other people relative to the user. In this study, consumers receive "moral utility" from energy conservation. Under the traditional utility model, people do not consider the behavior of others, but under a utility model that includes moral utility, consumers can receive utility from behavior that is good for society [61]. Social comparison allows people to understand whether their behavior good or bad for society. Use of social comparison can contribute to electricity conservation. In their study, Allcott [6] found that in a social comparison trial (by Opower), electricity usage was decreased by between 1.4 and 3.3%. Other studies have also shown an interaction effect between social comparison schemes and consumer preferences [13]. These new schemes that consider human behavioral factors of electricity demand can contribute to solving other problems for electrical systems. For example, Ito et al. [47] showed that a request for electricity conservation made to consumers decreased electricity consumption in the short run. In addition, they showed that information provision schemes are a cost-effective way of encouraging electricity conservation. From these findings, the application of psychological factors in policy-making is worth noting as a way to increase each policy's cost-effectiveness.

New Economic Analyses for the Future Energy Sector
Many researchers who study energy economics have contributed to actual policy-making, but rapid technological advancement and changing situations require further improvement to economic models of energy. Notably, climate change problems are now becoming more severe. We need to construct sustainable long-term energy systems that consider the economy, technological constraints, and mitigation of greenhouse gas emissions. To build such a system, the existing economic model must be supplemented with constraints that reflect new technology and economic behavior. Of course, such models will necessarily include new parameters and settings that may not yet be well understood. Several types of empirical analysis (natural experiments, field experiments, laboratory experiments, etc.) are of help in setting the details of the models.

1.3.3   Control and Optimization Perspective

Activities Toward Development of Smart-Grid Control
Control technologies are permeating every aspect of society. To solve social issues, which change over time, systems and control researchers have developed historically significant control theories as well as applications made possible by the science and technology available when those theories and applications were developed. Through state-of-the-art modeling, analysis, and design and synthesis tools of control systems, systems and control researchers have increased understanding of the underlying cyber-physical phenomena and solved corresponding engineering problems. Power systems are typical large-scale complex engineered systems and have not been an exception to this trend over the last half century. Traditional control in power systems is described in Sect. 1.3.1 and [57]. Advanced control mechanisms for power systems are described in [43–45, 105]. The central issue in such systems is *how system operators stabilize large-scale dynamical systems*. To solve large-scale control problems by computer with high compression quality and low computational load, decomposition techniques that allow computation in a decentralized (parallel) manner and methods of simplifying models have been developed [88].

Since the turn of the twenty-first century, it has become an urgent global issue to develop highly innovative technological solutions for cooperative distributed EMSs so as to fulfill multiple social criteria: global decarbonization; electric grid modernization; electricity demand growth; high penetration of VRES; and a mechanism for dynamic integration of electric grids to empower consumers despite a variety of sources, loads, and control systems. The vision report of the Institute of Electrical and Electronics Engineers (IEEE) [7] encourages huge investments in smart-grid R&D worldwide and proposes related policies. In 2010, IEEE Smart Grid was founded as a joint initiative of 14 IEEE societies; the breadth of its research scope is described in [42]. Around the same time, to realize the strategic objective of a next-generation EMS set by the Japanese government after the Great East Japan Earthquake and Tsunami of 2011, a cross-disciplinary research project—the JST

CREST EMS Project—centered on the systems and control field is being conducted from 2012 to 2020 [106].

Achieving system-level objectives is a universal requirement for all control systems. From the viewpoint of an innovative smart-grid structure, in particular, the desired control mechanism is *cooperative individual decision-making by spatially distributed components with limited local communication*, which is called distributed (cooperative) control of multi-agent systems (MASs). Distributed control is very different from the traditional decentralized approach proposed in [88] due to the decision-making process allowing only *limited* exchange of information among agents. During the last several decades, computation, communication, and sensing technology have become inexpensive and ubiquitous. As a consequence, distributed control policies running over (mesh) networks can be implemented in dynamic infrastructure systems that include both physical components and cyber networks; such systems are known as *cyber-physical systems* (CPSs). Many researchers in the field of network control systems have investigated fundamental theories of distributed and cooperative control. From 2007, when network control systems were first featured in a leading journal of the IEEE [8], community members have rigorously elucidated both fundamental theories for network control solutions and integration mechanisms to achieve system-level objectives alongside applications in networked robotics and networked sensor systems [35, 63, 66, 81, 90]. These control schemes are typically based on distributed optimization or distributed learning in games.

Recent Focus of Systems and Control Researchers

The rich mathematical tools offered by traditional control theory and state-of-the-art networked control theory have facilitated technological development of smart-grid control. The book [20] edited by Chakrabortty and Ilić in 2011 was the first book to comprehensively summarize feasible control and optimization technologies for smart grids and discuss related open problems. Later vision reports [7, 86] and the book [92] highlighted the role and research challenges of smart-grid controls, such as forecasting, prediction, and control of VRES combined with solar radiation estimation [96] from data provided by weather satellites and a light detection and ranging (LIDAR) system [73]; improvement of advanced power electronic devices, including flexible alternating current transmission system (FACTS) and high-voltage direct current (HVDC) transmission systems; improvement of ICT instruments including supervisory control and data acquisition (SCADA) systems and phasor measurement units (PMUs); and strengthening of cyber-physical security and resilience. Due to the wide range of topics, we limit our focus to three urgent research topics in systems, control, and optimization problems: distributed control, wide-area control, and electricity market-centric control. These are discussed in later chapters of this book.

Technological developments related to demand-side management (DSM) have enabled controllers to maintain supply–demand balance and regulate both voltage and frequency. Many papers have described progress in R&D related to direct load control and automated DR (ADR) [87] of novel distributed energy resources, including combined heat and power (CHP) plants; heating, ventilation, and air-conditioning (HVAC) systems [3, 33]; thermostatically controlled loads (TCLs) [34]; and plug-in (hybrid) electric vehicles (PEV/PHEV) [94]. These components and subsystems connect to power transmission and distribution systems due to redundant power exchange. Wide-area control of transmission and distribution networks is essential for network stabilization and ancillary services. The optimal power flow (OPF) problem can be reduced to a second-order cone program for radial networks [30] and a semi-definite program for general networks [14] via a convexification approach in a centralized fashion. Other methods of implementing wide-area control have been presented, notably in [20, 86, 92] and references therein. However, challenges remain, including how to synthesize an optimal distributed control for stochastic and nonlinear dynamical networks while extending the current consensus protocols of MAS.

Along another line, to integrate heterogeneous noncooperative subsystems in real time, it is absolutely necessary to start competitive electricity markets or reform existing markets. The economical operation of standard electricity markets is summarized in [89]. Veen and Hakvoort [102] discuss the design and analysis of an unbundled balancing power market, using examples of actual European energy markets. From a control viewpoint, the optimal EDC problem for electricity markets has been studied for many decades [44, 89, 105]. The traditional EDC problem is an optimal resource allocation problem of a market system operator under the assumption of *rational* market participants (agents). However, a more desirable approach is to analyze and synthesize the electricity market operation and a dynamic integration mechanism based on agents' *strategic behavior*. A promising concept for realizing such a market and control policy is called transactive energy and transactive control [9]. Kiani et al. [53] and Wasa et al. [103] presented a transactive control that integrated the pricing mechanism into a three-layer market architecture based on timescale control decomposition; these layers correspond to GF, LFC, and EDC, respectively (see Sect. 1.3.1). Dörfler et al. [23] proposed a distributed control for each timescale. However, all of these papers assume non-strategic agents.

When considering strategic agents, we must design a market mechanism that will satisfy all agents participating in the electricity market, allowing them to exchange information and goods honestly and to minimize their own cost [16]. Khargonekar, in [92], provides a big picture perspective, showing that future operational performance to support the infrastructure will be much quicker than using current practices. As a result, any *real-time* market mechanism needs to prevent the occurrence of moral hazard and adverse selection problems [16]. There are few papers addressing the dynamic mechanism design problem, despite it being an urgent research topic. For example, in one type of system it is critical to combine the standard (physical) dynamical control theory with a non-tatônnement model (a one-shot (economical) market-clearing mechanism with full information) and machine learning (ML) [52]. Other research directions in the field of systems and control are shown in Sect. 1.5.2. To address the issues with such systems, it is valuable to employ interdisciplinary modeling, which will be discussed in the next section.

1.3.4 Interdisciplinary Viewpoint

To establish economically enabled EMSs, it is essential to methodically consider the hierarchical and complex EMSs, which involve humans and an uncontrolled environment that includes physical networks, social networks, and economic networks, as shown in Fig. 1.1. In other words, to clarify human-/machine-cooperative control and decision-making schemes for each subsystem (component) in a dynamical CPHS, we need to newly prepare a common *mathematical* framework for system-level modeling, analysis, and synthesis, based on the networked integration of the subsystems. Interdisciplinary analysis of power systems was conducted in the 1990s [44]. In that research, Ilić et al. [44] analyzed the potential for real-time operation by the electricity market and industry, collaborating with power control engineers and economists. They noted that one of the aspects that was needed but yet to be achieved was the novel market mechanism described in Sect. 1.3.3 and emphasized the importance of systematically modeled transactions and model-based analysis. In this section, we discuss multidisciplinary representation via mathematical models and simulation-based analysis.

Models: Differences in Interpretation
We first discuss the model representation for systems and decision-making process.

As described in [24], system modeling to discover the principles, phenomena, and causality of real systems is divided into ideal *physical/economic* models and ideal *mathematical* models (Fig. 1.5). The ideal physical/economic models aim at precisely simulating the original real systems with mathematical tools, whereas the mathematical (theoretical) models are built by rigorously constructing a scientific theory from axioms and empirical laws. As Rosenblueth and Wiener [83] and Doyle et al. [24] have pointed out, no ideal mathematical (theoretical) model can perfectly realize all the universality of the real systems in their entirety. Physical systems tend to obey well-known universal laws and principles, connecting theoretical mathematical models and experimental data obtained by the actual systems, because replication in natural sciences is essential and implemented everywhere. Meanwhile, because real systems in social sciences include many unmodeled and uncertain effects, it is difficult to confirm the replicability of economic models. Therefore, classical economics (until the 1980s) and the corresponding models are built on the basis of general equilibrium models and behavioral hypotheses about *Homo economicus*, a perfectly rational economic actor who maximizes their own utility. Since the 1990s, agent-based analysis has become more common in economics, with researchers striving to understand behavioral economics by considering partly irrational decision-making by actual humans. In the decision-making process, so-called bounded rationality and bias are assumed, in contrast to the assumption of perfect rationality in Homo economicus. As shown in Sect. 1.3.3, the distributed control schemes of MAS have also been developed during the last decades. Therefore, the component-based network structure of a CPHS is convenient for both control engineers and economists who wish to analyze and synthesize the control problems of an EMS.

Fig. 1.5 Three models for integrating heterogeneous physical and economic systems

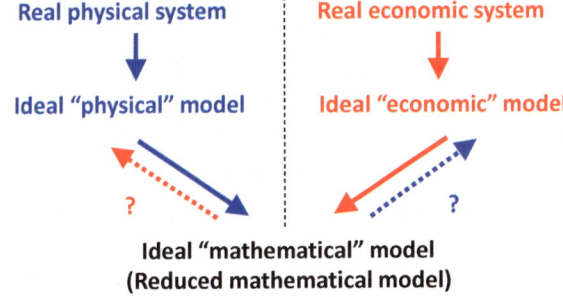

Our ultimate objective is to prepare a mechanism that integrates human-made strategic decisions with controls into a system that achieves pluralistic social welfare. From the viewpoint of social infrastructure development, modeling human-centric economic systems requires both economic models and physical models. Modeling the control systems in engineering is the same. We emphasize here that although we use a common mathematical model to reflect the decision-making mechanism, the understanding and viewpoint of economists are very different from that of control engineers. Economists are chiefly concerned with causal inference in real systems, with the economic interpretation of definite policy schemes considering heuristics (approximation) and nudges (incentives) for specific economic models. In contrast, control engineers are interested in admissible, implementable, and optimal control strategies for generic system models and in finding objective functions that are as unconstrained as possible. Because of this difference, the ideal (reduced) mathematical models and parameters that result from one physical (resp., economic) system may not be meaningful in the other economic (resp., physical) system (see the dashed arrows in Fig. 1.5). Enabling interdisciplinary interpretation is an important research topic.

Lastly, we focus on modeling human agents' strategic decision and control behavior. In traditional control theory, the decision model is mostly regarded as involving rational agents and optimization-based control. In ML, an intelligent agent that imitates human decisions is trained [85]. In the book [85], the decision-making process of intelligent agents is divided into four classes: simple reflex agents, model-based reflex agents, goal-based agents, and utility-based agents. Utility-based agents as defined in [85] are similar to the agent model found by causal inference in behavioral economics. Another type of classification is described in Sect. 1.5.2. Modeling human decision-making is still difficult; therefore, empirical social research is of great importance. Even if we obtain a rigorous human decision model, the consideration of economic interests in MAS (see Fig. 1.6), integration, and aggregation is a fresh research direction for control engineers. For instance, in CPHS, each DSM behaves so as to strengthen the price control authority of oligopolistic markets, striving to act in ways that provide financial benefit and advantages to themselves. Hence, there exist many unsolved theoretical challenges, such as how to balance the trade-off

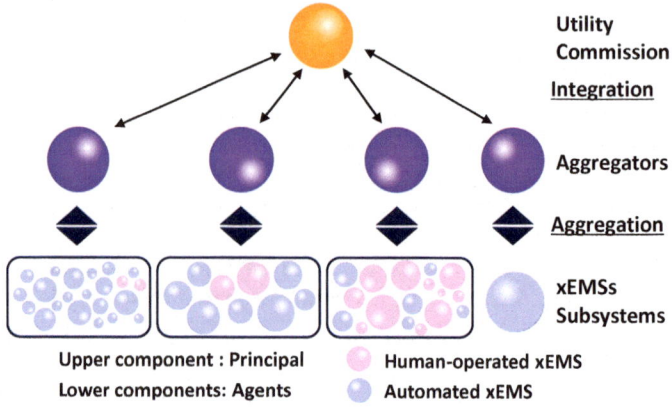

Fig. 1.6 Sketch of hierarchical decision-making scheme

between micro-level decision-making and macro-level phenomena in economically hierarchical CPHSs.

Virtual Test Bed Development of Smart Grids for Evaluation

A promising approach to bridging theory and practice/application is to use simulation-based analysis with ideal physical/economic models. In natural sciences, especially systems engineering in industry, model-based design/development (MBD) and model-based systems engineering (MBSE) have become popular due to the savings they allow on the cost of developing a social implementation. Detailed and precise models enable implementation of a novel control mechanism in a virtual world. Relative to field experiments, simulation-based analysis with rich data is easy, safe, and inexpensive. Additionally, it is relatively difficult to perform economic analysis and test novel controllers in social infrastructure systems, so the development of virtual CPS/CPHS simulators is very important. Chakrabortty and Bose [19] summarize the state-of-the-art simulation environment in smart grids. Hayashi et al. [37] have developed a real CPS test bed that imitates DSM for a distribution network. The most important work in building an all-purpose CPS/CPHS test bed is the system identification technique used to obtain precise physical/economic models while distinguishing among multiple spatiotemporal scales [65, 92].

1.4 Organization of the Book

To provide insight into the issues mentioned in the previous sections, this book includes contributions from a multidisciplinary research team comprising control engineering and economics researchers. This team was formed to address a central interdisciplinary social issue: economically enabled energy management. The research team is just one of the five super research teams organized for the JST

CREST EMS Project [106]. Several strong motives drive the research projects of our research team members. These are as follows:

- Construction of energy consumption models based on human behavior, analysis, optimization of energy supply–demand balance, and ADR promotion strategy;
- Construction and analysis of energy-economic models for consumer behavior, energy markets, and long-term energy policy;
- Principle-based design and evaluation of economic efficiency of dynamic integration mechanisms that combine strategic decision-making with control of dynamical (selfish) supply/demand agents to achieve public benefit, decentralized algorithms, fast algorithms, and reliable energy service; and
- An integration mechanism and control strategy based on the integration of economic and physical models for renewable energy.

This book's chapters introduce the main parts of the technical contributions to realize these four research objectives. In addition to this introductory chapter, this book contains eleven technical chapters written by leading researchers in power, economics, and control. The technical chapters are divided into four broad parts.

The first part covers fundamental viewpoints in power, economics, and control.

Chapter 2, "Supply and Demand Balance Control Based on Balancing Power Market," written by T. Tsuji, H. Bae, and J. Qi, focuses on balancing power markets to regulate frequency and power balance from the viewpoint of power engineering. Full installation of VRES would cause power imbalances at each node and power flow congestion on transmission lines. To avoid physical problems, the authors propose novel mechanisms for a balancing power market and advanced frequency control. The challenges discussed in the chapter include evaluating the impact and risk management of VRES and balancing power market operation on AGC through a detailed simulation of the Japanese power grid, which was provided by the Institute of Electrical Engineers of Japan (IEEJ).

Chapter 3, entitled "Resolving Discrepancies in Problem Formulations for Electricity Pricing by Control Engineers and Economists," by Y. Kawano and E. Sawada, who are a control researcher and an economist, respectively, aims at identifying fundamental differences between model-based approaches in control engineering and in economics from the viewpoint of budget constraints of consumers. To promote interdisciplinary research between economics and other engineering fields on the topic of smart grids, they point out that the economic model defined by non-economists often ignores the budget constraints of consumers, whose pricing model is a central topic in economics. Fundamental gaps caused by differences in the problem formulations appear in a very simple power grid model and economic model.

The three chapters after these, composing the second part, provide economics-oriented approaches to the subject.

Chapter 4, "Effectiveness of Feed-In Tariff and Renewable Portfolio Standard Under Strategic Pricing in Network Access," authored by Y. Kurakawa and A. Hibiki, analyzes the economic impact of policy schemes in the presence of strategic pricing with network access, with the aim of achieving a high penetration of VRES. In particular, they compare three policy schemes: fixed-price FIT, premium-price FIT, and renewable portfolio standards (RPSs). From the viewpoint of the theory of industrial organization, this chapter focuses on the effectiveness of these three policy schemes as arising from the influence of vertical structure, that is, vertical integration and separation. By this approach, they show that RPS is potentially more effective than FIT policies under vertical integration, and vertical separation improves the effectiveness of both FIT policies more than RPS does.

Chapter 5, entitled "The Welfare Effects of Environmental Taxation and Subsidization on Renewable Energy Sources in an Oligopolistic Electricity Market," by I. Matsukawa investigates the impact of environmental taxation and subsidization on non-polluting renewable power plants in a wholesale electricity market in which oligopolistic firms participate. After presenting a model to examine the welfare impact of a tax on emissions from fossil fuel power plants and a subsidy for renewable power plants, the author compares the welfare resulting from an emission tax on fossil fuel use against a subsidy for renewable power plants and uses a simulation to analyze how these environmental policies affect the wholesale electricity market.

Chapter 6, "Behavioral Study of Demand Response: Web-Based Survey, Field Experiment, and Laboratory Experiment," authored by T. Ida, Y. Ushifusa, K. Tanaka, K. Murakami, and T. Ishihara, analyzes customers' electricity conservation behavior for DR requests through valuable web-based survey and field and laboratory experiments performed in collaboration with major construction companies. These experiments provide particularly quantitative evaluation on residential electricity plan choice and residential and building energy conservation from the viewpoint of behavioral economics.

The third part consists of the following three chapters and addresses the modeling, analysis, and synthesis of optimal power markets integrating both physical and economic models.

Chapter 7, "Economic Impact and Market Power of Strategic Aggregators in Energy Demand Networks," written by Y. Okajima, K. Hirata, and V. Gupta, focuses on the function of demand-side strategic aggregators as intermediates between a wholesale electricity market and a massive number of consumers. The market pricing and the decision-making of both the strategic aggregators and the consumers, who maximize their own benefit individually, can be solved in a hierarchical distributed optimization method. Using the pricing mechanism, they also employ numerical examples to evaluate the economic impact of the market power of the aggregator in the (oligopolistic) wholesale market, the social welfare, the total operation, cost, and the installed battery capacity.

Chapter 8, "Incentive-Based Economic and Physical Integration for Dynamic Power Networks," authored by Y. Wasa, T. Murao, and K.-I. Akao, addresses a novel balancing power market with incentives (a so-called incentivizing market) and a policy agreement between the public utility commission, which is sometimes regarded as an independent system operator, and agents with individual dynamics in order to guarantee that physically ancillary services maximize social welfare economically. The critical issue in developing the incentivizing market is to prevent rational agents from selecting strategic bidding that minimizes their costs while realizing the public objective. It is also essential to cope with not only dynamic mechanism designs and contract theory approaches in economics but also an optimal control theory in physically dynamic power networks. To overcome these issues, the authors propose two incentivizing market models: a dynamic moral hazard model based on the principal–agent problem of inducing agents to select their private controls so as to maximize social welfare and a dynamic adverse selection model with a mechanism design that collects the agents' truthful private information.

Chapter 9, "Distributed Dynamic Pricing in Electricity Market with Information Privacy," authored by T. Namerikawa and Y. Okawa, proposes dynamic pricing rules for balancing power markets by using distributed optimization techniques. One of the proposed mechanisms is to not exchange the private information of the (selfish and rational) market participants, who include generators and consumers. By using only the iteratively updated market prices and the bidding information of quantities without the exchange of private information, the social welfare of multiple regional areas, where an electricity market exists in each area, can be optimized theoretically. The authors also illustrate, by numerical simulation, that these methods enable us not only to derive the optimal electricity prices but also to improve their convergence speed.

The last three chapters of the book focus mainly on engineering aspects of next-generation energy management, especially those for state-of-the-art systems and control approaches.

Chapter 10, "Real-Time Pricing for Electric Power Systems by Nonlinear Model Predictive Control," written by T. Ohtsuka, Y. Kawano, T. Hashimoto, Y. Okajima, and K. Kashima, discusses a real-time pricing mechanism based on nonlinear model predictive control (NMPC) for achieving highly accurate supply–demand balance and load frequency regulation in physically coupled power systems. Because the price signals of electricity are regarded as control inputs to the physical systems, several types of optimization problem are formulated for different models of consumers, suppliers, and levels of VRES. The authors also clarify that inherent uncertainties in consumers' characteristics and VRES are taken into account by NMPC combined with nonlinear estimation and/or stochastic optimization. They finally verify by numerical simulation that the proposed NMPC framework is flexible and effective for real-time pricing with realistic computational costs.

Chapter 11, "Distributed Multi-agent Optimization Protocol over Energy Management Networks," authored by I. Masubuchi, T. Wada, Y. Fujisaki, and F. Dabbene, proposes a distributed multi-agent optimization protocol that functions over large-scale unbalanced networks for the application of distributed EMSs. The proposed protocol, which is based on linear consensus algorithms and exact penalty methods, achieves partial consensus on decision variables of agents at a Pareto optimal solution of a minimax problem. With this generalization, the protocol enjoys more freedom in the formulation of optimization problems than those that can solve only full consensus problems. They also verify the effectiveness of the proposed protocol through numerical examples on a DC optimal power flow problem.

The final chapter, entitled "A Passivity-Based Design of Cyber-Physical Building HVAC Energy Management Integrating Optimization and Physical Dynamics," by

T. Hatanaka, T. Ikawa, and N. Li, addresses CPS design for enhancing the energy efficiency of HVAC systems in a building with multiple zones. After introducing a thermodynamics model of the building and formulating a set-point optimization problem to balance the human comfort and energy saving, they design a CPS that integrates optimization dynamics with locally controlled physical dynamics. The resulting CPS is interpreted as an interconnection of passive systems, and the system stability and convergence properties of the room temperature optimization can be rigorously proved. The present CPS framework, as extended to multiple connected buildings, is finally demonstrated on a simulator developed through the combination of a variety of software.

1.5 Further Research Opportunities

Compared with the wide perspective described in Sect. 1.3, the topics handled in this book are restricted, and many other research directions remain. One such direction is to develop new markets whose transactions promote more stable and high-quality energy supply–demand balance; in Japan, the wholesale market (spot market) is a central player at present, a regulation market (balancing power market or/and real-time market) is planned, and derivative markets for forward trading, option trading, and so on will be developed as a financial market for the energy sector. This topic is related not only to economically and technically efficient energy supply–demand matching but also to long-term environmental issues arising from both ethical and political points of view related to energy supply and demands. Further discussion of the topic is beyond the scope of this book. This section takes up two specific research topics considered by our group: long-term economic model-based analyses for decarbonization and renewable energy promotion, and systems and control analysis on MASs.

1.5.1 Long-Term Economic Models

The long-term perspective of EMSs is built on long-term policies and predictions of economic growth and climate change. The integrated assessment model of the economy and the climate system (IAM) provides these predictions and suggests appropriate climate policies aimed at maximizing the time-aggregated social welfare or minimizing the total cost to achieve a given target, such as the two-degree target of the Paris Agreement. Future EMSs and energy mixes are deployed on the basis of these predictions. Typically, the so-called carbon budget imposes a ceiling on the use of fossil fuels.

In the IAM, there are two crucial factors that determine the optimal climate policy and the optimal economic growth. One is how we evaluate future generations' welfare, and the other is our expectations of long-term technological progress.

The former is treated as a choice of social discount rate. There has long been debate about what the social discount rate is [11, 38, 75, 91]. It can be understood from an ethical viewpoint, under which a zero discount rate is used. This prescriptive discount rate means no discounting. Under this view, intergenerational ethics suggest that we should treat all future generations with equal concern. Another viewpoint for understanding the social discount rate is the empirical viewpoint, which suggests using the discount rate found by empirical macroeconomic analysis. With this descriptive discount rate, the IAM can predict the well-being of future generations and thereby provide useful information for choosing an appropriate policy.

Shortcomings of the prescriptive discount rate are that it fails to describe the future economy and the recommended policies will be inefficient: Since a different discount rate will actually be used, there is an alternative that achieves the same policy target at a lower social cost [74]. The key problem of the descriptive discount rate is that the recommended policy might be too conservative to avoid the threat of climate change. See [75, Figs. 5–7], which shows the deviation of the optimal GHG emission path from the cost minimization path to achieve the two-degree target.

An attempt to overcome this problem with the descriptive discount rate is to justify the use of a lower discount rate for events in the far distant future. Such a rate is known as a declining discount rate [12]. Ongoing theoretical inquiry considers the huge uncertainty and catastrophic risks associated with climate change [31].

Future technological progress is another crucial factor that determines the optimal climate policy and thus future energy mix and EMS. Acemoglu et al. [1] showed in an endogenous growth model that without intentionally directing technological change toward environmental conservation, the economy may experience environmental catastrophe. Technological change is a result of R&D activities; therefore, investment in R&D is an important part of climate change policy. Yet, in the practice of IAM, technological progress is treated as a given scenario and the extrapolated progress rate is roughly chosen by referencing past experience.

Basically, an IAM is a central planning model expressed in the form of a single-agent optimization problem. A decentralized economy can mimic such a solution if the government implements appropriate policies that correct for market failures and remove market distortion. For global issues such as climate change, however, there is no government that can implement a global-scale policy. The situation is a game (in the sense of game theory). If all countries cooperated, the optimal policy could be pursued via grand coalition. Pessimistically, cooperative game theory finds, in the context of climate change, that only a few countries will agree to a coalition [15]. In the framework of a noncooperative differential game, it is known that there are multiple equilibria in the class of Markov strategies. Dockner and Long [22] illustrate that, in the context of international pollution control, players in the game can choose the most environmentally friendly equilibrium through communication in advance of choice, and the steady state converges to the optimal steady state when the discount rate approaches zero. Recently, Akao et al. [5] strengthened these results: The equilibrium strategy is payoff-dominant when the pollution stock is larger than the steady-state level and the steady state is asymptotically stable worldwide. This result supports increased confidence in the implementation of the Paris Agreement and also supports the deployment of long-term EMSs based on international commitments.

We next give a brief economic analysis of renewable energy and its promotion, based on another long-term economic model called the computable general equilibrium model. Renewable energy is a key to mitigate GHG emissions resulting from power generation and to promote energy security. However, geographical imbalance of renewable energy potential across Japan poses a challenge for its promotion. For example, the island of Hokkaido has a large potential amount of wind power, but the highest demand for electricity is on the main island of Honshu, as shown in Fig. 1.2. Because the number of transmission lines between these two islands is limited, the potential of Hokkaido as a wind power producer is not fully realized. A promising model for evaluating this issue is a long-term computable general equilibrium (CGE) model. Specifically, we disaggregate the power generation sector by including solar and wind power generation and construct a CGE model that accounts for the limitations of the inter-island transmission lines. We investigate how investment in transmission lines between the islands would affect the penetration of renewable energy and, consequently, influence the Japanese economy and GHG emissions [10, 95].

1.5.2 Systems and Control Analysis for Multi-agent Systems

As shown in Sects. 1.3.3 and 1.3.4, synthesis of hierarchical and complex EMSs can essentially be reduced to control problems in MASs. Hence, the creation of a fundamental theory to establish MAS with the structures and functions required for both the present and future eras, aiming toward social applications, is an essential research direction.

Multi-agent consensus should be investigated from several theoretical aspects. In one line of inquiry, the convergence properties of synchronous consensus algorithms under noisy communication are established in [68]. The results are extended to two-layered networks in [70], with two-layered networks allowing both cooperative and antagonistic interactions to be described. The existence of antagonistic agents can also be modeled as a signed graph and/or a bounded confidence update rule, and asynchronous consensus/gossip algorithms of these models are investigated in [71]. Another important model is the network-of-network model, where subsystems of a MAS are themselves a MAS. A convergence property is derived for a class of network-of-networks via algebraic connectivity in [59]. Energy-management-related information technology is also investigated, with secure access to energy service data [51] and an efficient measurement–aggregation of distribution systems [56] established.

In the context of the market economy, research on learning-based decision-making by MAS has been tackled in agent-based computational economics (ACE) [21, 29, 44, 100]. The book [100, Chap. 18] classifies the learning models used in ACE. The learning models include traditional psychology-based models, rationality-based models, and belief learning-based models alongside models based on state-of-the-art artificial intelligence (AI), ML, and biology. ACE as applied to problems in energy markets is excellently summarized in [82, 99]. Although learning algorithms can be applicable in a variety of smart-grid scenarios (see, e.g., [82, Table 1]), the variety of the learning algorithms is very limited. The so-called Roth–Erev reinforcement learning algorithm [27, 84] seems likely to be the most valuable and popular (and possibly only) learning model that has been extensively studied in the smart-grid economics literature. The Roth–Erev algorithm is premised on psychology-based reinforcement learning. This agrees with investigation of transient and/or terminal performance by *human-based* strategic bidders [72], which should be one of the main interests of economists.

From the recent trends discussed in this chapter, the existence of multifarious market participants is seen to be inevitable, and the analysis of the economic and physical impact of this on energy markets is an urgent issue. Similar to the general market economy, an energy market with heterogeneous agents that integrates AI and ML is a promising research direction. From the viewpoint of behavioral economics and computational and memory capacity, both human agents and computer agents behave strategically while satisfying bounded rationality conditions. In contrast to the generalization of human-based bidding behavioral models, the modeling and analyses of specific market agents with marked individuality in oligopolistic energy markets are necessary parts of market power analysis and mitigation in the future. For instance, the reinforcement learning rule that reflects individuality is intended to slightly modify a *standard* objective function for the participants. As mentioned in Sect. 1.3.4, many engineers and natural scientists may think that specifically shaping individual cost functions is a meaningless theoretical discussion in terms of stability and convergence analysis. However, if the modified cost function is treated as a specific bidding behavior of market participants in the real world, creating a novel interpretation that integrates the ideas of behavioral economics and control theory is a valuable and essential research direction for interdisciplinary study.

The redesign of energy markets to include heterogeneous agents and maintain network stability is also a critical research topic. Tesfatsion [99] mentions that promising approaches toward improving the efficiency of the energy market under fundamental limitations are *incentive-based*, *price-based*, and *transactive-based* mechanisms. As shown in the book outline of Sect. 1.4, this book will also present some novel schemes to realize the mechanisms. One of the other options is to use a game-theoretic learning in distributed and cooperative control, an approach pioneered by Marden, Arslan, and Shamma [62] and developed in the systems and control field over the last decade. The fundamental idea in [62] is that the cooperative control problem in MAS can be reduced to the distinctively game-theoretic framework of so-called potential games. The game-theoretic approach offers rich mathematical tools, and potential games can be completely divided into problem formulation and rule learning, which is excellently summarized in [63]. In fact, many useful game-theoretic learning rules are presented (e.g., [36, 63] and the references therein). Cai et al. [18] apply the framework of potential games to the electricity market.

The fundamental theories and technologies presented in this book will also play a central role in cost-effective rebuilding of infrastructure systems to integrate human and social systems (i.e., so-called smart cities) due to the commonality of the economically enabled and engineered networked systems.

Acknowledgements The authors would like to thank Dr. Ken-Ichi Akao, Dr. Shiro Takeda, Dr. Toshi H. Arimura, Dr. Yasumasa Fujisaki, and Dr. Kenji Hirata for invaluable comments and advice on further research opportunities.

References

1. Acemoglu D, Aghion P, Bursztyn L, Hemous D (2012) The environment and directed technical change. Amer Econ Rev 102(1):131–166
2. Ackermann T (2012) Wind power in power systems. Wiley Ltd
3. Afram A, Janabi-Sharif F (2014) Theory and applications of HVAC control systems—a review of model predictive control (MPC). Build Environ 72:343–355
4. Agency for Natural Resources and Energy. Survey of electric power statistics. https://www.enecho.meti.go.jp/statistics/electric_power/ep002/results.html. Accessed 16 Aug 2019
5. Akao K, Uchida K, Wasa Y (2018) International environmental agreement as an equilibrium choice in a differential game. In: Proceedings of the 6th world congress of environmental and resource economists
6. Allcott H (2011) Social norms and energy conservation. J Publ Econ 95(9–10):1082–1095
7. Amin M, Annaswamy AM, DeMarco CL, Samad T (eds) (2013) IEEE vision for smart grid controls: 2030 and beyond. IEEE Press
8. Antsaklis PJ, Baillieul J (eds) (2007) Special issue on the technology of networked control systems. Proc IEEE 95(1)
9. Architecture Council (2015) GridWise trasactive energy framework version 1.0
10. Arimura TH, Takeda S, Onuma H (2018) Double dividend of carbon tax: to solve environmental and economic issues together. Rev Environ Econ Policy Stud 11(2):73–78 (in Japanese)
11. Arrow KJ, Cline WR, Mäler K-G, Munasinghe M, Squitieri R, Stiglitz JE (1996) Intertemporal equity, discounting, and economic efficiency. In: Bruce JP, Lee H, Haites EF (eds) Climate change 1995: economic and social dimensions of climate change: contribution of working group III to the second assessment report of the intergovernmental panel on climate change. Cambridge University Press, pp 125–144
12. Arrow KJ, Cropper M, Gollier C, Groom B, Heal G, Newell R, Nordhaus WD, Pindyck RS, Pizer W, Portney PR, Sterner T, Tol RSJ, Weitzman ML (2013) Determining benefits and costs for future generations. Science 341(6144):349–350
13. Asensio OI, Delmas MA (2015) Nonprice incentives and energy conservation. Proc Nat Acad Sci 112(6):E510–E515
14. Bai X, Wei H, Fujisawa K, Wang Y (2008) Semidefinite programming for optimal power flow problems. Int J Electr Power Energy Syst 30(6–7):383–392
15. Barrett S (2005) The theory of international environmental agreements. In: Maler K-G, Vincent JR (eds) Handbook of environmental economics, vol 3. Elsevier, pp 1457–1516
16. Bolton P, Dewatripont M (2005) Contract theory. The MIT Press
17. Borenstein S, Bushnell J, Stoft S (2000) The competitive effects of transmission capacity in a deregulated electricity industry. Rand J Econ 31(2):294–325
18. Cai D, Bose S, Wierman A (2019) On the role of a market maker in networked Cournot competition. Math Oper Res 44(3):1122–1144
19. Chakrabortty A, Bose A (2017) Smart grid simulations and their supporting implementation methods. Proc IEEE 105(11):2220–2243

20. Chakrabortty A, Ilić M (eds) (2011) Control and optimization methods for electric smart grids, Springer Science & Business Media
21. Chen S-H (2016) Agent-based computational economics—how the idea originated and where it is going. Routledge
22. Dockner EJ, Long NV (1993) International pollution control: cooperative versus noncooperative strategies. J Environ Econ Manage 25(1):13–29
23. Dörfler F, Simpson-Porco JW, Bullo F (2016) Breaking the hierarchy: distributed control and economic optimality in microgrids. IEEE Trans Control Netw Syst 3(3):241–253
24. Doyle JC, Francis BA, Tannenbaum AR (1990) Feedback control theory. Macmillan Publishing Co
25. EirGrid Electricity System Operator for Northern Ireland (SONI) (2018) Annual renewable energy constraint and curtailment report 2017. http://www.eirgridgroup.com/Annual-Renewable-Constraint-and-Curtailment-Report-2017-V1.pdf. Accessed 4 May 2019
26. Electricity and Gas Market Surveillance Commission (EGC) The 33rd technical report. https://www.emsc.meti.go.jp/activity/emsc_system/pdf/033_06_00.pdf. Accessed 13 Apr 2019 (in Japanese)
27. Erev I, Roth AE (1998) Predicting how people play games: reinforcement learning in experimental games with unique, mixed strategy equilibria. Amer Econ Assoc 88(4):848–881
28. Faruqui A, George S (2005) Quantifying customer response to dynamic pricing. Electr J 18(4):53–63
29. Frantz R, Marsh L (eds) (2016) Minds, models and milieux: commemorating the centennial of the birth of Herbert Simon. Palgrave Macmillan
30. Gan L, Li N, Topcu U, Low SH (2015) Exact convex relaxation of optimal power flow in radial networks. IEEE Trans Autom Control 60(1):72–87
31. Gollier C (2008) Discounting with fat-tailed economic growth. J Risk Uncertain 37(2–3):171–186
32. Gomez T, Herrero I, Rodilla P, Escobar R, Lanza S, de la Fuente I, Llorens ML, Junco P (2019) European union electricity markets, current practice and future view. IEEE Power Energy Mag 17(1):20–31
33. Hao H, Lin Y, Kowli AS, Barooah P, Meyn S (2014) Ancillary service to the grid through control of fans in commercial building HVAC systems. IEEE Trans Smart Grid 5(4):2066–2074
34. Hao H, Sanandaji BM, Poolla K, Vincent TL (2015) Aggregate flexibility of thermostatically controlled loads. IEEE Trans Power Syst 30(1):189–198
35. Hatanaka T, Chopra N, Fujita M, Spong MW (2015) Passivity-based control and estimation in networked robotics. Springer
36. Hatanaka T, Wasa Y, Funada R, Charalambides A, Fujita M (2016) A payoff-based learning approach to cooperative environmental monitoring for PTZ visual sensor networks. IEEE Trans Autom Control 61(3):709–724
37. Hayashi Y, Fujimoto Y, Ishii H, Takenobu Y, Kikusato H, Yoshizawa S, Amano Y, Tanabe S-I, Yamaguchi Y, Shimoda Y, Yoshinaga J, Watanabe M, Sasaki S, Koike T, Jacobsen H-A, Tomsovic K (2018) Versatile modeling platform for cooperative energy management systems in smart cities. Proc IEEE 106(4):594–612
38. Heal G (2017) The economics of climate. J Econ Lit 55:1046–1063
39. Herter K (2007) Residential implementation of critical-peak pricing of electricity. Energy Policy 35(4):2121–2130
40. Hobbs BF, Helman U (2004) Complementarity-based equilibrium modeling for electric power markets. In: Bunn DW (ed) Modeling prices in competitive electricity markets. Wiley, pp 69–98
41. IEA Wind Technology Collaboration Programme (2017) Annual report 2017
42. IEEE Smart Grid Domains & Sub-domains. https://smartgrid.ieee.org/domains. Accessed 30 July 2019
43. Ilić MD (2016) Toward a unified modeling and control for sustainable and resilient electric energy systems. Found Trends Electric Energy Syst 1(1–2):1–141

44. Ilić MD, Galiana F, Fink L (eds) (1998) Power systems restructuring: engineering and economics. Springer Science & Business Media
45. Ilić MD, Zaborszky J (2000) Dynamics and control of large electric power systems. Wiley
46. International Renewable Energy Agency (IRENA) (2018) Renewable capacity statistics 2018. https://irena.org/publications/2018/Mar/Renewable-Capacity-Statistics-2018. Accessed 16 Aug 2019
47. Ito K, Ida T, Tanaka M (2018) Moral suasion and economic incentives: field experimental evidence from energy demand. Amer Econ J Econ Policy 10(1):240–67
48. Japan Electric Power eXchange (JEPX). http://jepx.org/english/index.html. Accessed 13 Apr 2019
49. Jessoe K, Rapson D (2014) Knowledge is (less) power: experimental evidence from residential energy use. Amer Econ Rev 104(4):1417–1438
50. Kadokura S, Nohara D, Ohba M, Hashimoto A, Nakao K, Hattori Y, Watanabe T, Hirakuchi H (2018) Japan's R&D project of ramp forecasting technology: deterministic forecast with post-processing using real-time monitoring data. In: Proceedings of 17th international wind integration workshop, 3C-3
51. Kawada Y, Yano K, Mizuno Y, Tsuchiya T, Fujisaki Y (2017) Data access control for energyrelated services in smart public infrastructures. Comput Ind 88:35–43
52. Khargonekar PP, Dahleh MA (2018) Advancing systems and control research in the era of ML and AI. Annu Rev Control 45:1–4
53. Kiani A, Annaswamy A, Samad T (2014) A hierachical transactive control architecture for renewables integration in smart grids: analytical modeling and stability. IEEE Trans Smart Grid 5(4):2054–2065
54. Kleit AN, Terrell D (2001) Measuring potential efficiency gains from deregulation of electricity generations: a Bayesian approach. Rev Econ Stat 83(3):523–530
55. Klobasa M, Haendel M, Pfluger L (2018) Negative market prices and market premium support schemes—impacts on wind integration in the German electricity market. In: Proceedings of 17th wind integration workshop
56. Kojima H, Tsuchiya T, Fujisaki Y (2018) The aggregation point placement problem for power distribution systems. IEICE Trans Fundam Electron Commun Comput Sci E101.A(7):1074–1082
57. Kundur P (1994) Power system stability and control. McGraw-Hill Professional
58. Kuwahata R, Merk P (2017) German paradox demystified: Why is need for balancing reserves reducing despite increasing VRE penetration? In: Proceedings of 16th wind integration workshop
59. Lee H, Nguyen LTH, Fujisaki Y (2017) Algebraic connectivity of network-of-networks having a graph product structure. Syst Control Lett 104:15–20
60. Leibenstein H (1966) Allocative efficiency versus "X-efficiency". Amer Econ Rev 56(3):392–415
61. Levitt SD, List JA (2007) What do laboratory experiments measuring social preferences reveal about the real world? J Econ Perspect 21(2):153–174
62. Marden JR, Arslan G, Shamma JS (2009) Cooperative control and potential games. IEEE Trans Syst Man Cybern 39(6):1393–1407
63. Marden JR, Shamma JS (2018) Game-theoretic learning in distributed control. In: Basar T and Zaccour (eds) Handbook of dynamic game theory. Springer International Publishing, pp 511–546
64. Matsukawa I, Asano H, Kakimoto H (2000) Household response to incentive payments for load shifting: a Japanese time-of-day electricity pricing experiment. Energy J 21(1):73–86
65. Meier A von (2011) Integration of renewable generation in California; coordination challenges in time and space. In: Proceedings of 11th international conference on electrical power quality and utilisation, pp 1–6
66. Mesbahi M, Egerstedt M (2010) Graph theoretic methods in multiagent networks. Princeton University Press

67. Ministry of Economy, Trade and Industry (METI) (2018) Cabinet decision on the new strategic energy plan. https://www.meti.go.jp/english/press/2018/0703_002.html. Accessed 1 Aug 2019
68. Morita R, Wada T, Masubuchi I, Asai T, Fujisaki Y (2016) Multi-agent consensus with noisy communication: stopping rules based on network graphs. IEEE Trans Control Netw Syst 3(4):358–364
69. Nakano M, Managi S (2008) Regulatory reforms and productivity: an empirical analysis of the Japanese electricity industry. Energy Policy 36(1):201–209
70. Nguyen LTH, Wada T, Masubuchi I, Asai T, Fujisaki Y (2018) A consensus protocol over noisy two-layerd networks with cooperative and antagonistic interactions. Asian J Control 20(1):548–557
71. Nguyen LTH, Wada T, Masubuchi I, Asai T, Fujisaki Y (2019) Bounded confidence gossip algorithms for opinion formation and data clustering. IEEE Trans Autom Control 64(3):1150–1155
72. Nicolaisen J, Petrov V, Tesfatsion L (2001) Market power and efficiency in a computational electricity market with discriminatory double-auction pricing. IEEE Trans Evolut Comput 5(5):504–523
73. Njiri JG, Söffker, (2016) State-of-the-art in wind turbine control: trends and challenges. Renew Sustain Energy Rev 60:377–393
74. Nordhaus WD (1999) Discounting and public policies that affect the distant future. In: Weyant JP, Portney PR (ed) Discounting and intergenerational equity, resource for the future, pp 145–162
75. Nordhaus WD (2008) A question of balance: weighing the options on global warming policies. Yale University Press
76. Nudell TR, Annaswamy AM, Lian J, Kalsi K, D'Achiardi D (2018) Electricity markets in the United States: a brief history, current operations, and trends. In: Stoustrup J, Annaswamy AM, Chakrabortty A, Qu Z (eds) Smart gird control: overview and research opportunities. Springer, pp 3–27
77. Organization for Cross-regional Coordination in Transmission system Operators (OCCTO). https://www.occto.or.jp/en/index.html. Accessed 13 Apr 2019
78. OCCTO, The Investigation Committee on the Major Blackout by the 2018 Hokkaido Eastern Iburi Earthquake (2018) The investigation committee on the major blackout by the 2018 Hokkaido Eastern Iburi earthquake, final report
79. Pollitt MG, Shaorshadze I (2011) The role of behavioral economics in energy and climate policy. EPRG working paper 1130
80. Price WC, Weyman-Jones T (1996) Malmquist indices of productivity change in the UK gas industry before and after privatization. Appl Econ 28(1):29–39
81. Ren W, Beard RW (2008) Distributed consensus in multi-vehicle cooperative control. Springer
82. Ringler P, Keles D, Fichtner W (2016) Agent-based modelling and simulation of smart electricity grids and market—a literature review. Renew Sustain Energy Rev 57:205–215
83. Rosenblueth A, Wiener N (1945) The role of models in science. Philos Sci 12(4):316–321
84. Roth AE, Erev I (1995) Learning in extensive-form games: experimental data and simple dynamic models in the intermediate term. Games Econ Behav 8(1):164–212
85. Russel S, Norvig P (2009) Artificial intelligence: a modern approach, 3rd edn. Prentice Hall
86. Samad T, Annaswamy AM (2017) Controls for smart grids: architectures and applications. Proc IEEE 105(11):2244–2261
87. Samad T, Koch E, Stluka P (2016) Automated demand response for smart buildings and microgrids: the state of the practice and research challenges. Proc IEEE 104(4):726–744
88. Sandell NR Jr, Varaiya P, Athans M, Safonov MG (1978) Survey of decentralized control methods for large scale systems. IEEE Trans Autom Control AC-23(2):108–128
89. Shahidehpour M, Yamin H, Li Z (2003) Market operations in electric power systems: forecasting, scheduling, and risk management. Wiley
90. Shamma JS (ed) (2007) Cooperative control of distributed multi-agent systems. Wiley

91. Stern N (2007) The economics of climate change: the Stern review. Cambridge University Press
92. Stoustrup J, Annaswamy AM, Chakrabortty A, Qu Z (eds) (2018) Smart gird control: overview and research opportunities. Springer
93. Suga T, Hayasaki N, Ogimoto K (2018) Japan's R&D project of ramp forecasting technology: project overview. In: Proceedings of 17th international wind integration workshop, 3C-1
94. Suzuki T, Inagaki S, Susuki Y, Tuan A (in press) Design and analysis of distributed energy management systems—integration of EMS, EV, and ICT. Springer
95. Takeda S, Arimura TH, Sugino M (2019) Labor market distortions and welfare-decreasing international emissions trading. Environ Res Econ. https://doi.org/10.1007/s10640-018-00317-4
96. Takenaka H, Nakajima TY, Higurashi A, Higuchi A, Takamura T, Pinker RT, Nakajima T (2011) Estimation of solar radiation using a neural network based on radiative transfer. J Geophys Res 116:D08215
97. Tanaka K, Managi S (2013) Measuring productivity gains from deregulation of the Japanese urban gas industry. Energy J 34(4):181–198
98. Tanaka M (2006) Transmission-constrained oligopoly in the Japanese electricity market. Energy Econ 31(5):690–701
99. Tesfatsion L (2017) Electric power markets in transition: agent-based modeling tools for transactive energy support. In: Handbook of computational economics, vol 4, pp 715–766
100. Tesfatsion L, Judd KL (eds) (2006) Handbook of computational economics: volume 2—agent-based computational economics. Elsevier
101. Thiesen H, Jauch C, Gloe A (2016) Design of a system substituting today's inherent inertia in the European continental synchronous area. Energies 9(8):582
102. van der Veen RAC, Hakvoort RA (2016) The electricity balancing market: exploring the design challenge. Utilities Policy 43-B:186–194
103. Wasa Y, Sakata K, Hirata K, Uchida K (2017) Differential game-based load frequency control for power networks and its integration with electricity market mechanisms. In: Proceedings of 1st IEEE conference on control technology and applications, pp 1044–1049
104. Winston C (1993) Economic deregulation: days of reckoning for microeconomists. J Econ Lit 31(3):1263–1289
105. Wood AJ, Wollenberg BF, Sheblé GB (2013) Power generation, operation, and control, 3rd edn. Wiley
106. Creation of Fundamental Theory and Technology to Establish a Cooperative Distributed Energy Management System and Integration of Technologies Across Broad Disciplines Toward Social Application. https://www.jst.go.jp/kisoken/crest/en/research_area/ongoing/areah24-1.html. Accessed 30 July 2019

Chapter 2
Supply and Demand Balance Control Based on Balancing Power Market

Takao Tsuji, Hyangryul Bae and Jingting Qi

Abstract The supply and demand balance control with frequency regulation has been more and more important in power systems as the integration of variable renewable energy (VRE) is going on. Nowadays, both procurement and operation of control reserves to meet the supply and demand have been traded through the market in many countries in order to economically manage the balancing power with utilizing not only conventional generators but also new resources including the demand response. On the other hand, the power flow congestion caused by the VRE integration is another serious issue. This power flow issue should be also taken into account in activating the balancing power procured at the market. From the above perspectives, the two control strategies are introduced in this chapter after the information about the balancing power market in Japan. First, an optimal operation method of the balancing power based on the market mechanism is given considering the constraints of the power flow congestion. Second, a frequency control by using new resources whose control reserves are not symmetric is introduced. The specific performance of these control strategies is shown by numerical simulations.

2.1 Introduction

The restructuring of the power system has been rapidly progressed in Japan after the Great East Japan Earthquake followed by the nuclear power accident. In order to promote the liberalization in the electric power industry, transmission and distribution sectors are being separated and various new power markets which treat ancillary services have been developed under the national government accordingly.

T. Tsuji (✉) · H. Bae · J. Qi
Graduate School of Engineering, Yokohama National University, 79-5, Tokiwadai, Hodogaya, Yokohama 240-8501, Japan
e-mail: t-tsuji@ynu.ac.jp

H. Bae
e-mail: hyangryul.bae.cx@hitachi-systems.com

J. Qi
e-mail: qi-jingting-kd@ynu.jp

© Springer Nature Singapore Pte Ltd. 2020
T. Hatanaka et al. (eds.), *Economically Enabled Energy Management*,
https://doi.org/10.1007/978-981-15-3576-5_2

Specifically, as of July 2019, it is planned that the centralized capacity market and the balancing power market will be newly started from 2024 to 2021, respectively. In particular, since the balancing power market to adjust supply and demand balance in various time cycles is closely related to this CREST-EMS project, the overview of the balancing power market in Japan is explained in this section.

In many countries, the balancing power market to procure control reserves and energy consists of multiple different time scales. They are typically divided into primary, secondary, and tertiary controls, for example, in Germany [12]. The reasonable procurement of the balancing power can be realized by this multi-time scale market mechanism because conventional generators and other control resources have different controllability. For example, control resources whose load or generation can be controlled rapidly are suitable for primary or secondary control, while control resources whose generation costs are low should be used for tertiary control even if their control speeds are low. Considering the ongoing framework of the supply and demand balance control, it is planned that the balancing power market in Japan consists of five products as shown in Table 1.1 in Chap. 1. The tentative schedule to start the balancing power market in Japan is shown in [17]. The balancing power market for procurement will be started with Replacement Reserve for Feed-In-Tariff (RR-FIT) whose response time is the shortest in 2021. The markets for the other control reserves will be started step by step after the RR-FIT. It was decided that cross-regional trades would be implemented based on the balancing power market as for Frequency Restoration Reserve (FRR), Replacement Reserve (RR), and RR-FIT, while the availability of this cross-regional trade is now still being discussed as for the other balancing power because Frequency Containment Reserve (FCR) and Synchronized Frequency Restoration Reserve (S-FRR) are important to keep frequency stability in the case of the power system separation. Namely, the risk of insufficient control reserves for governor-free or load frequency control has to be carefully taken into account when a connection between control areas is lost if a large part of the control reserves is located in a certain control area. Here, it should be noted that sufficient margins on cross-regional interconnection lines are needed to realize the cross-regional procurement of the balancing control reserves. It is still being discussed how much the margin should remain for the balancing control reserves since it affects the cross-regional power trade at spot markets.

2.2 Power Trade and Locational Marginal Price

As the VRE integration is going on, the power flow condition changes accordingly. The entire power network in Japan has been conventionally divided into 10 areas, and the zonal pricing method has been applied considering the power flow congestion on cross-regional interconnection lines. In particular, when a large amount of VRE is introduced to a certain area, the power flow congestion often becomes a serious problem on the interconnection lines. Specifically, in Japan, the cumulative amount of installed photovoltaics (PV) has already reached 50 GW as of 2018, and the rate

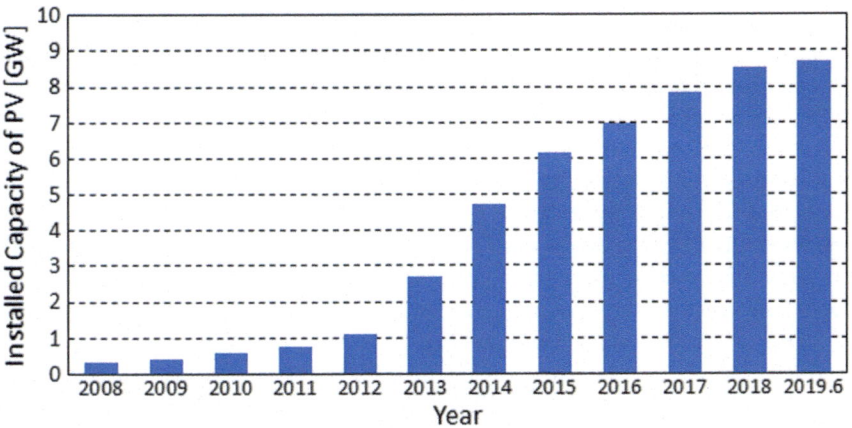

Fig. 2.1 Cumulative installed capacity of PV in Kyushu area [13]

of the PV capacity to the peak demand is already critical values in several areas mainly in westernmost Kyushu and Shikoku areas. Figure 2.1 shows the cumulative installed capacity of PV in Kyushu area. The installed capacity has reached 8.7 GW as of June 2019 although the off-peak demand is around 8 GW. As the capacity of the interconnection between Kyushu and Chugoku areas is around 2 GW, it becomes a more challenging issue to meet the supply and demand even with exporting excess power under the available transfer capability constraint. Also, in [9], Kyushu area is categorized to phase 3, "flexibility is key," in the integration study of VRE by International Energy Agency (IEA).

Consequently, the first case of PV output curtailment in Japan other than remote islands was conducted in Kyushu area in October 2018 [8]. As of 2019, it is ruled in Japan that Transmission System Operator (TSO) is able to send the curtailment signal to VRE only when the oversupply cannot be avoided by the other countermeasures [18, Article 174]. The outline of the specific procedure is as follows:

Step 1 Based on the forecasted demand and VRE output, the generation schedule is decided the day before.
Step 2 When the oversupply is forecasted the day before, the scheduled outputs of conventional generators are reduced to resolve the oversupply. Here, the control reserve in downward direction has to be taken into account.
Step 3 If the oversupply has not been resolved yet with the minimum output of conventional generators, pump-up hydro-generators are operated with pump-up operation mode to increase the demand as high as possible. Moreover, the excess power is exported to neighboring areas through cross-regional interconnection lines as much as possible.
Step 4 If the oversupply has not been resolved with the above countermeasures yet, TSO is able to send the control order for curtailment to VRE.

As shown in this example about the curtailment of VRE output, non-uniform installations of VRE may cause power flow congestions on interconnection lines. On the other hand, the wind power (WP) integration is also going on in Japan. Compared to PV, WP is introduced to specific locations with good wind energy resources. Due to the penetration of both PV and WP, the power flow congestion becomes a serious problem even inside the control areas. For example, in Tohoku area, power flow in the north-south direction has been increasing, and the control area is sometimes divided into two areas due to the congestion on corresponding transmission lines. As shown in this case, it is expected that more power flow congestion cases will appear inside the control areas in the near future with much more VRE. One of the countermeasures to avoid the overloading on transmission lines is to apply the locational marginal price (LMP) to power trade as already applied in PJM interconnection, one of regional transmission organizations in the USA. Given the different marginal prices in the different locations, the power consumption in the area with high price is naturally reduced and vice versa. As a result, the power flow could be indirectly controlled.

The LMP has been applied to the day-ahead or intraday spot markets [21, 14]. For example, the effectiveness of the LMP on the congestion management was discussed in [15] supposing the demand response works based on the LMP. Moreover, as mentioned in the previous section, the balancing power market is now being developed in Japan, and the congestion issue should be properly taken into account in operating the balancing power. Hence, in this chapter, a research activity on the application of LMP to the balancing power operation is introduced based on [3].[1]

2.3 Supply and Demand Balance Control Considering Power Flow Congestion

It is assumed that a large part of power trades has been decided through day-ahead and intraday spot markets. Generators whose marginal costs are low are generally cleared first, and their generation output are often fixed to the rated output without control reserves. On the other hand, generators with higher marginal costs are not necessarily operated at the rated generation output, and they might have control reserves. The main aim of this paper is how to activate those control reserves together with the demand response for the tertiary balancing control supposing the control reserves were procured in the balancing power market in terms of capacity in advance. If the power trade at the spot markets were conducted ideally, the marginal costs of generators which have control reserves are the same to minimize the total generation cost although their fuel cost functions are different from each other. Then, supposing that all the balancing power providers have the same marginal cost with different marginal cost functions, an optimal operation method of the balancing power in terms of energy based on the LMP is discussed here as shown in Fig. 2.2.

---

[1]This section is based on the Ref. [2] which was presented in IEEJ Transactions on Power and Energy.

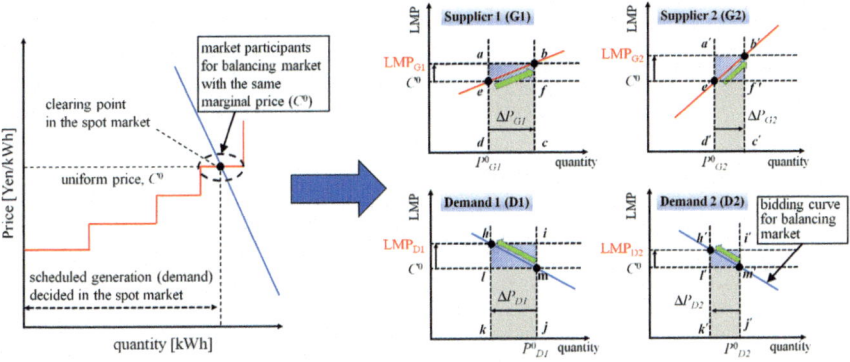

Fig. 2.2 Concept of the proposed balancing power market [3]

LMPs at all the nodes are regarded as decision variables in the proposed method. Given the LMP, the dispatch for the generation and demand response is automatically decided based on their bidding curves as shown in Eqs. (2.1–2.3).

$$\mathbf{C}^k = [C_{Gi}^k \dots C_{Di}^k \dots] \quad Gi \in \mathbf{N}_G, \, Di \in \mathbf{N}_D \tag{2.1}$$

$$P_{GSi}^k(C_{Gi}^k) = \frac{C_{Gi}^k - b_{Gi}}{a_{Gi}} \tag{2.2}$$

$$P_{DSi}^k(C_{Di}^k) = \frac{C_{Di}^k - b_{Di}}{a_{Di}} \tag{2.3}$$

here,

$\mathbf{C}^k$	Vector of LMP at step k
C_{Gi}^k, C_{Di}^k	LMP at generator or demand bus i at step k
$\mathbf{N}_G, \mathbf{N}_D$	Set of balancing power suppliers and customers
P_{GSi}^k, P_{DSi}^k	Generation or load dispatch for player i at step k
$a_{Gi}, a_{Di}, b_{Gi}, b_{Di}$	Price elasticity coefficients of player i ($a_{Gi} > 0$, $a_{Di} < 0$).

Based on this relationship between LMP and generation or demand dispatch, the desired operation of the balancing power can be formulated as the following optimization problem:

$$\min \sum_i^{N_G} \Delta C_{Gi}^k \Delta P_{GSi}^k + \sum_i^{N_D} \Delta C_{Di}^k \Delta P_{DSi}^k$$
$$\Delta C_{Gi}^k = C_{Gi}^k - C_{Gi}^{k0}$$
$$\Delta C_{Di}^k = C_{Di}^k - C_{Di}^{k0} \tag{2.4}$$
$$\Delta P_{GSi}^k = P_{GSi}^k - P_{GSi}^{k0}$$
$$\Delta P_{DSi}^k = P_{DSi}^k - P_{DSi}^{k0}$$

subject to

$$\sum_{i}^{N_G} P_{GSi}^k (C_{Gi}^k) + \sum_{i}^{N_D} P_{DSi}^k (C_{Di}^k) + \sum_{i}^{N_W} P_{Wi}^k = 0 \tag{2.5}$$

$$F_{l,\min} \leq \sum_{i}^{N_G} A_{l,Ni} P_{GSi}^k (C_{Gi}^t) + \sum_{i}^{N_D} A_{l,Ni} P_{DSi}^k (C_{Di}^k) + \sum_{i}^{N_W} A_{l,Ni} P_{Wi}^k \leq F_{l,\max} \tag{2.6}$$

$$\sum_{i}^{N_G} SR_{20,\text{up},i} \geq SR_{20,\text{up},\min} \tag{2.7a}$$

$$SR_{20,\text{up},i} = \begin{cases} \text{ramp}_{20,i} & \text{if } P_{GSi}^k + \text{ramp}_{20,i} \leq P_{Gi,\max} \\ P_{Gi,\max} - P_{GSi}^k & \text{otherwise} \end{cases} \tag{2.7b}$$

$$\sum_{i}^{N_G} SR_{20,\text{down},i} \geq SR_{20,\text{down},\min} \tag{2.8a}$$

$$SR_{20,\text{down},i} = \begin{cases} \text{ramp}_{20,i} & \text{if } P_{GSi}^k - \text{ramp}_{20,i} \geq P_{Gi,\min} \\ P_{GSi}^k - P_{Gi,\min} & \text{otherwise} \end{cases} \tag{2.8b}$$

here,

N_G, N_D	Number of players as generator and demand response in the balancing power market
$\Delta C_{Gi}^k, \Delta C_{Dj}^k$	Variation of LMP from the original value at node i with a generator and demand response at step k
P_{GSi}^0, P_{DSi}^0	Initial value of dispatch to generator or demand i
P_{Wi}^k	Generation output of wind power plant i
N_W	Number of wind power plants in the system
$F_{l,\max}, F_{l,\min}$	Upper and lower limits of power flow on transmission line l
N_G, N_D, N_W	Generator, load, and wind power plant
$A_{l,Ni}$	Sensitivity of power flow on line l to nodal power at node Ni
$SR_{20,\text{up},i}, SR_{20,\text{down},i}$	Secondary control reserve to increase and decrease generation by generator i
$SR_{20,\text{up},\min}, SR_{20,\text{down},\min}$	Upper and lower limits of secondary control reserve to increase and decrease generation by generator i
$P_{Gi,\max}, P_{Gi,\min}$	upper and lower limits of generation output of generator i
$\text{ramp}_{20,i}$	Ramp rate limit in 20 min of generator i.

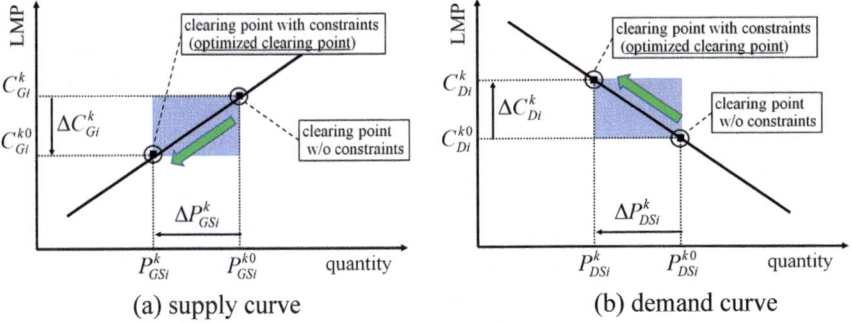

Fig. 2.3 Objective function [3]

Equation (2.4) represents the objective function in this optimization problem, the total change of surplus of generators and loads. Supposing the bidding prices by generators or demands are continuous straight lines just for simplicity and clearing prices at different locations are separately decided, namely Pay-as-Bid auction method is applied from a viewpoint of the balancing control in the entire grid, $\Delta C_{Gi}^k \Delta P_{GSi}^k$ and $\Delta C_{Di}^k \Delta P_{DSi}^k$ represent the payment to the market player i as shown in Fig. 2.3. Therefore, to minimize the total payment to all the players is defined as the objective function. If it is possible to minimize this equation without any other constraints other than the supply and demand balance, the imbalance can be compensated based on the merit-order list of all the available generators and demand response providers.

Equations (2.5)–(2.8a) are constraints in this optimization problem. First, Eq. (2.5) represents the supply and demand balance constraint. Reference dispatch signals for generation and demand have to be decided to meet the supply and demand balance including the wind power output. Equation (2.6) represents the congestion constraint based on the linearized power flow calculation. Given the dispatch signals for generators and demands, the power flow on transmission lines is automatically determined. Equations (2.7a) and (2.8a) are constraints on secondary control reserves in upward and downward directions, respectively. Since it is supposed in this chapter that the secondary control reserves are provided by generators only, the control reserves have to still remain after the generation dispatch decided by this optimal tertiary control has been realized as shown in Fig. 2.4. Here, the new generation dispatch, P_{Gi}^{k+1}, is calculated at time t, and the control reserve of this generator after this control has been finished at time $t + \Delta t$ is defined as $P_{Gi,\max} - P_{Gi}^{k+1}$. If this control reserve is big enough, the control reserve for the secondary control is decided mainly by the ramp rate limit as shown in the figure. Supposing the secondary control has to be activated within 20 min, the secondary control reserve provided by this generator is defined as the maximum variation of generation output in 20 min, $SR_{20,\text{up},i}$ or $SR_{20,\text{down},i}$. The total amount of these reserves provided by all the generators is defined as the secondary control reserve in this network.

This optimization is mainly based on the nonlinear objective function with the linear constraints with upper and lower limits. As it is not so hard to solve this problem,

Fig. 2.4 Secondary control
reserve constraint [3]

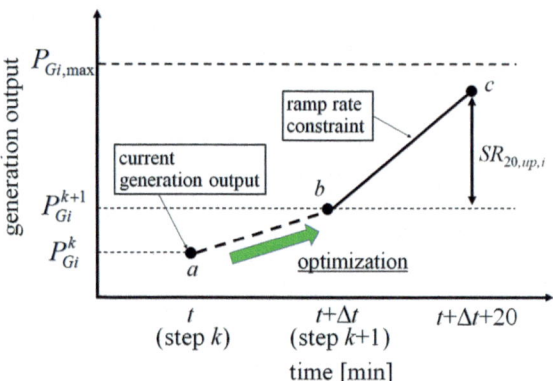

we assumed that reference dispatch signals are updated every minute and sent to all
the generators and demand response providers from the TSO.

In order to simulate both the system frequency and the power flow on transmission
lines at the same time, the combined simulation model consisting of the dynamic frequency control and the power flow calculation was developed. Thus, Fig. 2.5 shows
the simplified frequency control model combined with the power flow calculation
used in this study. The frequency control part consists of primary, secondary, and
tertiary controls. The generation dispatch calculated by the proposed method is used
as the reference signal of generation output, P_{MSi}^{ref}, corresponding to the conventional economic dispatch control (EDC). The supply and demand imbalance can be
compensated by the primary and secondary controls which were modeled as the proportional and integral logics with first-order lag elements and the rate limiter as
shown in Fig. 2.5. The generation output from each generator is calculated based on
the phase angle difference between buses by using the power flow calculation. Here,
Direct Current (DC) method was applied to the power flow calculation because only
active power flow is the scope of this analysis and the voltage–reactive power control
can be ignored just for simplicity. When the mechanical input does not meet with

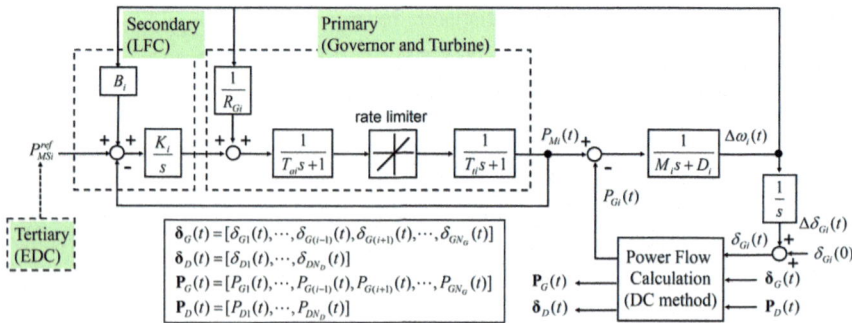

Fig. 2.5 Simplified frequency control model with power flow calculation [3]

the electrical output, the rotational speed of the generator changes due to the acceleration or deceleration energy. For example, when the mechanical input is increased, the generator is accelerated by the acceleration energy and the phase angle of the generator bus also increases. As the electrical output normally increases when the phase angle of the bus increases, the acceleration of the generator gradually converges to a new equilibrium point. While the above dynamic characteristic is modeled as for generators, the load control by demand response providers is simply modeled as the first-order lag element only for the tertiary control supposing the demand response does not contribute to the primary and secondary controls. Here, it should be noted that demand response providers have already contributed to the primary and secondary controls in many real cases, however, the primary and secondary controls by the demand response were ignored for simplicity in this chapter. Figure 2.6 shows the flowchart of the entire simulation. The dynamic frequency simulation is working with the step size of 0.1 s, while the LMP is updated every minute based on the proposed optimization technique.

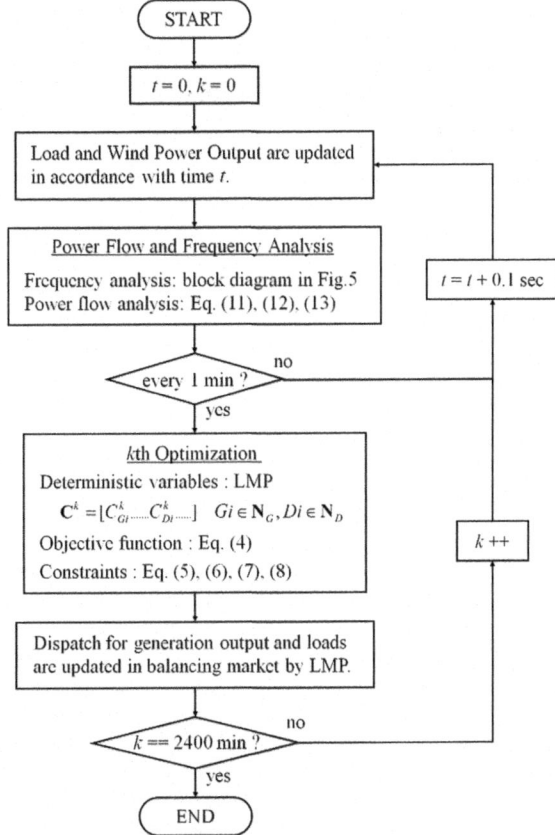

Fig. 2.6 Flowchart of the proposed method [3]

Fig. 2.7 IEEE 39 bus system model with wind power plants [3]

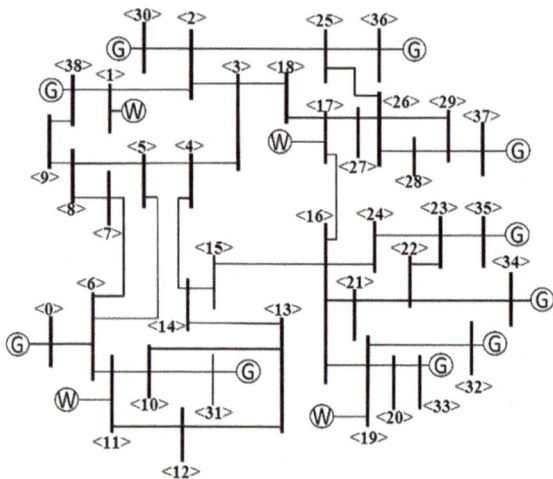

IEEE-39 New England System model [7] with four wind power plants shown in Fig. 2.7 was used for the numerical simulation. The entire simulation time is set to 40 h, and wind power fluctuation data opened at the Web site of 50 Hertz in Germany [23] are used for the four wind power plants as shown in Fig. 2.8. Although the time interval of the original data was 15 min, the short time cycle component was superposed to this data based on the proposed method in [11]. The aggregated

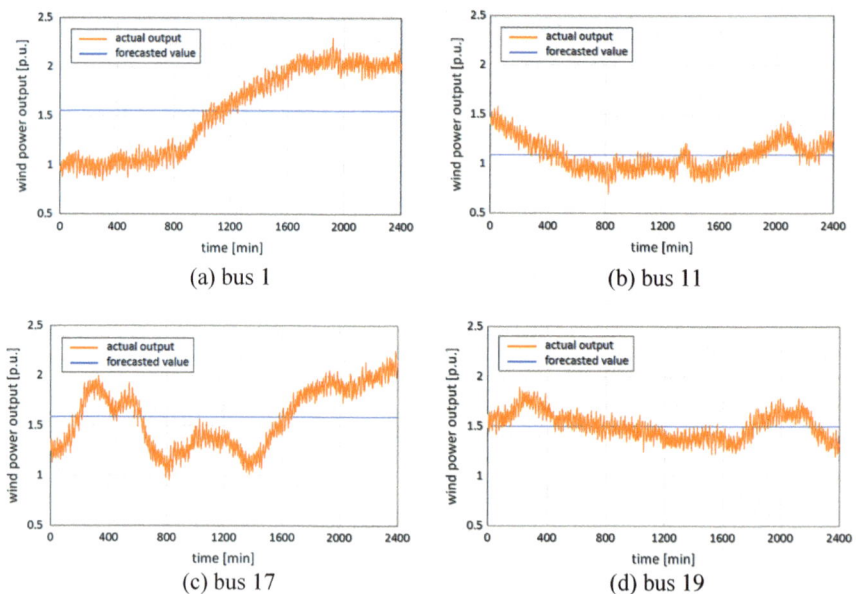

Fig. 2.8 Wind power fluctuation at bus 1, 11, 17, and 19 [3]

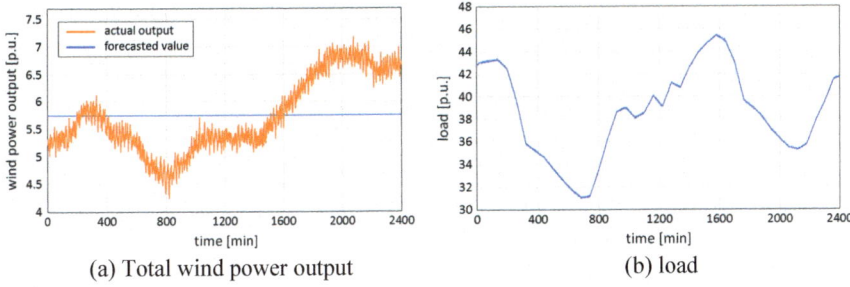

Fig. 2.9 Total wind power output and load [3]

total wind power output and the entire load change are shown in Fig. 2.9. Here, this load data is cited from the actually measured demand curve opened at the Web site of Hokkaido Electric Power Company in Japan [6] and that is assigned to all the load buses in proportional to their capacities. Although details about the daily operation such as unit commitment have to be decided to simulate realistic daily operations, the outputs of all the conventional generators are adjusted to follow the imbalance correctly in terms of the daily operation whose control cycle is longer than that of the tertiary control supposing the perfect forecasting is available for simplicity.

Figure 2.10 shows the simulation result about output of two conventional generators. Their generation output were adjusted to compensate the imbalance, and some irregular changes are seen to satisfy all the constraints. Around this daily generation curve, the high-frequency component of generation output can be seen due to the primary and secondary controls to regulate the system frequency. As a result, as shown in Fig. 2.11, the frequency fluctuation caused by the high-frequency component of the wind power fluctuation was properly controlled around 50 Hz with the maximum frequency deviation from 50 Hz less than 0.15 Hz.

The power flow changes on transmission lines are shown in Fig. 2.12 by focusing on the two important lines only where the congestion was often caused by the wind power fluctuation. When the power flow congestion occurs, the LMP was adjusted as shown in Fig. 2.13 in order to avoid the overloading. Consequently, the congestions on

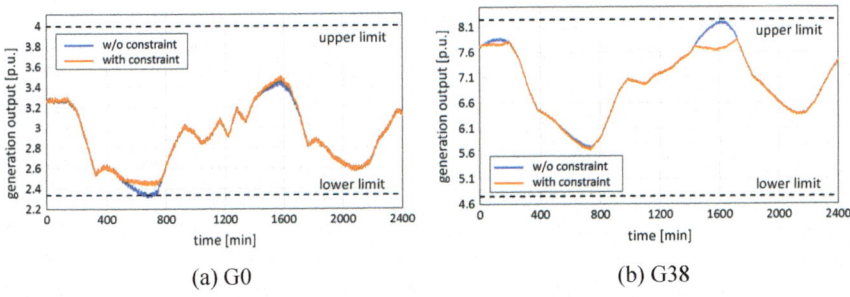

Fig. 2.10 Generation output [3]

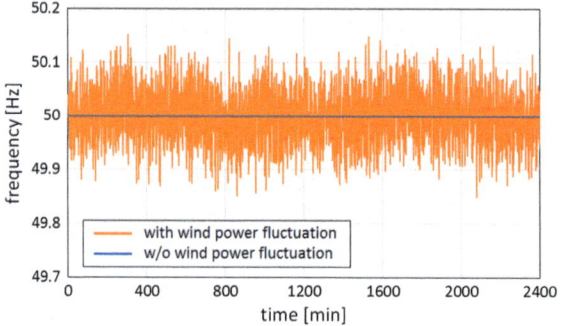

Fig. 2.11 Frequency fluctuation [3]

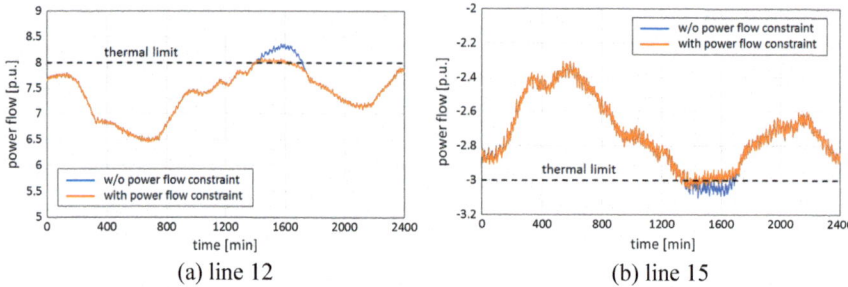

(a) line 12   (b) line 15

Fig. 2.12 Power flow on transmission lines [3]

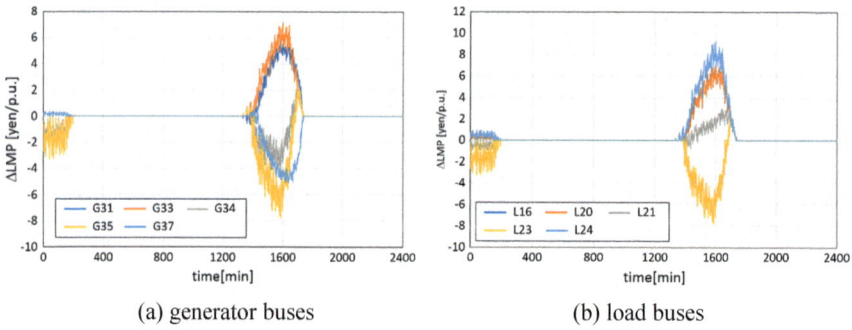

(a) generator buses   (b) load buses

Fig. 2.13 LMP adjusted by the proposed method [3]

those two lines were successfully avoided as shown in the figure. Finally, Fig. 2.14 shows the total of the secondary control reserves in both upward and downward directions. Here, the lower limit of the secondary control reserve was set to 10% of the total demand on trial. The proposed method worked effectively and the LMPs at the multiple buses were properly adjusted so that the secondary reserve constraint

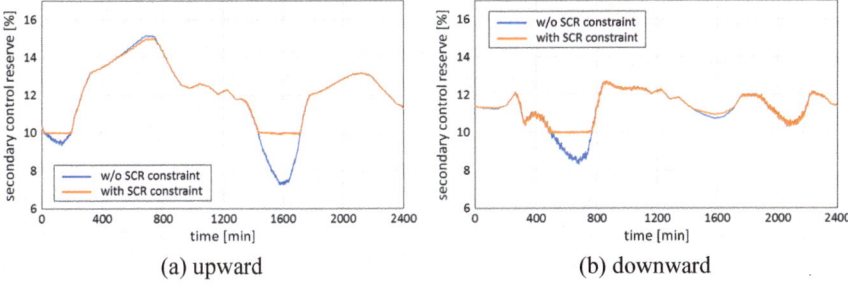

(a) upward   (b) downward

Fig. 2.14 Total secondary control reserves [3]

was successfully satisfied. Please see [3] for further details about this simulation results.

As above, the balancing power operation method based on the LMP considering the congestion management on transmission lines was introduced and tested through the numerical simulation. The tertiary control is the main scope of this paper, and the LMPs at all the nodes are precisely determined by using the optimization technique. Here, it is also taken into account to procure the secondary control reserve in order to properly maintain the system frequency in addition to the thermal limit constraints of the transmission lines. The areas of the future works are as follows:

- Not only the operation but also the procurement of the balancing control reserves should be considered for more economic operations. The coordinated procurement and operation of balancing power should be discussed.
- The power flow is affected by the primary and secondary controls in addition to the tertiary control which was mainly analyzed in this chapter. In particular, the primary and secondary controls are directly related to the frequency stability of the system, and the sufficient reserve capacity has to be secured for those controls. The proposed method should be enhanced in order to optimize the power flow based on all the balancing power.
- The congestion management becomes a more serious issue when any transmission line tripped due to any faults. The proposed method should be updated so that the power flow change caused by the fault can be considered.
- The supply and demand curves in the market are modeled as continuous straight lines in this chapter. However, they are generally discontinuous stepwise curves in the real markets. The simulation model has to be more realistic to obtain more credible results.

2.4   Asymmetric Procurement of Balancing Control Reserves

It was assumed in the previous section that both conventional generators and demand response providers could offer both upward and downward balancing control reserves. As the integration of VRE is going on, the VREs have participated in the balancing power market as players [1]. In many balancing power markets, the upward and downward balancing control reserves are separately treated. For example, in Japan, only the upward control reserve will be traded for the time being because many generators are operated around the rated output and sufficient downward control reserves could be procured consequently. In this section, a combined frequency control strategy consisting of multiple asymmetric balancing control reserves is introduced based on [20].[2]

2.4.1   Balancing Control Reserves by VRE

(a)   Photovoltaic

Generation output from VRE basically depends on weather conditions. However, both PV and WP can reduce their output arbitrarily, namely, the active power curtailment is available. PV is generally connected with the grid via a power conditioning subsystem (PCS) consisting of a DC/DC converter and a DC/AC inverter. Here, the DC voltage imposed on the PV module could be controlled by the DC/DC converter, and the maximum power point tracking (MPPT) control is realized by this DC voltage control based on the I-V characteristic of the PV panel. The active power curtailment is available with shifting the operating point rightward along with the P-V curve from the optimal point found by the MPPT control. Based on this active power curtailment, there are various ways to contribute to the balancing control.

A fundamental way is to reduce the PV output in the case of the oversupply. This concept can be applied to various time scales from the primary to the tertiary controls. In the case of the application to the tertiary control, TSO decides the amount and timing of the curtailment when the oversupply or any other issue (overloading in transmission line and so on) is forecasted in advance. PV systems will apply the curtailment when they receive the control signal from the TSO. On the other hand, in the case of the primary control, the active power curtailment could be applied based on the locally measured frequency as each PV could roughly understand the supply and demand balance in the entire AC network from the local frequency information. When the active power curtailment is applied beforehand, the PV could contribute to not only the over-frequency but also under-frequency cases since the PV has upward

---

[2]This section is based on the Ref. [20] which was presented in 16th Wind Integration Workshop organized by Energynautics.

control reserve [19]. Also, this control reserve can be used to smooth the rapid change of the PV output when the weather condition changes so fast.

(b) Wind Turbine

There are mainly four types of connection methods of wind turbines, which are defined in [2, Chap. 4]. In the conventional connection type I, an induction generator was directly connected to the grid resulting in no controllability in the rotational speed. It is well known that the power coefficient depends on the rotational speed of the turbine. Therefore, the rotational speed becomes controllable for the MPPT control in the case of modern wind turbines such as Doubly Fed Induction Generator (DFIG) or DC link, corresponding to type III and IV. The active power curtailment is available by shifting the rotational speed from the optimal value as well as the pitch angle control of the blades.

As the wind turbine has a rotating generator and blades, the kinetic energy is normally stored in the rotating mass. Consequently, wind turbines are able to increase their generation output temporarily by slowing down the rotational speed even when the wind speed does not change. Based on this principle, the wind turbines are able to contribute to mitigating sudden frequency drops in short time by increasing the generation output. This control concept is called temporal power surge, one of the inertial response controls [5].

Two control strategies were taken into account in this section. One is the Delta control based on a droop characteristic [2, Chap. 11]. WP output is reduced by a certain amount, Δ, in advance and is controlled to compensate frequency fluctuations based on the droop characteristic within the control margin. It is desired to reduce the amount of Δ as much as possible in order to save the active power curtailment. In this section, the control response of this Delta control is modeled as the first-order lag element considering the mechanical delay supposing this control strategy is realized by the pitch control. The other is the temporal power surge as one of the inertial response controls. Figure 2.15 shows a block diagram for this control that was developed by authors based on [20]. Since the wind turbines have the kinetic energy in the rotating mass, it is possible to increase their output by reducing their rotational speed, namely, consuming the stored kinetic energy [22]. The magnitude of the power surge depends on the droop characteristic and the system frequency, and a dead zone is set around the normal frequency. Consequently, only when the system frequency becomes smaller than the threshold value, the temporal power surge works to increase the system frequency. Here, it should be noted that the control mode can be shifted into the recovery mode when the rotational speed reaches the threshold value. After the recovery control starts, the temporal power surge does not work until the rotational speed is almost restored to the optimal rotational speed.

(c) Electric Vehicle

It is expected that rechargeable batteries could compensate for the supply and demand imbalance by charging and discharging controls. There are various projects in the world to install large-scale rechargeable batteries to verify their effectiveness. From

Fig. 2.15 Control block for temporal power surge of wind turbine

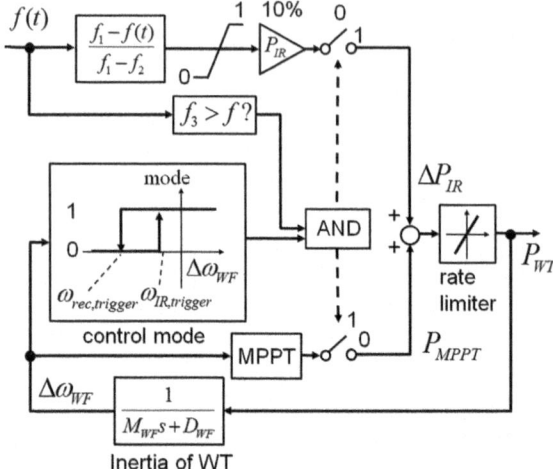

Inertia of WT

a viewpoint of the capital investment cost, the utilization of rechargeable batteries of EVs is effective supposing EV owners contribute to the balancing control with receiving proper economic incentives. Here, the State of Charge (SOC) of EV has to be carefully taken into account in the charging and discharging controls. For example, it is natural that the EV owners dislike that the SOC is reduced by the frequency control when the EV is being charged. Also, the overcharging has to be avoided in order to protect the rechargeable battery. From the above perspectives, it is assumed that EVs can be classified into the following three groups according to the SOC.

Group 1   SOC is over 90%
Group 2   SOC is from 80 to 90%
Group 3   SOC is less than 80%.

Supposing the desirable range of the SOC is from 80 to 90% to its rated capacity, EV owners belonging to Group 1 could cooperate to discharging controls because the SOC becomes closer to the desirable range. Similarly, it is assumed that Group 2 could agree with charging control signals with improving its SOC, while Group 2 contributes to both the charging and discharging controls. Hence, in this study, it is supposed that the discharging and charging control signals can be given to "Group 1 and 2" and "Group 2 and 3," respectively. However, only when the frequency fluctuation is beyond another threshold value that is bigger than the others, it is supposed to be mandatory to contribute to the frequency regulation regardless of the SOC. In other words, all the groups contribute to both the charging and discharging controls only in this emergent condition with the large frequency deviation.

Specific control signals for EVs consist of primary and secondary control signals. The primary control signal is generated by a droop characteristic. Here, it is assumed that 50% of the rated capacity is available for this primary control at maximum. Here, the control signal is calculated by each EV with observing the frequency change at

Fig. 2.16 Coordination of multiple balancing power

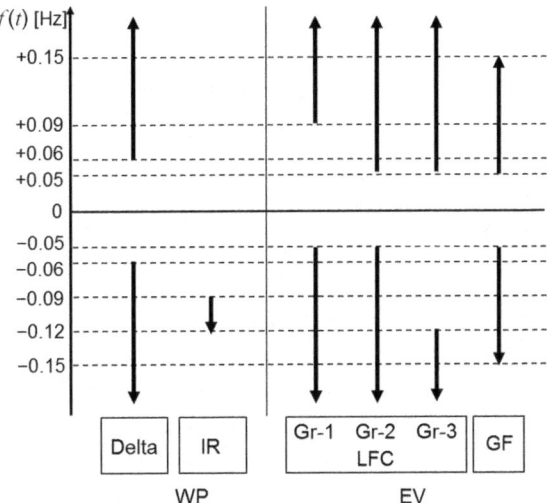

its terminal bus, namely decentralized control. On the other hand, load frequency control (LFC) signals are calculated and sent to EVs by the TSO like centralized operation based on the high-frequency component of the imbalance as the secondary control signal.

(d) Coordinated Control Strategy

It is needed to put priorities to all the balancing control reserves in order to obtain a good control performance in terms of the minimization of the active power curtailment and the SOC management. To this end, the different dead zones are set for the above control reserves as shown in Fig. 2.16. The charging and discharging control of EV other than the emergency mode works first to improve both the frequency and the SOC because the utility of EV owner does not decrease since the SOC is expected to be gradually improved throught the control strategy mentioned above. Here, the impact of the frequent charging control on its lifetime is ignored. The second priority is the droop control strategy of WP based on the Delta control supposing the active power curtailment was applied in advance to procure the upward control reserve. It is expected that the required amount of the active power curtailment for this Delta control can be reduced by giving lower priority compared to the charging and discharging control of EV. The third one is the temporal power surge by WP because the duration time of the control is limited to ten or several ten seconds. Finally, when the frequency deviation from the normal value becomes more significant, the primary and secondary control signals can be sent to another EV group that has not worked in order not to get the SOC worse. This option should be avoided as much as possible; however, it is expected that the SOC does not deteriorate even with this emergent control because the implementation of this control is activated only when the frequency fluctuation is particularly big and the SOC will be improved on average throughout the entire simulation time.

2.4.2 Simulation Model and Results

(a) IEEJ AGC30 Model and Simulation Conditions

IEEJ, the Institute of Electrical Engineers of Japan, newly developed a recommended practice for automatic generation control, AGC30, in 2016 to provide a common base of the simulation model for the supply and demand balance control with the frequency regulation because this simulation has been more and more important these days due to the penetration of VRE. Figure 2.17 shows an overview of the AGC30 model which consists of control block diagrams, algorithms, and typical time-domain fluctuation data of load and VRE output. The block diagrams cover conventional generation plants, load frequency control (LFC), system inertia, and tie-line power flow. Economic dispatch control (EDC) is represented as the algorithm based on the lambda-iteration method [4]. It is possible to verify or compare various control methods to stabilize the frequency fluctuation based on this model. The proposed coordinated control strategy was tested by using this model with the following simulation conditions (see [10] for details about this model).

As shown in Table 2.1, the three hours' data (from 11 to 14) of load curve in an off-peak season included in this model was used for numerical simulations because the frequency regulation issue becomes more serious in this season with a smaller amount of the LFC capacity. The maximum load is around 11 GW, while 3.2 GW of PV and 2.8 GW of WP are installed in this system. Figure 2.18 shows the fluctuation

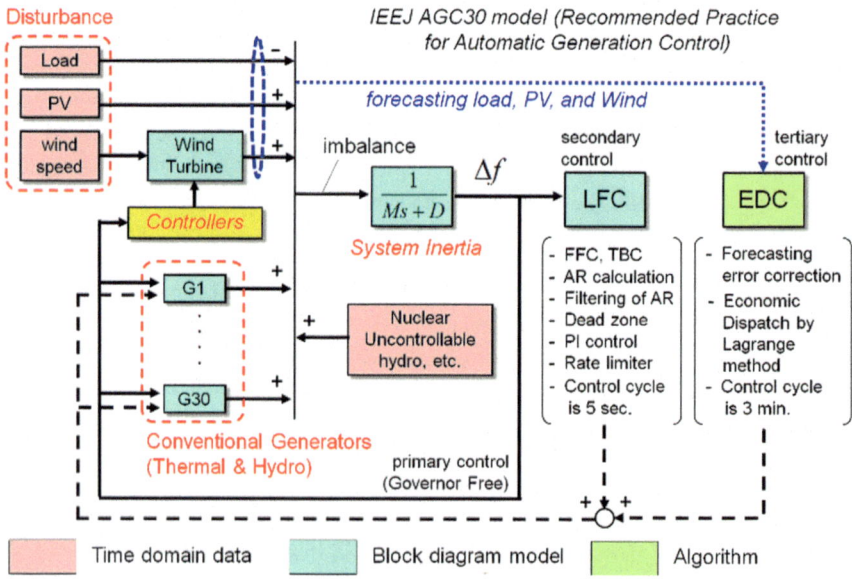

Fig. 2.17 Overview of AGC30 model with controllers for wind power and EV

Table 2.1 Simulation conditions

Simulation time	11:00–14:00 in off-peak season
Demand	Light load pattern in AGC30
Load prediction error	3%
PV	Ramp-down case in AGC30
PV prediction error	Before 12 a.m.: 9%, Other: 37% (MAE%)
Wind power	Modified ramp-down case in AGC30
WP prediction error	RMSE10%
Reserve capacity	15%
LFC capacity	2%
GF capacity	3%

data of load and VRE used in this simulation. Here, the original data of WP output included in the AGC30 was proportionally increased to 2.8 GW without considering the smoothing effect [16] because the degree of the smoothing effect depends on various geographical characteristics and is not easily predicted.

The original AGC30 model consists of 37 generators: nuclear, uncontrollable hydro, coal, LNG, combined cycle (GTCC), and oil. The list of available generators is shown in Table 2.2. The scheduled generation output is decided in advance to meet the supply and demand based on the forecasted load and VRE output. However, the imbalance caused by forecasting error and short-period fluctuations of the loads and VRE output is inevitable, and the frequency change can be calculated with the imbalance and the inertia model in AGC30 model. The governor-free control of conventional generators works first based on locally measured frequency to mitigate the frequency change. Also, three generating plants, one LNG and two GTCC, work for LFC (around 200 MW of LFC capacity is secured by these generators), and other thermal generating plants are following EDC control signals as shown in Table 2.2. Area Requirement (AR) is calculated for the LFC based on the frequency deviation from the normal value, system constant, low-pass filter, and PI controller. The AR is distributed as the LFC signal to the LFC generators in proportional to their control speeds (see [10] for more details). A part of this LFC signal will be assigned to EVs as well as the LFC generators in the case of the proposed method.

(b) Simulation Results

The numerical simulation was conducted by using MATLAB/Simulink in the following two cases. The control performance is evaluated by using "sojourn rate," the ratio of the duration time when the frequency is within ±0.1 Hz to the entire simulation time. In 60 Hz network in Japan, the requirement of the frequency regulation is to keep the sojourn rate at least 95% with the maximum frequency deviation from the normal value being less than 0.2 Hz.

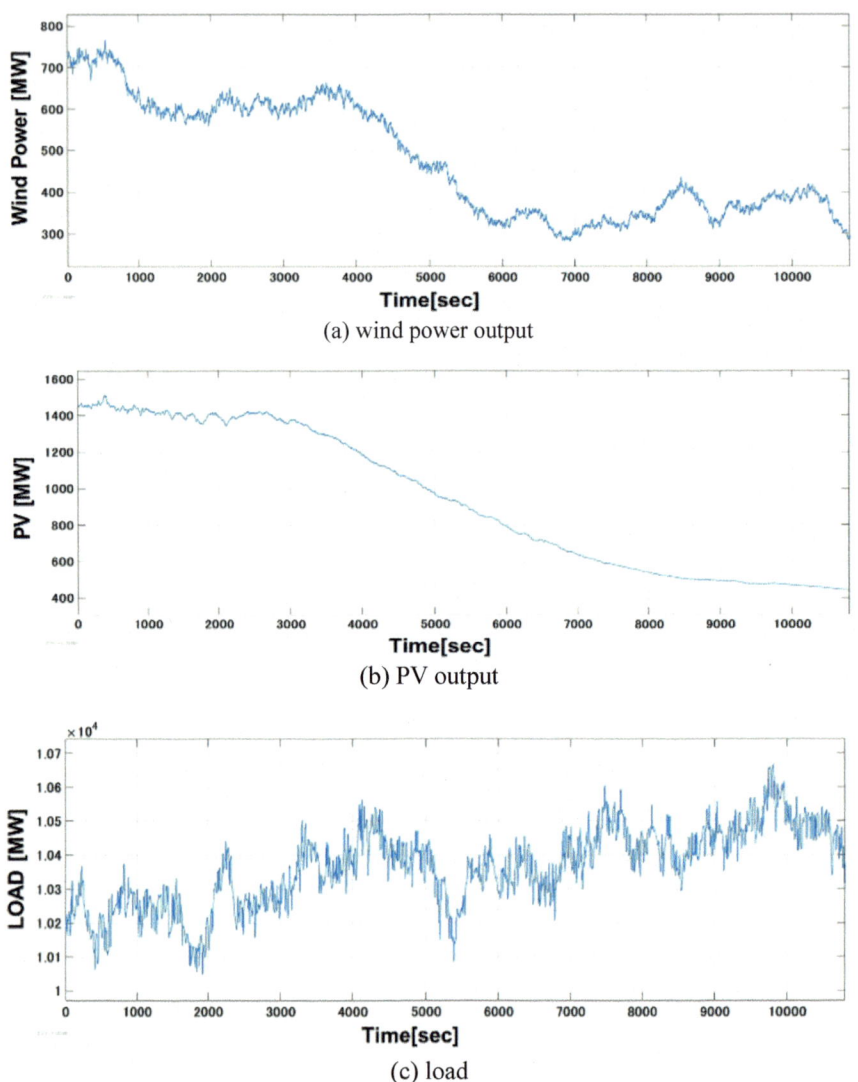

Fig. 2.18 VREs output and load

(a) Case 1: frequency regulation without any contribution from VRE and EV

The frequency fluctuation in this case is shown in Fig. 2.19. Since the automatic generation control by conventional generators was not enough in terms of both control capacity and ramp rate limit, the frequency deviation sometimes reached 0.2 Hz, the

Table 2.2 LFC and EDC
generators

Type	Rated output (MW)	LFC	EDC
Coal1	3400 in total of 4	–	O
LNG1	700	–	–
LNG2	700	O	–
GTCC1	1500 in total of 6	–	O
GTCC2	500 in total of 2	O	–
Hydro	3415	–	–
Nuclear	2500 in total of 4	–	–

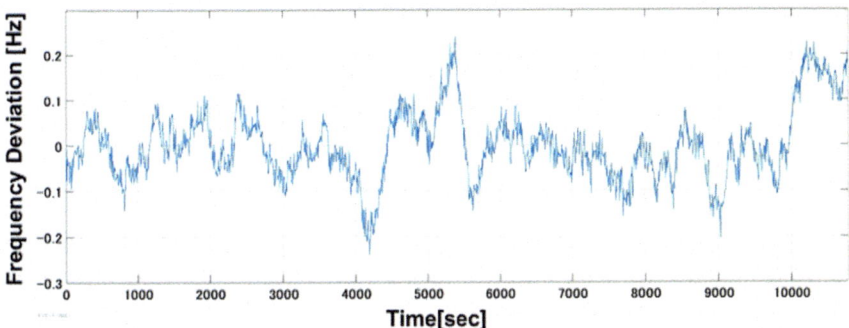

Fig. 2.19 Frequency fluctuation in Case 1

boundary of the proper range. Besides, the sojourn rate was 82.4% which does not
satisfy the requirement. Figure 2.20 shows the generation output of the LNG which

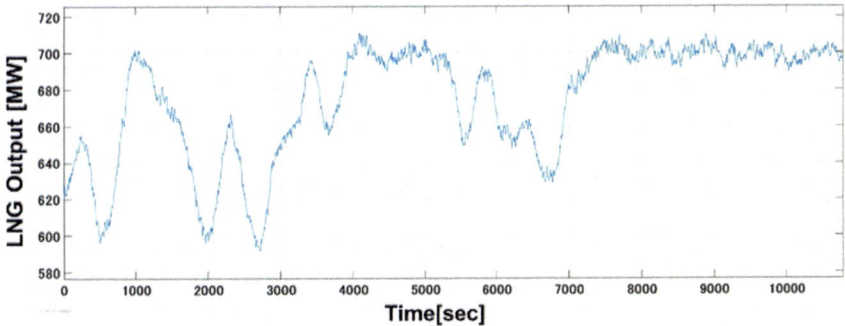

Fig. 2.20 LNG output in Case 1

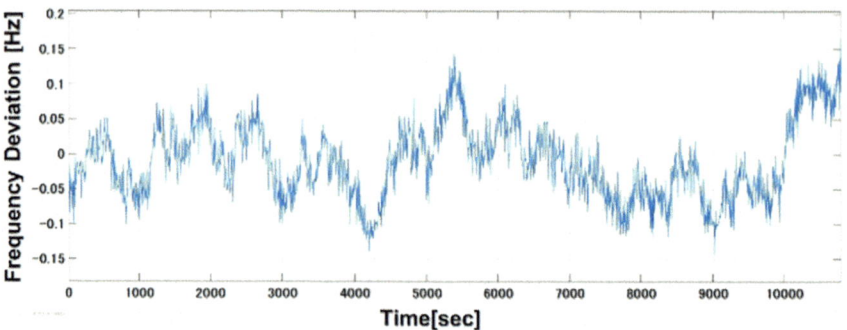

Fig. 2.21 Frequency fluctuation in Case 2

is one of the LFC generators. The control response of the conventional generator is not so fast that the frequency sometimes fluctuated largely.

(b) Case 2: frequency regulation with the proposed method using VRE and EV

The frequency fluctuation is shown in Fig. 2.21. The maximum frequency deviation from the normal value is maintained within ±0.15 Hz due to the proposed method in this case. Moreover, the sojourn rate was 94.7% which has almost satisfied the requirement. To achieve this requirement, the Delta control was applied to WP as shown in Fig. 2.22. As 40 MW of the power curtailment was applied in advance, the maximum upward balancing power is 40 MW, accordingly. Figure 2.23 shows the temporal power surge of WP which sometimes worked to increase the total power supply in order to mitigate frequency drops.

Figure 2.24 shows the charging and discharging control of three EV groups. Group 1 and 3 contribute to only discharging and charging controls since their SOC are higher and lower than the targeted range, respectively. On the other hand, Group 2 contributes to both upward and downward balancing controls as its SOC is inside of the targeted range. As the frequency deviation becomes bigger than the threshold

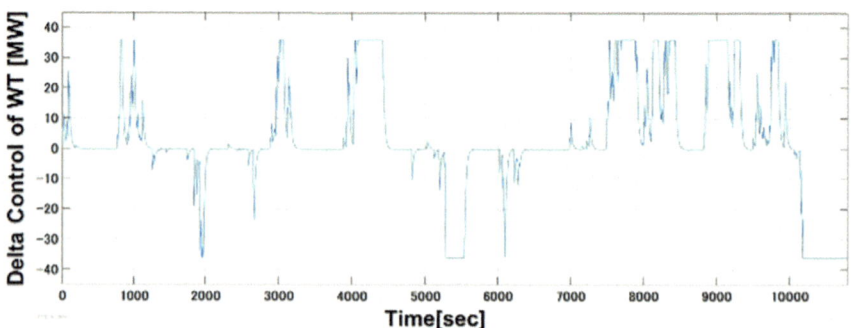

Fig. 2.22 Delta control by wind turbines

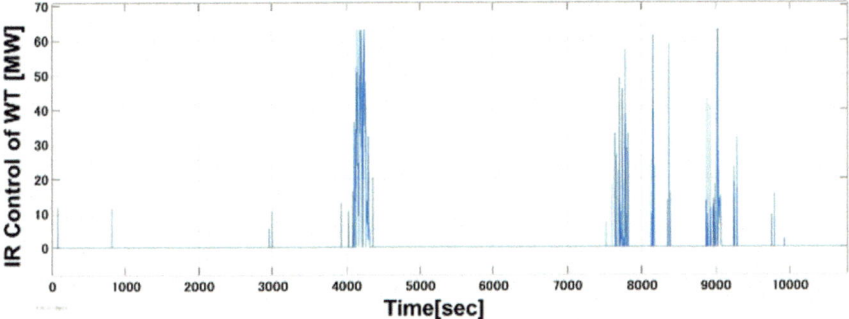

Fig. 2.23 Temporal power surge by wind turbines

value at around 5500 s and after 10,200 s, all the groups including Group 1 applied the charging control to mitigate the over frequency regardless of their SOC. Here, Group 3 hardly contributed to the discharging control because the inertial response control by WP mainly works in the case of the under-frequency based on the asymmetric control strategy as shown in Fig. 2.16.

As shown above, the control performance of the frequency regulation was improved by combining various control reserves including asymmetric control reserves such as the temporal power surge of WP or the charging control of EV. The areas of future works about this topic are as follows.

- Not only WP but also PV should be included in the coordinated control strategy.
- The priority or width of the dead zone for each control has to be carefully redesigned to maximize the control performance.
- The control capacity of EV in each group has to be dynamically changed depending on the SOC.

2.5   Conclusion

It is needed for TSO to procure the balancing control reserves in order to successfully maintain the frequency in power systems with VREs. Balancing power markets have been developed in many countries for economically procuring the balancing control reserves, and various new resources such as VREs or EVs are expected to join the markets, providing the flexibility to the power system. Supposing that more VREs are integrated in power systems in the near future, a LMP-based supply and demand balance control method and a frequency control method based on coordinations between EVs and VREs were introduced in this chapter. It is highly desired that distributed control resources are integrated and activated effectively based on state-of-the-art control techniques.

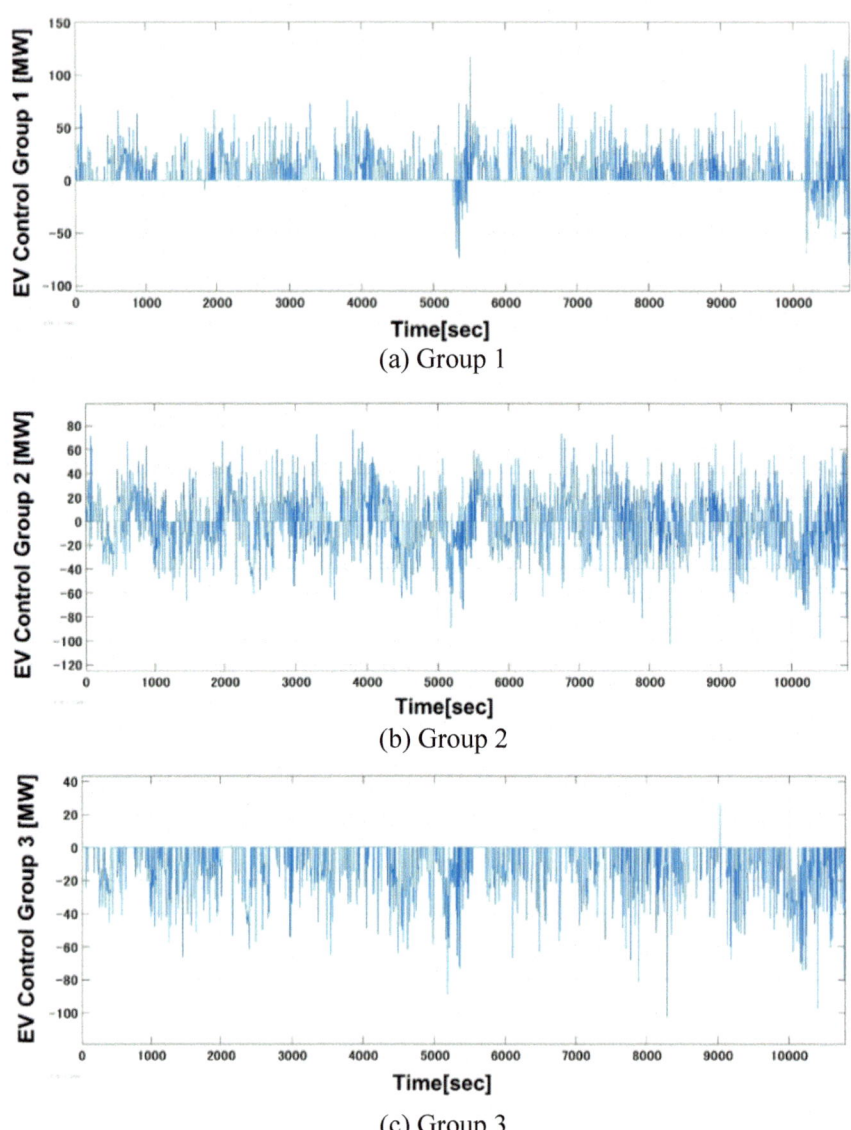

(a) Group 1

(b) Group 2

(c) Group 3

Fig. 2.24 Charging and discharging controls by EVs

References

1. ACCIONA (2016) ACCIONA Energía, pioneer in providing electric power system adjustment only using wind power. https://www.acciona.com/pressroom/news/2016/march/acciona-energia-pioneer-providing-electric-power-system-adjustment-only-using-wind-power/. Accessed 28 Aug 2019
2. Ackermann T (ed) (2013) Wind power in power systems, 2nd edn. Wiley
3. Bae H, Tsuji T, Oyama T, Uchida K (2018) Supply and demand balance control and congestion management by locational marginal price based on balancing market in power system with wind power integration. IEEJ Trans Power Energy 138(8):671–684
4. Chen P-H, Chang H-C (1995) Large-scale economic dispatch by genetic algorithm. IEEE Trans Power Syst 10(4):1919–1926
5. de Vyver JV, de Kooning JDM, Meersman B, Vandevelde L, Vandoorn TL (2016) Droop control as an alternative inertial response strategy for the synthetic inertia on wind turbines. IEEE Trans Power Syst 31(2):1129–1138
6. Hokkaido Electric Power Company. http://denkiyoho.hepco.co.jp/area_forecast.html. Accessed 28 Aug 2019 (in Japanese)
7. IEEE 39-Bus System. https://icseg.iti.illinois.edu/ieee-39-bus-system/. Accessed 28 Aug 2019
8. Institute for Sustainable Energy Plicies (2019) Share of renewable energy power in Japan 2018. https://www.isep.or.jp/en/717/. Accessed 17 Aug 2019
9. International Energy Agency (2018) System integration of renewables
10. Investigating R&D Committee (2016) Recommended practice for simulation models for automatic generation control. In IEEJ Technical Report 1386
11. Kazuya Y, Kurimoto M, Sujuoki Y, Manabe Y, Funabashi T, Kato T (2015) Variation characteristics of spatial average irradiance in time domain of electricity supply and demand/frequency control according to daily pattern classification. In: Proceedings of IEEJ joint technical meeting on PE and PSE. PE-15-014/PSE-15-036
12. Kuwahata R, Merk P (2017) German paradox demystified: why is need for balancing reserves reducing despite increasing VRE penetration? In: Proceedings of 16th international wind integration workshop, 5A-4
13. Kyushu Electric Power Company. http://www.kyuden.co.jp/effort_renewable-energy_application.html. Accessed 17 Aug 2019 (in Japanese)
14. Litvinov E (2010) Design and operation of the locational marginal prices-based electricity markets. IET Gener Transm Distrib 4(2):315–323
15. Liu W, Wu Q, Wen F, Østergaard J (2014) Day-ahead congestion management in distribution system through household demand response and distribution congestion prices. IEEE Trans Smart Grid 5(6):2739–2747
16. Nanahara T, Asari M, Sato T, Yamaguchi K, Shibata M, Maejima T (2004) Smoothing effects of distributed wind turbines. Part 1. Coherence and smoothing effects at a wind farm. Wind Energy 7(2):61–74
17. Organization for Cross-regional Coordination in Transmission system Operators (OCCTO) (2018) Expectations for new resources in balancing power markets. https://www.occto.or.jp/iinkai/chouseiryoku/files/jukyuchousei_dsr.pdf. Accessed 17 Aug 2019 (in Japanese)
18. Organization for Cross-Regional Coordination in Transmission System Operators (OCCTO) (2018) Network Codes https://www.occto.or.jp/en/about_occto/articles/files/Network_Codes_1810.pdf. Accessed 28 Aug 2019
19. Pourbeik P, Soni S, Gaikwad A, Chadliev V (2017) Providing primary frequency response from photovoltaic power plants. In: Proceedings of CIGRE symposium 2017, 022
20. Qi J, Tsuji T (2017) Coordinated frequency control by inertial response of wind power and electric vehicles in Japanese power system. In: Proceedings of solar & wind integration workshop, WIW17-205

21. Singh K, Padhy NP, Sharma J (2011) Influence of price responsive demand shifting bidding on congestion LMP in pool-based day-ahead electricity markets. IEEE Trans Power Syst 26(2):886–896
22. Ye H, Pei W, Qi Z (2016) Analytical modeling of inertial and droop responses from a wind farm for short-term frequency regulation in power systems. IEEE Trans Power Syst 31(5):3414–3423
23. 50Herts (2019) http://www.50hertz.com/en/Grid-Data/Wind-power. Accessed 28 Aug 2019

Chapter 3
Resolving Discrepancies in Problem Formulations for Electricity Pricing by Control Engineers and Economists

Yu Kawano and Eiji Sawada

Abstract Electricity pricing mechanisms have attracted significant attention in recent years from researchers in different fields including systems and control. Many papers have been published in each field exemplifying different experts' insights. These investigations have been partly based on knowledge of and techniques in economics, where pricing is a central topic. However, it seems that several discrepancies exist between problem formulations for electricity pricing by economists and non-economists. In this chapter, through the joint work of a control engineer and an economist, we investigate one such discrepancy, focusing on the analysis of peak shifts of electricity consumption.

3.1 Introduction

In keeping with the intense societal scrutiny of energy and environmental problems in recent decades, the mechanisms of power systems are being reformed. For instance, in Japan, the liberalization of electricity retail sales began in April, 2016 [1]. Because of this and other major events, the design of electricity pricing mechanisms has attracted significant research attention, not only in economics but also in other fields including systems and control. A significant body of research has developed which combines the knowledge of these fields with economics. An early example of a paper combining the knowledge of control engineering and economics was [2]. In this reference, a pricing mechanism was proposed to simultaneously achieve stabilization of frequency deviations of power generators and maximization of social welfare in terms of the utility of consumers and the profit of firms. From the perspective of

Y. Kawano (✉)
Hiroshima University, Kagamiyama 1-4-1, Higashi,
Hiroshima 739-8527, Japan
e-mail: ykawano@hiroshima-u.ac.jp

E. Sawada
Kyushu Sangyo University, 2-3-1 Matsukadai Higashi-ku,
Fukuoka 813-8503, Japan
e-mail: e.sawada@ip.kyusan-u.ac.jp

© Springer Nature Singapore Pte Ltd. 2020
T. Hatanaka et al. (eds.), *Economically Enabled Energy Management*,
https://doi.org/10.1007/978-981-15-3576-5_3

control engineering, the problem is formulated as an optimal control problem, a well-known problem in the field, and thus, the paper opened the door for control researchers to study pricing problems within the systems and control framework.

However, some discrepancies between problem formulations in economics and other research fields can be observed. In this chapter, we focus on methodological differences between control engineering and economics. Although [2] presents an optimal control problem formulation from the systems and control perspective, the maximization of social welfare is formulated as a static problem. That is, consumers are supposed to decide their consumption based on electricity prices at a specific point in time. The problem formulation does not enable consideration of the time profiles of electricity prices. Because of this, it is difficult to evaluate the effects of incentives such as negawatt trading based on time profiles of electricity consumption. In other words, it is difficult to analyze peak shifts.

To fix this problem, we utilize insights from economics to improve the objective functions. New objective functions can be defined for the analysis of peak shifts by using two problem formulations: (1) social welfare maximization problems in one area; (2) social welfare maximization problems in two areas connected by a transmission line. The improved objective functions are described as functions of time integrals (or summations) of some functions. This sort of problem structure does not appear in systems and control in general. Therefore, to examine these structures, a new framework for optimal control problems must to be developed. This is an interesting challenge for future work in a branch of research which combines systems and control with economics.

3.2   Motivating Examples

In this section, we consider a simple electricity pricing problem to effectively illustrate the discrepancy in the problem formulations of control engineers and economists. First, we summarize a problem formulation of electricity pricing based on the systems and control viewpoint proposed in [2]. In this formulation, the objective is to dynamically determine the electricity price at which social welfare is maximized and the frequency deviation of the power generator is reduced. It is noteworthy that the definition of social welfare which is implied by this approach is different from the standard definition in economics. Because of this discrepancy, peak shifts caused by negawatt trading may not be properly represented if one uses the social welfare definition of [2]. First, we demonstrate this error and then suggest another way to evaluate social welfare.

3.2.1 Control Problems

In this subsection, we provide a simplified version of the control problem in [2]. Suppose that there is only one area where one representative consumer, firm, and generator exist. The objective is to decide the electricity price at each time point such that the two aforementioned objectives are achieved.

A swing equation of the generator in the power system is given by

$$\begin{cases} \dfrac{d\theta(t)}{dt} = \omega(t), \\ \dfrac{d\omega(t)}{dt} = -D(t) + S(t), \end{cases} \tag{3.1}$$

where $\theta(t) \in \mathbb{R}$ and $\omega(t) \in \mathbb{R}$ denote the phase and frequency of the generator, respectively. Furthermore, $D(t) \in \mathbb{R}$ and $S(t) \in \mathbb{R}$ represent power demand by the consumer and power supply by the firm, respectively.

As standing assumptions, suppose that the consumer (w.r.t. the firm) has the utility function $u(x)$ (w.r.t. the cost function $c(x)$) that represents utility from using x units of electricity (w.r.t. the minimized cost required to generate x units of electricity), and the consumer (w.r.t. the firm) decides their consumption amount (w.r.t. their power generation amount) in response to a given electricity price so as to maximize their utility. Moreover, suppose that the utility (w.r.t. cost) function is a C^2 function, strictly increasing, and strictly concave (or convex). Under these assumptions, the electricity demand D and supply S are, respectively, decided by

$$D = \arg\max_x \{u(x) - px\}, \tag{3.2}$$

$$S = \arg\max_x \{px - c(x)\}, \tag{3.3}$$

where $p \in \mathbb{R}$ denotes the given electricity price. Under the assumptions, $D(p)$ and $S(p)$ are obtained by, respectively, solving

$$\frac{du(x)}{dx} - p = 0, \tag{3.4}$$

$$p - \frac{dc(x)}{dx} = 0 \tag{3.5}$$

with respect to x. By solving them, it follows that

$$D(p) = \left(\frac{du}{dx}\right)^{-1}(p), \tag{3.6}$$

$$S(p) = \left(\frac{dc}{dx}\right)^{-1}(p), \tag{3.7}$$

where the inverse means the inverse function.

By substituting (3.6) and (3.7) into (3.1), we have

$$
\begin{cases}
\dfrac{d\theta(t)}{dt} = \omega(t), \\[2mm]
\dfrac{d\omega(t)}{dt} = -\left(\dfrac{du}{dx}\right)^{-1}(p(t)) + \left(\dfrac{dc}{dx}\right)^{-1}(p(t)).
\end{cases}
\tag{3.8}
$$

Therefore, if the price $p(t)$ is changed appropriately at each time point, the frequency deviation can be lessened, which is one of the objectives of the price mechanism design. The other objective is to maximize the utility of the consumer and the profit of the firm. To achieve these two objectives simultaneously, the following objective function is considered.

$$
J = qJ_g - \alpha J_s,
\tag{3.9}
$$

$$
J_g := \int_0^T \omega^2(t)\,dt,
$$

$$
J_s := \int_0^T (u(x) - p(t)x)\Big|_{x=\left(\frac{du}{dx}\right)^{-1}(p(t))} + (p(t)x - c(x))\Big|_{x=\left(\frac{dc}{dx}\right)^{-1}(p(t))} \, dt,
$$

where $T > 0$ represents the time interval considered to design the price, and $q > 0$ and $\alpha > 0$ are weights. The price $p(t)$ has to be designed dynamically such that the cost function (3.9) is minimized under the constraint (3.8). Indeed, if J_g becomes small, the frequency deviation becomes small. Moreover, if $-J_s$ becomes small, social welfare J_s becomes large.

3.2.2 Posing Problems

Hereafter, we focus on social welfare J_s in the objective function (3.9) and aim to find the electricity price $p(t)$ to maximize it. That is, consider the following maximization problem

$$
\max_{p(t);0\le t\le T} J_s = \int_0^T \max_{p(t)} \bigg((u(x) - p(t)x)\Big|_{x=\left(\frac{du}{dx}\right)^{-1}(p(t))}
$$

$$
+ (p(t)x - c(x))\Big|_{x=\left(\frac{dc}{dx}\right)^{-1}(p(t))} \bigg) dt.
\tag{3.10}
$$

This equality means that the maximization problem of J_s can be solved by simply solving the maximization problem of $(u(x) - px) + (px - c(x))$. That is, the problem itself is static. To clarify, considering that $p(t) = p_1$ for $0 \le t \le T/2$

and $p(t) = p_2$ for $T/2 < t \leq T$, (3.10) becomes

$$
\max_{p_1,p_2} J_s = \max_{p_1} \left((u(x_1) - p_1 x_1) \Big|_{x_1 = \left(\frac{du}{dx}\right)^{-1}(p_1)} + (p_1 x_1 - c(x_1)) \Big|_{x_1 = \left(\frac{dc}{dx}\right)^{-1}(p_1)} \right)
$$
$$
+ \max_{p_2} \left((u(x_2) - p_2 x_2) \Big|_{x_2 = \left(\frac{du}{dx}\right)^{-1}(p_2)} + (px_2 - c(x_2)) \Big|_{x_2 = \left(\frac{dc}{dx}\right)^{-1}(p_2)} \right).
$$

(3.11)

Again, it is emphasized that the problem can be solved by computing p_1 and p_2 separately, which implies that there is no causality for the optimal electricity prices p_1^* and p_2^* or for the optimal electricity consumption levels x_1^* and x_2^*. Moreover, in this case, it follows that $p_1^* = p_2^*$ and $x_1^* = x_2^*$, since the utility function $u(x)$ and the cost function $c(x)$ are independent of time. Therefore, problem (3.11) does not always correspond to a realistic situation.

3.2.3 The Economist's Approach

In this subsection, we provide a problem formulation to address the causality problem of electricity consumption based on insights from economics. A simple way of doing so is to improve the assumption for the behavior of the consumer.

To compare the problem formulation here with (3.11), we consider two time periods. In (3.2), it is assumed that the consumer decides their electricity consumption based on the electricity price at each time point. However, some consumers prefer to minimize total electricity expenditure overtime. In this case, the consumer decides electricity consumption as follows:

$$
\max_{\bar{x}_1,\bar{x}_2} (\bar{u}(\bar{x}_1, \bar{x}_2) - \bar{p}_1 \bar{x}_1 - \bar{p}_2 \bar{x}_2),
$$

(3.12)

where $\bar{x}_i$ and $\bar{p}_i$ denote electricity consumption and price in each time period, respectively, and $\bar{u}(\bar{x}_1, \bar{x}_2)$ denotes an improved utility function, which is assumed to be C^2, strictly increasing, and is also strictly concave. This maximization problem means that the consumer determines a schedule of consumption depending on the time profile of the price; for example, a washing machine is used when the price of electricity is low.

Consumption levels $\bar{x}_1$ and $\bar{x}_2$, which maximize (3.12), can be obtained as functions of $\bar{p}_1$ and $\bar{p}_2$ by solving a set of two algebraic equations,

$$
\begin{cases}
\dfrac{\partial \bar{u}(\bar{x}_1, \bar{x}_2)}{\partial \bar{x}_1} - \bar{p}_1 = 0, \\[2ex]
\dfrac{\partial \bar{u}(\bar{x}_1, \bar{x}_2)}{\partial \bar{x}_2} - \bar{p}_2 = 0.
\end{cases}
$$

(3.13)

Solutions $\bar{x}_1$ and $\bar{x}_2$ are functions of both $(\bar{p}_1, \bar{p}_2)$ in general. This implies that the causality naturally appears by considering the utility function in (3.12). This is illustrated by an example below. To simplify the discussion, these levels of consumption are denoted by $\bar{D}_1(\bar{p}_1, \bar{p}_2)$ and $\bar{D}_2(\bar{p}_1, \bar{p}_2)$, respectively.

By contrast, it is possible to assume that the electricity firm determines electricity supply in each time period as in (3.3). The reason for this is that the profit of the firm in each time period is independently determined by the monetary terms, so there is no problem to add them. In summary, an improved formulation of the social welfare maximization problem is

$$\max_{\bar{p}_1, \bar{p}_2} \bar{J}_s \tag{3.14}$$

$$\bar{J}_s := \left. (\bar{u}(\bar{x}_1, \bar{x}_2) - \bar{p}_1\bar{x}_1 - \bar{p}_2\bar{x}_2) \right|_{\bar{x}_1 = \bar{D}_1(\bar{p}_1, \bar{p}_2), \bar{x}_2 = \bar{D}_2(\bar{p}_1, \bar{p}_2)}$$

$$+ \left. (\bar{p}_1\bar{x}_1 - c(\bar{x}_1)) \right|_{\bar{x}_1 = \left(\frac{dc}{dx}\right)^{-1}(\bar{p}_1)} + \left. (\bar{p}_2\bar{x}_2 - c(\bar{x}_2)) \right|_{\bar{x}_2 = \left(\frac{dc}{dx}\right)^{-1}(\bar{p}_2)}.$$

By solving this problem with respect to $(\bar{p}_1, \bar{p}_2)$, the optimal electricity prices in each time period $\bar{p}_1^*$ and $\bar{p}_2^*$ are obtained. Then, by substituting $(\bar{p}_1^*, \bar{p}_2^*)$ into $\bar{D}_1(\bar{p}_1, \bar{p}_2)$ and $\bar{D}_2(\bar{p}_1, \bar{p}_2)$, the optimal electricity consumption levels $\bar{x}_1^* := \bar{D}_1(\bar{p}_1^*, \bar{p}_2^*)$ and $\bar{x}_2^* := \bar{D}_2(\bar{p}_1^*, \bar{p}_2^*)$ are computed.

Remark It is known that the optimization problem (3.14) can be solved in another way, as presented by [4]. This methodology proceeds as follows: First, $\bar{D}_1(\bar{p}_1, \bar{p}_2)$, $\bar{D}_2(\bar{p}_1, \bar{p}_2)$, $\bar{x}_1 = S(\bar{p}_1)$, and $\bar{x}_2 = S(\bar{p}_2)$ are computed. Then, because consumption and supply in each time period must be balanced, the following equations hold:

$$\begin{cases} \bar{D}_1(\bar{p}_1, \bar{p}_2) = S(\bar{p}_1), \\ \bar{D}_2(\bar{p}_1, \bar{p}_2) = S(\bar{p}_2). \end{cases} \tag{3.15}$$

By solving these equations with respect to $(\bar{p}_1, \bar{p}_2)$, the optimal electricity price in each time period, $\bar{p}_1^*$ and $\bar{p}_2^*$, is also obtained. In other word, one only have to solve a set of algebraic equations consisting of (3.5), (3.13), and (3.15),

$$\begin{cases} \bar{p}_1 - \dfrac{dc(\bar{x}_1)}{d\bar{x}_1} = 0, \\ \bar{p}_2 - \dfrac{dc(\bar{x}_2)}{d\bar{x}_2} = 0, \\ \dfrac{\partial \bar{u}(\bar{x}_1, \bar{x}_2)}{\partial \bar{x}_1} - \bar{p}_1 = 0, \\ \dfrac{\partial \bar{u}(\bar{x}_1, \bar{x}_2)}{\partial \bar{x}_2} - \bar{p}_2 = 0. \end{cases} \tag{3.16}$$

Then, the optimal electricity prices and consumption levels are obtained simultaneously. A similar approach can be used for the problem in (3.11). That is, a solution to (3.11) is obtained by solving

$$
\begin{cases}
p_1 - \dfrac{dc(x_1)}{dx_1} = 0, \\
p_2 - \dfrac{dc(x_2)}{dx_2} = 0, \\
\dfrac{du(x_1)}{dx_1} - p_1 = 0, \\
\dfrac{du(x_2)}{dx_2} - p_2 = 0.
\end{cases}
\tag{3.17}
$$

From the above equation, for the problem in (3.11), it is also clear that (p_1^*, x_1^*) and (p_2^*, x_2^*) are independent from each other.

Now, we compare social welfare in (3.11) and (3.14).

Example First, consider the social welfare maximization problem in (3.11) with the following utility and cost functions:

$$
u(x) = 2\alpha x^{1/2},
$$

$$
c(x) = \frac{\beta}{2} x^2,
$$

where α and β are positive real numbers. From (3.17), the optimal price and electricity consumption in each time period are obtained by solving

$$
\begin{cases}
\alpha x_i^{-1/2} - p_i = 0, & i = 1, 2, \\
p_i - \beta x_i = 0, & i = 1, 2.
\end{cases}
$$

The optimal price in each time period p_i^* is

$$
p_i^* = \left(\beta \alpha^2\right)^{1/3}, \quad i = 1, 2.
\tag{3.18}
$$

Moreover, the optimal electricity consumption in each time period x_i^* is

$$
x_i^* = \left(\frac{\alpha}{\beta}\right)^{2/3}, \quad i = 1, 2.
\tag{3.19}
$$

Now, we consider how to evaluate the effect of a negawatt trading. Suppose that, to save consumption in time period 2, an incentive is provided to encourage the consumption of electricity in time period 1. For instance, this can be represented by changing the utility function at time 1 from $u(x) = 2\alpha x^{1/2}$ to

$$
u_1(x) = 2(\alpha + \alpha_0) x^{1/2}, \quad \alpha_0 > 0,
$$

where $\alpha_0 x^{1/2}$ represents an increasing amount of profit for the consumer resulting from negawatt incentives. Note that the utility function at time 2 is the same as in the previous period, that is,

$$u_2(x) = u(x) = 2\alpha x^{1/2}.$$

After introducing the negawatt incentive, the optimal price in each time period changes to

$$p_{n,1}^* = \left(\beta(\alpha + \alpha_0)^2\right)^{1/3},$$
$$p_{n,2}^* = \left(\beta\alpha^2\right)^{1/3}.$$

Furthermore, the optimal electricity consumption in each time period changes to

$$x_{n,1}^* = \left(\frac{\alpha + \alpha_0}{\beta}\right)^{2/3},$$
$$x_{n,2}^* = \left(\frac{\alpha}{\beta}\right)^{2/3}.$$

While the price at time 1 increases, electricity consumption at time 1 increases because of the incentive. In contrast, electricity consumption and price at time 2 do not change. This is illustrated in Fig. 3.1. Therefore, the causality of electricity consumption is still difficult to capture when social welfare is considered using (3.11).

Next, the social welfare maximization problem in (3.14) must be considered based on the following utility function,

$$\bar{u}(\bar{x}_1, \bar{x}_2) = 2(\alpha_1 \bar{x}_1 + \alpha_2 \bar{x}_2)^{1/2},$$

where α_1 and α_2 are positive real numbers, and we use the same cost function, that is,

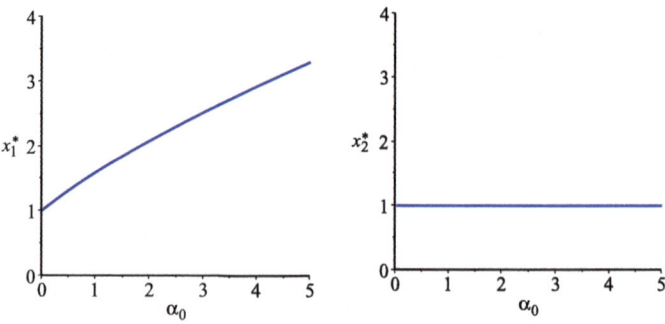

Fig. 3.1 Optimal electricity consumptions versus α_0 for social welfare in (3.11) when $\alpha = \beta = 1$

$$c(x) = \frac{\beta}{2}x^2.$$

From (3.16), the optimal price and electricity consumption in each time period are obtained by solving

$$\begin{cases} \bar{p}_1 - \beta\bar{x}_1 = 0, \\ \bar{p}_2 - \beta\bar{x}_2 = 0, \\ \dfrac{\alpha_1}{(\alpha_1\bar{x}_1 + \alpha_2\bar{x}_2)^{1/2}} - \bar{p}_1 = 0, \\ \dfrac{\alpha_2}{(\alpha_1\bar{x}_1 + \alpha_2\bar{x}_2)^{1/2}} - \bar{p}_2 = 0. \end{cases} \tag{3.20}$$

The optimal price in each time period $\bar{p}_i^*$ is

$$\bar{p}_1^* = \alpha_1 \left(\frac{\beta}{\alpha_1^2 + \alpha_2^2} \right)^{1/3},$$

$$\bar{p}_2^* = \alpha_2 \left(\frac{\beta}{\alpha_1^2 + \alpha_2^2} \right)^{1/3}.$$

The optimal electricity consumption in each time period $\bar{x}_i^*$ is

$$\bar{x}_1^* = \frac{\alpha_1}{\beta} \left(\frac{\beta}{\alpha_1^2 + \alpha_2^2} \right)^{1/3},$$

$$\bar{x}_2^* = \frac{\alpha_2}{\beta} \left(\frac{\beta}{\alpha_1^2 + \alpha_2^2} \right)^{1/3}.$$

To compare $\bar{p}_i^*$ and $\bar{x}_i^*$ with p_i^* in (3.18) and x_i^* in (3.19), consider the case when $\alpha_1 = \alpha_2 = \alpha$. Then, we have

$$\bar{p}_i^* = (\alpha\beta/2)^{1/3}, \ i = 1, 2,$$

$$\bar{x}_i^* = \left(\frac{4\alpha}{\beta^2} \right)^{1/3}, \ i = 1, 2.$$

Finally, to evaluate an effect of the negawatt trading, consider the case when $\alpha_1 = \alpha + \alpha_0$ and $\alpha_2 = \alpha$. Then, the optimal price in each time period changes to

$$\bar{p}_1^* = (\alpha + \alpha_0) \left(\frac{\beta}{\alpha^2 + (\alpha + \alpha_0)^2} \right)^{1/3},$$

$$\bar{p}_2^* = \alpha \left(\frac{\beta}{\alpha^2 + (\alpha + \alpha_0)^2} \right)^{1/3}.$$

68   Y. Kawano and E. Sawada

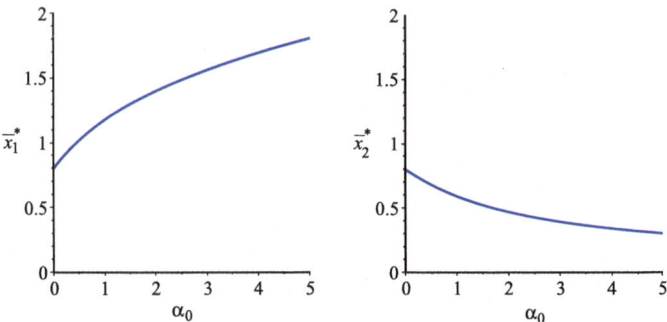

Fig. 3.2 Optimal electricity consumption levels versus α_0 for social welfare in (3.14) when $\alpha = \beta = 1$

Moreover, the optimal electricity consumption in each time period changes to

$$\bar{x}_1^* = \frac{\alpha + \alpha_0}{\beta} \left(\frac{\beta}{\alpha^2 + (\alpha + \alpha_0)^2} \right)^{1/3},$$

$$\bar{x}_2^* = \frac{\alpha}{\beta} \left(\frac{\beta}{\alpha^2 + (\alpha + \alpha_0)^2} \right)^{1/3}.$$

It is clear that the electricity consumption in time 1 increases, but to balance electricity consumption across the whole time interval, consumptions decreases at time 2, as shown in Fig. 3.2. Therefore, the social welfare definition from (3.14) can be used to describe the causality of consumption caused by negawatt trading.

3.2.4 Budget Constraints

In economics, the utility maximization problem of consumers is considered under a budget constraint. However, in some situations, it is possible to relax this constraint so the existence of a budget constraint is not an integral consideration in terms of the discrepancies in problem formulations.

Under a budget constraint, problem (3.12) can be replaced with

$$\max_{\bar{x}_1, \bar{x}_2} \bar{u}(\bar{x}_1, \bar{x}_2),$$

$$\text{s.t. } \bar{p}_1 \bar{x}_1 - \bar{p}_2 \bar{x}_2 = I,$$

where $I > 0$ is the budget of the consumer. In general, the consumer uses their budget for several consumption goods, and not just electricity. Hence, let y represent consumption goods other than electricity, the price of which can be normalized by 1. Then, y can be directly regarded as the cost of consumption. Furthermore, the following type of utility function can be assumed:

$$\tilde{u}(y, \bar{x}_1, \bar{x}_2) = y + \bar{u}(\bar{x}_1, \bar{x}_2). \tag{3.21}$$

This is called the "quasi-linear utility function"; this function can be specified without any loss of generality if the expenditure for $\bar{x}_1$ and $\bar{x}_2$ is only a small share of the total income. Then, the generalized utility maximization problem becomes

$$\max_{y, \bar{x}_1, \bar{x}_2} \tilde{u}(y, \bar{x}_1, \bar{x}_2) \tag{3.22}$$

under the budget constraint

$$y + \bar{p}_1 \bar{x}_1 - \bar{p}_2 \bar{x}_2 = \bar{I}, \tag{3.23}$$

where $\bar{I} > 0$. By substituting y obtained by solving (3.23) into (3.22), it follows that

$$\max_{\bar{x}_1, \bar{x}_2}(\bar{u}(\bar{x}_1, \bar{x}_2) - \bar{p}_1 \bar{x}_1 - \bar{p}_2 \bar{x}_2) + \bar{I}. \tag{3.24}$$

Since $\bar{I}$ is constant, this problem is equivalent to (3.12). Therefore, the budget constraint does not affect electricity consumption decisions so long as $\bar{p}_1^* \bar{x}_1^* + \bar{p}_2^* \bar{x}_2^*$ are small enough. In reality, the electricity expenditure is small (c.f. in Japan, electricity accounted for only around 3.5% of consumption expenditures from 2010 to 2017 [3]).

3.2.5 Improved Formulations of the Control Problem

In this subsection, we apply our observations to update the objective function in (3.9). This improvement leads to a new direction for research in the systems and control literature.

According to Sect. 3.2.3, to investigate the causality of electricity consumption, one possible approach is to change the evaluation of social welfare J_s from (3.9) as follows,

$$J_s := u \left(\int_0^T v(x(t), t)dt \right) - \int_0^T p(t)x(t)dt + \int_0^T (p(t)x(t) - c(x(t)))dt.$$

However, this type of objective function is not typically used in the systems and control discipline. That is, minimization problems consisting of functions of integrals do not appear in problem formulations in systems and control. Therefore, to integrate problems of control and economics, it would be interesting for future studies to develop an optimal control framework, which can be applied to this type of objective function.

3.3 Pricing Problems Between Two Connected Areas

3.3.1 Problem Formulation

In this section, we further illustrate the effectiveness of the improved utility function in (3.12) by considering a more general problem. In the previous section, we study a pricing problem of electricity in a single area. Now, we consider the scenario where there are two areas connected by a single transmission line and investigate the interactions between the two areas.

Again, we consider two time periods. Suppose that in each area, a unique representative consumer and firm exist. The consumer in each area determines their electricity consumption to maximize their respective utility, given by

$$\max_{\bar{x}_{c,1}, \bar{x}_{c,2}} (\bar{u}_x(\bar{x}_{c,1}, \bar{x}_{c,2}) - \bar{p}_{x,1}\bar{x}_{c,1} - \bar{p}_{x,2}\bar{x}_{c,2}), \tag{3.25}$$

$$\max_{\bar{y}_{c,1}, \bar{y}_{c,2}} (\bar{u}_y(\bar{y}_{c,1}, \bar{y}_{c,2}) - \bar{p}_{y,1}\bar{y}_{c,1} - \bar{p}_{y,2}\bar{y}_{c,2}), \tag{3.26}$$

where $\bar{x}_{c,i}$ and $\bar{y}_{c,i}$ denote electricity consumption levels of areas x and y in time period i, respectively, and $\bar{p}_{x,i}$ and $\bar{p}_{y,i}$ denote the electricity prices of areas x and y in time period i, respectively. Furthermore, $\bar{u}_x$ and $\bar{u}_y$ denote the utility functions of areas x and y, respectively.

Next, the firm in each area decides electricity supply to maximize their respective profit, given by

$$\max_{\bar{x}_{s,1}, \bar{x}_{s,2}} (\bar{p}_{x,1}\bar{x}_{s,1} + \bar{p}_{x,2}\bar{x}_{s,2} - \bar{c}_x(\bar{x}_{s,1}) - \bar{c}_x(\bar{x}_{s,2})), \tag{3.27}$$

$$\max_{\bar{y}_{s,1}, \bar{y}_{s,2}} (\bar{p}_{y,1}\bar{y}_{s,1} + \bar{p}_{y,2}\bar{y}_{s,2} - \bar{c}_y(\bar{y}_{s,1}) - \bar{c}_y(\bar{y}_{s,2})), \tag{3.28}$$

where $\bar{x}_{s,i}$ and $\bar{y}_{s,i}$ denote the electricity supplies of areas x and y in time period i, respectively, and $\bar{c}_x$ and $\bar{c}_y$ denote the cost functions of areas x and y, respectively.

Suppose that the two areas are connected by a power line of a single transmission company. That is, the transmission company earns a profit by sending $\bar{z}$ units of electricity from area x to area y, where $\bar{z}$ takes a minus sign if electricity is sent from area y to area x. The transmission company faces the cost function $\bar{v}(\bar{z})$ that represents the cost of transmitting $\bar{z}$ units of electricity, including maintenance cost. The transmission company determines the amount of transmission to maximize its profit. Furthermore, suppose that $\bar{v}(\bar{z})$ is strictly increasing and strictly convex, as defined by C^2, and $\bar{v}(0) = 0$. Under these assumptions, the amount of transmission is decided to maximize

$$\max_{\bar{z}_1, \bar{z}_2}((\bar{p}_{y,1} - \bar{p}_{x,1})\bar{z}_1 + (\bar{p}_{y,2} - \bar{p}_{x,2})\bar{z}_2 - \bar{v}(\bar{z}_1) - \bar{v}(\bar{z}_2)), \qquad (3.29)$$

where $\bar{z}_i$ denote the amount of electricity transmission from area x to area y in time period i.

In this scenario, electricity consumption and supply in each area do not need to be balanced. It suffices that consumption levels and supplies are balanced in the whole area. Therefore, we impose

$$\bar{x}_{s,1} - \bar{x}_{c,1} = \bar{z}_1, \qquad (3.30)$$
$$\bar{y}_{c,1} - \bar{y}_{s,1} = \bar{z}_1, \qquad (3.31)$$
$$\bar{x}_{s,2} - \bar{x}_{c,2} = \bar{z}_2, \qquad (3.32)$$
$$\bar{y}_{c,2} - \bar{y}_{s,2} = \bar{z}_2. \qquad (3.33)$$

In summary, the pricing problem of the two connected areas can be solved by finding the prices $\bar{p}_{x,i}$ and $\bar{p}_{y,i}$, $i = 1, 2$ balancing (3.25)–(3.29) while satisfying constraints (3.30)–(3.33).

3.3.2 Examples

3.3.2.1 Problems

This analysis can be extended to specific utility and cost functions. Consider the following utility and cost functions:

$$\bar{u}_x(\bar{x}_{c,1}, \bar{x}_{c,2}) = 2(\alpha_{x,1}\bar{x}_{c,1} + \alpha_{x,2}\bar{x}_{c,2})^{1/2},$$
$$\bar{u}_y(\bar{y}_{c,1}, \bar{y}_{c,2}) = 2(\alpha_{y,1}\bar{y}_{c,1} + \alpha_{y,2}\bar{y}_{c,2})^{1/2},$$
$$\bar{c}_x(\bar{x}) = \frac{\beta_x}{2}\bar{x}^2,$$
$$\bar{c}_y(\bar{y}) = \frac{\beta_y}{2}\bar{y}^2,$$
$$\bar{v}(\bar{z}) = \frac{\gamma}{2}\bar{z}^2,$$

where $\alpha_{x,i}, \alpha_{y,i}, \beta_x, \beta_y$, and γ are positive, real numbers. Then, the optimal electricity prices, consumption levels, and supplies are determined by solving the following set of algebraic equations:

$$
\left\{
\begin{array}{l}
\dfrac{\alpha_{x,1}}{(\alpha_{x,1}\bar{x}_{c,1}+\alpha_{x,2}\bar{x}_{c,2})^{1/2}} - \bar{p}_{x,1} = 0, \\[2mm]
\dfrac{\alpha_{x,2}}{(\alpha_{x,1}\bar{x}_{c,1}+\alpha_{x,2}\bar{x}_{c,2})^{1/2}} - \bar{p}_{x,2} = 0, \\[2mm]
\dfrac{\alpha_{y,1}}{(\alpha_{y,1}\bar{y}_{c,1}+\alpha_{y,2}\bar{y}_{c,2})^{1/2}} - \bar{p}_{y,1} = 0, \\[2mm]
\dfrac{\alpha_{y,2}}{(\alpha_{y,1}\bar{y}_{c,1}+\alpha_{y,2}\bar{y}_{c,2})^{1/2}} - \bar{p}_{y,2} = 0, \\[2mm]
\bar{p}_{x,1} - \beta_x \bar{x}_{s,1} = 0, \\
\bar{p}_{x,2} - \beta_x \bar{x}_{s,2} = 0, \\
\bar{p}_{y,1} - \beta_y \bar{y}_{s,1} = 0, \\
\bar{p}_{y,2} - \beta_y \bar{y}_{s,2} = 0, \\
\bar{p}_{y,1} - \bar{p}_{x,1} - \gamma \bar{z}_1 = 0, \\
\bar{p}_{y,2} - \bar{p}_{x,2} - \gamma \bar{z}_2 = 0, \\
\bar{x}_{s,1} - \bar{x}_{c,1} = \bar{z}_1, \\
\bar{y}_{c,1} - \bar{y}_{s,1} = \bar{z}_1, \\
\bar{x}_{s,2} - \bar{x}_{c,2} = \bar{z}_2, \\
\bar{y}_{c,2} - \bar{y}_{s,2} = \bar{z}_2.
\end{array}
\right.
\tag{3.34}
$$

It is not clear if this set of equations can be solved analytically. Therefore, in the following, for specific choices of parameters, we numerically compute the optimal electricity prices, consumption levels, and supplies to illustrate our problem formulation.

3.3.2.2 Scenario 1

First, consider the scenario where every parameter is 1, that is, $\alpha_{x,1} = \alpha_{x,2} = \alpha_{y,1} = \alpha_{y,2} = \beta_x = \beta_y = \gamma = 1$. In this scenario, by solving (3.34), we have

$$
\begin{aligned}
& \bar{p}_{x,1} = \bar{p}_{x,2} = \bar{p}_{y,1} = \bar{p}_{y,2} = 0.794, \\
& \bar{x}_{c,1} = \bar{x}_{s,1} = \bar{x}_{c,2} = \bar{x}_{s,2} = \bar{y}_{c,1} = \bar{y}_{s,1} = \bar{y}_{c,2} = \bar{y}_{s,2} = 0.794, \\
& \bar{z}_1 = \bar{z}_2 = 0.
\end{aligned}
$$

A key fact is that there is no transmission. This is natural because utility and cost functions in both areas are identical, which means that there is no advantage for consumers (w.r.t. firms) to use (w.r.t. sell) electricity in the other area. Moreover, it is clear that there is no difference in behaviors between time 1 and 2, which is also natural, since utility and cost functions are the same in both time periods.

3.3.2.3 Scenario 2

For the sake of further discussion, consider the scenario where $\beta_x = \beta_y = 2$, and the other parameters are 1. That is, it costs more to generate electricity in both areas x and y than in the previous example. In this scenario, by solving (3.34), we have

$$\bar{p}_{x,1} = \bar{p}_{x,2} = \bar{p}_{y,1} = \bar{p}_{y,2} = 1.0,$$
$$\bar{x}_{c,1} = \bar{x}_{s,1} = \bar{x}_{c,2} = \bar{x}_{s,2} = \bar{y}_{c,1} = \bar{y}_{s,1} = \bar{y}_{c,2} = \bar{y}_{s,2} = 0.5,$$
$$\bar{z}_1 = \bar{z}_2 = 0.$$

The difference from the previous scenario is that, since it costs more to generate electricity, prices increase, and consequently, consumption levels decrease.

3.3.2.4 Scenario 3—Including Transmission

To evaluate the effect of electricity transmission, we consider the scenario where $\beta_y = 2$ and the other parameters are 1. That is, in area y, it costs more to generate electricity than in the first scenario. Note that, if there is no electricity transmission line between two areas, then the price and consumption in area x or area y are the same as those in scenario 1 or scenario 2, respectively.

In this scenario, by solving (3.34), we have

$$\bar{p}_{x,1} = \bar{p}_{x,2} = 0.830,$$
$$\bar{p}_{y,1} = \bar{p}_{y,2} = 0.935,$$
$$\bar{x}_{c,1} = \bar{x}_{c,2} = 0.725,$$
$$\bar{x}_{s,1} = \bar{x}_{s,2} = 0.830,$$
$$\bar{y}_{c,1} = \bar{y}_{c,2} = 0.572,$$
$$\bar{y}_{s,1} = \bar{y}_{s,2} = 0.467,$$
$$\bar{z}_1 = \bar{z}_2 = 0.307.$$

The main difference from the previous two scenarios is that there is electricity transmission from area x to area y. Since $\beta_x = 1$ and $\beta_y = 2$, it costs more for the firm in area y to generate electricity than for the firm in area x. Therefore, the price in area y tends to be higher than in area x.

Since there is the transmission line, the consumer in area y can also use cheaper electricity from area x. Because of this, the demand for electricity generated in area y decreases, and the demand for electricity in area x increases at time 1. This makes consumption in area y higher than in scenario 2 (where there is no transmission line). In contrast, since electricity is sent from area x, supply in area y decreases as compared to scenario 2. On the other hand, since the demand for electricity in area x increases, the price in area x rises, and consequently, consumption in area x decreases as compared to scenario 1. Finally, since electricity is sent to area y, supply in area x increases. Figures 3.3, 3.4, and 3.5, respectively, show the difference in the price, consumption levels, and supply depending on the existence of the transmission line.

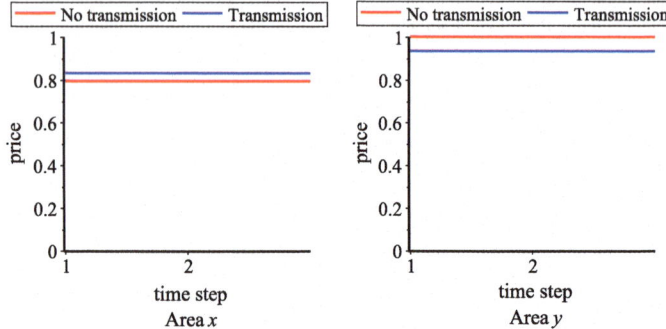

Fig. 3.3 Changes in optimal electricity prices with the transmission line

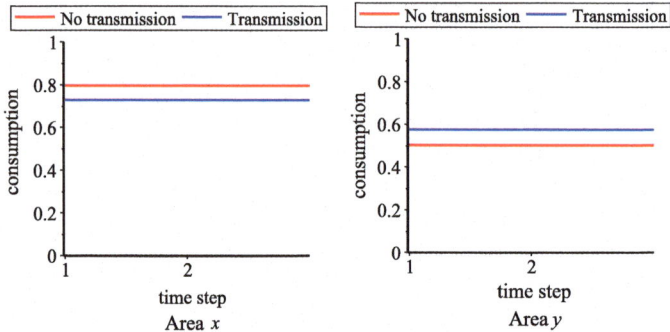

Fig. 3.4 Changes in optimal electricity consumption levels with the transmission line

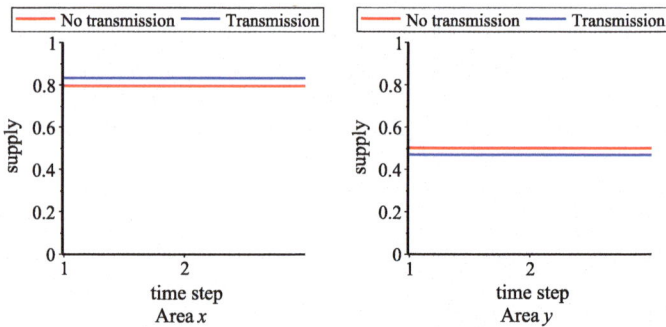

Fig. 3.5 Changes in optimal electricity supplies with the transmission line

3.3.2.5   Scenario 4—Including Causality and Transmission

To evaluate the effect of interaction between a negawatt of trading and transmission, we consider the scenario where $\alpha_{x,1} = 2$ and the other parameters are 1. That is, an incentive is provided in area x at time 1 to encourage electricity consumption at time 1. In Sect. 3.2.3, the causality of electricity consumption is demonstrated in the case of one area. Now, we investigate how negawatt trading affects electricity transmission.

In this scenario, by solving (3.34), we have

$$
\begin{aligned}
\bar{p}_{x,1} &= 1.12, & \bar{p}_{x,2} &= 0.560, \\
\bar{p}_{y,1} &= 0.805, & \bar{p}_{y,2} &= 0.805, \\
\bar{x}_{c,1} &= 1.44, & \bar{x}_{c,2} &= 0.315, \\
\bar{x}_{s,1} &= 1.12, & \bar{x}_{s,2} &= 0.560, \\
\bar{y}_{c,1} &= 0.491, & \bar{y}_{c,2} &= 1.05, \\
\bar{y}_{s,1} &= 0.805, & \bar{y}_{s,2} &= 0.805, \\
\bar{z}_1 &= -0.315, & \bar{z}_2 &= 0.245.
\end{aligned}
$$

First, we evaluate the effect of the incentive, and we compare these figures with scenario 1. Figures 3.6, 3.7, and 3.8, respectively, show consumption levels, supplies, and transmission levels of scenarios 1 and 4. As a result of the incentive, consumption in area x at time 1 is higher than that in scenario 1 (where there is no incentive). In contrast, consumption in area x at time 2 is less. Therefore, a peak shift occurs as a result of the incentive. Accordingly, since consumption in area x at time 1 (w.r.t. 2) increases (w.r.t. decreases), the price and supply in area x at time 1 (w.r.t. 2) increase (w.r.t. decrease).

In area y, the prices and supply are the same in both time periods, which implies that an incentive in area x does not affect power supply in area y. However, it does affect consumption in area y. Since consumption in area x at time 1 (w.r.t. 2) increases

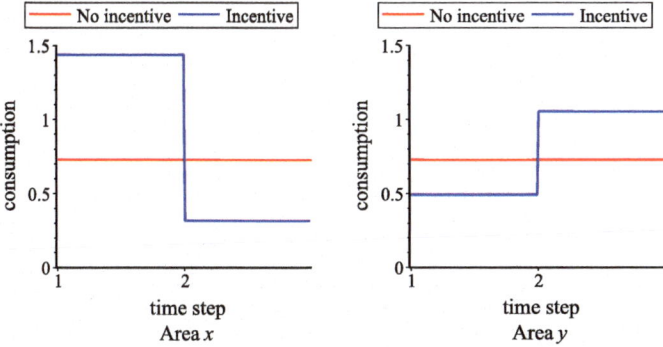

Fig. 3.6 Changes in optimal electricity consumption levels with an incentive

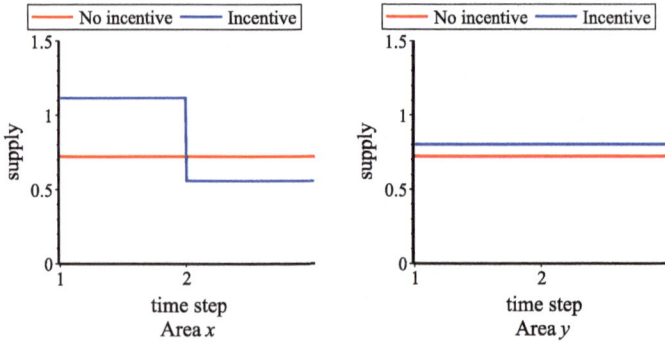

Fig. 3.7 Changes in optimal electricity supplies with an incentive

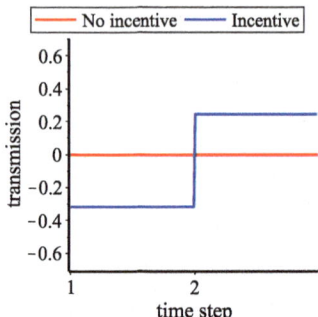

Fig. 3.8 Changes in optimal electricity transmission with an incentive

Table 3.1 Profits of scenario 1 and 4

	Scenario 1	Scenario 4
Utility of the consumer in area x	1.26	1.78
Utility of the consumer in area y	1.26	1.24
Profits of the firm in area x	0.630	1.00
Profits of the firm in area y	0.630	0.593
Profit of the transmission company	0	0.0797

(w.r.t. decreases), the amount of consumption in area y at time 1 (w.r.t. 2) becomes lower (w.r.t. higher).

Table 3.1 shows the profits of the consumer, firm, and transmission company in each area for scenarios 1 and 4. One notices that profits in scenario 4 are higher than in scenario 1, which implies that an incentive to area x is beneficial for the consumer and the firm in area x. In addition, since the incentive boosts the demand for transmission, this is also beneficial for the transmission company. However, the incentive is not beneficial for area y.

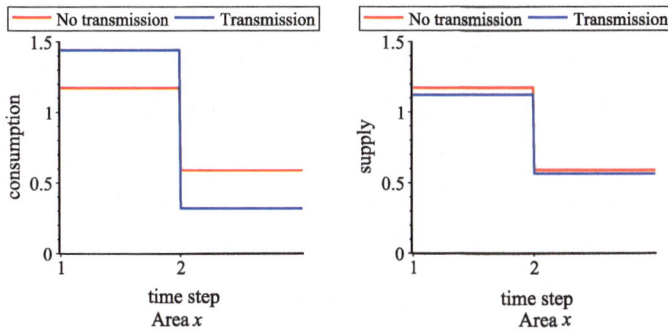

Fig. 3.9 Changes in optimal electricity consumption levels in area x with transmission

Table 3.2 Changes in profits with transmission

	(3.35)	Scenario 4
Utility of the consumer in area x	1.71	1.78
Profits of the firm in area x	0.854	1.00

Next, to evaluate the interaction between an incentive and transmission, a single area is considered, along with the problem formulation from Sect. 3.2.3. To coincide with scenario 4 here, we choose $\alpha_1 = 2$, $\alpha_2 = \beta = 1$ in (3.16). By solving (3.16), we have

$$
\begin{aligned}
\bar{p}_{x,1} &= 1.17, \\
\bar{p}_{x,2} &= 0.585, \\
\bar{x}_{c,1} = \bar{x}_{s,1} &= 1.17, \\
\bar{x}_{c,2} = \bar{x}_{s,2} &= 0.585.
\end{aligned}
\tag{3.35}
$$

Figure 3.9 shows the consumption levels and supplies in area x of (3.35) and in scenario 4. Because of transmission, consumption in area x at time 1 (w.r.t. 2) increases (w.r.t. decreases). Therefore, the transmission line encourages a peak shift. Also, Table 3.2 shows the utility of the consumer and the profit of the firm in area x increases due to the transmission line.

3.4 Conclusion

In this chapter, we developed a structure for the utility function for electricity pricing from the perspective of economics. First, we demonstrated that the proposed utility functions based on economic theory can help analyze the peak shifts in electricity markets caused by incentives. This approach makes it possible to analyze interactions among incentives and transmission lines. However, by employing the proposed utility functions in control problems, the objective functions become functions of time integrations (or summations) of some functions. This kind of objective functions is not well studied in the systems and control literature; thus, it could be a fruitful new research direction with respect to optimizing control in the systems and control field.

References

1. Agency for Natural Resources and Energy (2016). http://www.globus.org/toolkit/
2. Berger AW, Schweppe FC (1989) Real time pricing to assist in load frequency control. IEEE Trans Power Syst 4(3):920–926
3. e-Stat (2017). https://www.e-stat.go.jp/en
4. Jehle GA, Reny PJ (2011) Advanced microeconomic theory, 3rd edn. Pearson London

Chapter 4
Effectiveness of Feed-In Tariff and Renewable Portfolio Standard Under Strategic Pricing in Network Access

Yukihide Kurakawa and Akira Hibiki

Abstract Although some policy schemes are intended to promote production from renewable energy sources (RES), strategic pricing in network access possibly offsets the effectiveness of these policies. This study compares the effectiveness of fixed-price and premium-price feed-in tariffs (FIT) and renewable portfolio standards (RPS) for promoting production from RES, explicitly considering strategic pricing in network access. The effects of vertical structure, i.e., vertical integration and separation, are also investigated. An analytical model consists of a monopolist and a competitive fringe, where the fringe firm produces from RES. The vertically integrated monopolist is able to set the access price incurred by the fringe and its own output. Under vertical separation, in contrast, the access price is set by an independent operator in the network sector. It is shown that under vertical integration, the effectiveness of both FIT policies are fully offset by strategic pricing in network access, whereas RPS does not create an incentive for the manipulation. This is because a higher access price induces a higher cost for the vertically integrated monopolist to meet the purchase obligation under RPS. Consequently, RPS is potentially more effective than FIT policies under vertical integration. It is also shown that vertical separation improves the effectiveness of both FIT policies but adversely reduces that of RPS.

4.1 Introduction

Feed-in tariffs (FIT) and renewable portfolio standards (RPS) are mainstream policy schemes for promoting generation from renewable energy sources (RES). FIT is a price regulation that includes broadly divided fixed-price FIT and premium-price

---

Y. Kurakawa (✉)
Faculty of Economics, Kanazawa Seiryo University, Ishikawa 920-8620, Japan
e-mail: kurakawa@seiryo-u.ac.jp

A. Hibiki
Graduate School of Economics and Management, Tohoku University, Miyagi 980-8576, Japan
e-mail: hibiki@tohoku.ac.jp

FIT.[1] The former type fixes the price of renewable electricity, whereas the latter adds a premium set by a policymaker to the market equilibrium price. RPS mandates an electric utility to procure a certain percentage of the electricity that it sells from renewable energy sources.

Most theoretical studies on FIT and RPS focus on the performance of each scheme (e.g., [1, 2, 4, 6, 8, 10, 11]). Only a few studies, such as Tamás et al. [9] and Hibiki and Kurakawa [5], compare the effectiveness and efficiency of these policies. Tamás et al. [9] compare the market equilibria of premium-price FIT and RPS policy schemes by using a model that does not explicitly consider strategic pricing in network access.

Although some policy schemes are intended to promote production from RES, strategic pricing in network access possibly offsets the effectiveness of these policies. Ropenus and Jensen [7] theoretically show that the strategic manipulation of access charges by the monopolist partially offsets the effectiveness of fixed-price FIT for promoting renewable power production, and unbundling with perfect regulation on access charges eliminates the possibility of manipulation. Consequently, it is possible for the fixed-price FIT to work entirely.

The purpose of this study is to compare the effectiveness of fixed-price and premium price FIT and RPS for promoting production from RES, explicitly considering strategic pricing in network access. We also investigate the effects of vertical structure, i.e., vertical integration and separation.

Our study is most closely related to the work of Ropenus and Jensen [7], who investigate how the effectiveness of fixed-price feed-in tariffs depends on industry structure, i.e., vertical integration and separation. They assume a market structure with a monopolist and a competitive fringe, where the fringe produces from renewable energy sources. They show that under vertical integration, FIT induces the monopolist to raise access charges for the fringe, and effective unbundling with an externally set access charge does not create the possibility for the monopolist to extract part of the fringe's profit. Consequently, unbundling increases the effectiveness of the fixed-price feed-in tariff. Our study expands their analysis by considering RPS and premium-price FIT and showing comparative effectiveness between these policies under vertical integration and separation.

We analyze a model with a monopolist and a competitive fringe, where the monopolist determines the output from non-renewable energy sources and the competitive fringe determines the output from renewable energy sources. Both firms need to access a transmission network sector to sell the electricity that they produce. Under vertical integration, a vertically integrated monopolist is able to set the access price incurred by the fringe in addition to its own output. Under vertical separation, the access price is set by an independent operator in a network sector to maximize its own profit.

We show that under vertical integration, the market equilibria of the fixed-price and premium price feed-in tariff policy schemes are same as that of the benchmark case. This result indicates that the effectiveness of both FIT policy schemes is fully offset

---

[1]Premium-price FIT is also referred to as feed-in premium (FIP). See Couture et al. [3] for more details.

by strategic pricing in the network access under vertical integration. It also indicates that under these policy schemes, the monopolist practically faces the same decision making as the benchmark case; it is able to set a price for renewable electricity as a monopsony through manipulating the access price. In contrast to the FIT policy schemes, RPS does not create an incentive for the manipulation. This is because a higher access price shifts the inverse supply function upward, which in turn induces a higher cost for the vertically integrated monopolist to meet purchase obligations. Consequently, RPS is potentially more effective than FIT policy schemes under vertical integration.

We also show that vertical separation improves the effectiveness of both FIT policies but adversely reduces that of RPS. RPS under vertical separation gives the independent network operator room to increase its profit by raising the access price, whereas RPS under vertical integration does not create an incentive for access price manipulation. In the case of fixed-price and premium-price feed-in tariff policies, the unbundling increases renewable electricity production because it makes it impossible for the monopolist to set a renewable electricity price as a monopsony. The effect of vertical separation under RPS is a contrast to the cases of fixed-price and premium-price FIT policies, in which unbundling enables the policymaker to increase the output of the fringe.

The remainder of this paper is organized as follows: Sect. 4.2 sets out the model and derives conditions for market equilibria in the cases of the policies investigated. Section 4.3 investigates the effects of vertical separation comparing the outcomes between vertical integration and separation. Section 4.4 summarizes the analysis and concludes.

4.2 The Model

4.2.1 Outline

Consider a market structure with a monopolist and a competitive fringe, where the monopolist determines the output from non-renewable energy sources q_M, whereas the competitive fringe determines the output from renewable energy sources q_F. The fringe firm sells its output to the monopolist at price P_R in the renewable electricity market. The monopolist sells total electricity $Q = q_M + q_F$ to a final representative consumer in a retail market, where inverse demand is $P(Q)$. Both firms need to access a transmission network sector to sell the electricity they produce.

Throughout the analysis, the fringe firm determines its output q_F to maximize its profit π_F, taking the prices for renewable electricity and network access as given,

$$\max_{q_F} \pi_F, \quad \pi_F := P_R q_F - C_F(q_F) - a q_F, \tag{4.1}$$

where $C_F(\cdot)$ is the production-cost function with $C'_F > 0$ and $C''_F > 0$. The first-order condition of profit maximization $P_R - a = C'_F(q_F)$ yields renewable electricity inverse supply function $P_R(q_F, a) = a + C'_F(q_F)$ and supply function $q_F(P_R, a)$ with $\partial q_F(P_R, a)/\partial P_R = 1/C''_F > 0$ and $\partial q_F(P_R, a)/\partial a = -1/C''_F$.

Under vertical integration, a vertically integrated monopolist is able to set the access price a incurred by the fringe in addition to its own output. Under vertical separation, the access price is set by an independent operator in network sector to maximize its own profit.

The monopolist is obliged to purchase renewable electricity from the fringe firm under a policy scheme set by a policymaker in advance. We investigate three main-stream policy schemes: (a) the fixed-price feed-in tariff (FIT), (b) the premium-price feed-in tariff, and (c) renewable portfolio standard (RPS). Under the fixed-price FIT, the price for renewable electricity is fixed at $P_R = \overline{P}_R$, which is set by the policy-maker in advance. Under the premium-price FIT, the price for renewable electricity is set at $P_R = P + t$, where P is an equilibrium price in the retail market, and t is a premium set by the policymaker. RPS policy requires the monopolist to purchase a certain proportion of renewable electricity to its own output. Let β denote the proportion of the purchase obligation set by the policymaker. The monopolist is obliged to purchase at least $q_F = \beta q_M$ of electricity from renewable energy sources (RES).

We also consider a benchmark case in which no policy for promoting production from renewable energy sources is implemented. Under the benchmark case, the monopolist is able to set a price for renewable electricity as a monopsony.

4.2.2 Vertical Integration

Under vertical integration, the vertically integrated monopolist is able to set the access price incurred by the fringe in addition to its own output.

4.2.2.1 The Benchmark

In the case of the benchmark, the vertically integrated monopolist is able to set a price for renewable electricity P_R as a monopsony. The vertically integrated monopolist determines the prices for network access and renewable electricity, in addition to its own output,

$$\max_{a, P_R, q_M} \pi_M, \quad \pi_M := PQ - C_M(q_M) - (P_R + \theta - a)q_F - \theta q_M - F_T, \quad (4.2)$$

where F_T is a fixed cost at the network sector and θ is the constant marginal cost of transmission. Let $\Delta P_R := P_R - a$, and note that the first-order condition of profit maximization for the fringe can be written as $C'_F(q_F(\Delta P_R)) = \Delta P_R$. Then, we can rearrange (4.2) as the following expression:

$$\max_{\Delta P_R, q_M} \pi_M, \quad \pi_M := PQ - C_M(q_M) - (C_F' + \theta)q_F - \theta q_M \tag{4.3}$$

Differentiating with respect to ΔP_R and q_M, we obtain the first-order conditions of profit maximization for the monopolist:

$$\frac{\partial \pi_M}{\partial(\Delta P_R)} = [(P + P'Q) - (C_F''q_F + C_F' + \theta)]\frac{dq_F}{d(\Delta P_R)} = 0, \tag{4.4}$$

$$\frac{\partial \pi_M}{\partial q_M} = P + P'Q - C_M' - \theta = 0. \tag{4.5}$$

The conditions (4.4) and (4.5) determine the equilibrium outputs of the monopolist and the fringe (q_M^0, q_F^0), where the superscript 0 denotes the case of the benchmark.

4.2.2.2   Fixed-Price FIT

Under FIT, the policymaker sets a fixed price of renewable electricity, $\overline{P}_R$. The vertically integrated monopolist determines a price for network access incurred by the fringe and its own output, taking the price of renewable electricity as given,

$$\max_{q_M, a} PQ - C_M(q_M) - (\overline{P}_R + \theta - a)q_F(a, \overline{P}_R) - \theta q_M. \tag{4.6}$$

Substituting the first-order condition for profit maximization of the fringe $\overline{P}_R - a = C_F'$, we can rearrange the above expression as the following:

$$\max_{q_M, a} PQ - C_M(q_M) - (C_F' + \theta)q_F(a, \overline{P}_R) - \theta q_M. \tag{4.7}$$

Differentiating with respect to a and q_M, we obtain the first-order conditions of the profit maximization of the monopolist,

$$\frac{\partial \pi_M}{\partial a} = [(P + P'Q) - (C_F''q_F + C_F' + \theta)]\frac{\partial q_F}{\partial a} = 0, \tag{4.8}$$

$$\frac{\partial \pi_M}{\partial q_M} = P + P'Q - C_M' - \theta = 0. \tag{4.9}$$

The conditions (4.8) and (4.9) determine the equilibrium outputs of the fringe and the monopolist.

We can see that (4.8) and (4.9) are practically the same as (4.4) and (4.5), respectively. This result indicates that the equilibrium outputs of the monopolist and the fringe under fixed-price FIT equal those under benchmark (q_M^0, q_F^0). That is, the policy implementation does not have the effect of increasing output from renewable energy sources. This is because the vertically integrated monopolist is practically able to behave as a monopsony, as (4.8) indicates, which in turn eliminates the effectiveness of the policy. From the first-order condition of the profit maximization of the fringe, and noting that the fringe output is q_F^0, we obtain the access price set by the monopolist as $a(\overline{P}_R) = \overline{P}_R - C_F'(q_F^0)$. From $a'(P_R) = 1$, we can see that an increase in the price of renewable electricity is fully offset by an increase in the access price set by the vertically integrated monopolist. The following proposition summarizes the above discussion.

Proposition 1 *Under vertical integration, the market equilibrium realized in the fixed-price FIT is identical to that in the benchmark case, in which the monopolist is able to set a price of renewable electricity as a monopsony. That is, the implementation of the FIT policy fails to increase production from renewable energy sources.*

4.2.2.3   Premium-Price FIT

The policymaker sets a premium, t, added to the market price in the retail market. The fringe determines its output taking the price of renewable electricity, $P_R = P + t$, as given,

$$\max_{q_F} (P + t)q_F - C_F(q_F) - aq_F. \tag{4.10}$$

The first-order condition for profit maximization $P + t - a = C_F'(q_F)$ yields an output from renewable energy sources as $q_F = q_F(a, q_M, t)$. Note that under the premium-price FIT policy, the output of the monopolist affects the fringe output via the retail market price, in contrast to fixed-price FIT.

The vertically integrated monopolist determines its output and the network access price incurred by the fringe,

$$\max_{a, q_M} PQ - C_M(q_M) - (P + t + \theta - a)q_F - \theta q_M. \tag{4.11}$$

Substituting the first-order condition of the fringe yields

$$\max_{a, q_M} PQ - C_M(q_M) - (C_F' + \theta)q_F(a, q_M; t) - \theta q_M. \tag{4.12}$$

Differentiating with respect to a and q_M, we obtain the first-order conditions as follows:

$$\frac{\partial \pi_M}{\partial a} = [(P + P'Q) - (C_F'' q_F + C_F' + \theta)]\frac{\partial q_F(a, q_M)}{\partial a} = 0, \tag{4.13}$$

$$\frac{\partial \pi_M}{\partial q_M} = P + P'Q - C_M' - \theta + [(P + P'Q) - (C_F'' q_F + C_F' + \theta)]\frac{\partial q_F(a, q_M)}{\partial q_M} = 0. \tag{4.14}$$

Because $\partial q_F / \partial a = -1/C_F'' < 0$ from the first-order condition of the fringe, (4.13) is equivalent to the following expression:

$$(P + P'Q) - (C''_F q_F + C'_F + \theta) = 0. \tag{4.15}$$

Substituting (4.15) into (4.14), we obtain

$$P + P'Q - C'_M(q_M) - \theta = 0. \tag{4.16}$$

Equations (4.15) and (4.16) are, respectively, equivalent to (4.4) and (4.5), which are conditions for market equilibrium under the benchmark. That is, the premium-price FIT under vertical integration yields market equilibrium (q_M^0, q_F^0), which is same as that of the benchmark, irrespective of the level of premium t set by the policymaker. From the first-order condition of the fringe, and noting that equilibrium output of the fringe is q_F^0, the access price set by the monopolist is represented as $a = t + P^0 - C'_F(q_F^0)$, where P^0 denotes the equilibrium price in the retail market under the benchmark. This expression indicates that an increase in the premium is entirely offset by an increase in the access price set by the vertically integrated monopolist. The following proposition summarizes the above.

Proposition 2 *The market equilibrium realized in the premium-price FIT under vertical integration is the same as that of the benchmark (in which the monopolist in the retail market is able to set a price for renewable electricity); this result indicates that the policy does not have the effect of increasing production from renewable energy sources.*

4.2.2.4 RPS

RPS policy requires the monopolist to provide a certain proportion of electricity to its own production from renewable sources. The policymaker sets a mandatory proportion, β. The monopolist is required to satisfy $q_F \geq \beta q_M$, purchasing electricity from the fringe.

The monopolist determines its own output and a price for network access:

$$\max_{a, q_M} \pi_M, \quad \pi_M := PQ - C_M(q_M) - [P_R(q_F, a) + \theta - a]q_F - \theta q_M. \tag{4.17}$$

Recall that $P_R(q_F, a) = a + C'_F(q_F)$ is the inverse supply function of renewable electricity, which is derived from the first-order condition of profit maximization by the fringe. Differentiating (4.17) with respect to a yields

$$\frac{\partial \pi_M}{\partial a} = -\left(\frac{\partial P_R(q_F, a)}{\partial a} - 1\right) q_F. \tag{4.18}$$

Because $\partial P_R(q_F, a)/\partial a = 1$, the effect of the access price on the monopolist's profit is $\partial \pi_M/\partial a = 0$ for all $a \geq 0$. That is, RPS does not create any incentive for manipulation in access prices. This is because a higher access price shifts the inverse supply function upward, which in turn induces higher cost for the vertically inte-

grated monopolist to meet the purchase obligation, and vice versa. Then we obtain
the following proposition.

Proposition 3 *RPS policy under vertical integration does not motivate the vertically integrated monopolist to manipulate the network access price incurred by the fringe.*

The equilibrium output of the monopolist is determined by the following first-order condition with respect to q_M:

$$\frac{\partial \pi_M}{\partial q_M} = (P + P'Q - C'_M - \theta) + \left[P + P'Q - \left(\frac{\partial P_R}{\partial q_F} q_F + P_R + \theta - a \right) \right] \frac{dq_F}{dq_M} = 0.$$
(4.19)

Substituting $P_R = a + C'_F(q_F)$ and $q_F = \beta q_M$ and rearranging terms, we obtain the following expression:

$$(P + P'Q)(1 + \beta) = C'_M + (C'_F + C''_F \beta q_M)\beta + (1 + \beta)\theta.$$
(4.20)

The left-hand side of (4.20) is the marginal revenue of increasing production. The first term of the right-hand side is the marginal production cost, the second term is the marginal procurement cost of renewable electricity, and the third term is the marginal transmission cost.

4.2.3   Vertical Separation

In the case of vertical separation, the network sector is separated from the monopolist. An independent operator in the network sector sets a price for network access incurred by the monopolist and the fringe to maximize its profit, π_T,

$$\max_a \pi_T, \quad \pi_T := (a - \theta)Q - F_T.$$
(4.21)

From the first derivative with respect to a: $Q + (a - \theta)dQ/da$, we obtain

$$\frac{d\pi_T}{da}\bigg|_{a=\theta} = Q > 0.$$
(4.22)

This indicates that the independent network operator sets the access price higher than the marginal transmission cost irrespective of the policy implemented.[2]

---

[2]Recall that in the case of RPS under vertical integration, an increase in access price does not have a positive/negative effect on the profit of the vertically integrated monopolist.

4.2.3.1 The Benchmark

In the case of the benchmark, the monopolist is able to set a price for renewable electricity as a monopsony. In contrast to the case of vertical integration, the monopolist must absorb the network access price set by the independent operator. The maximization problem of the monopolist is represented as follows:

$$\max_{q_M, P_R} \pi_M, \quad \pi_M := PQ - C_M(q_M) - aq_M - P_R q_F. \tag{4.23}$$

Differentiating with respect to q_M and P_R yields first-order conditions of profit maximization,

$$\frac{\partial \pi_M}{\partial q_M} = P + P'Q - C'_M - a = 0, \tag{4.24}$$

$$\frac{\partial \pi_M}{\partial P_R} = (P + P'Q)\frac{dq_F}{dP_R} - \left(q_F + P_R \frac{dq_F}{dP_R}\right) = 0. \tag{4.25}$$

The conditions (4.24) and (4.25) determine the equilibrium outputs of the monopolist and the fringe as a function of a, and the total output can be represented as $Q(a) = q_M(a) + q_F(a)$. The independent operator in the network sector sets a price for network access a^*:

$$a^* = \arg\max_a (a - \theta)Q(a) - F_T. \tag{4.26}$$

4.2.3.2 Fixed-Price FIT

The policymaker sets a fixed price for renewable electricity, $\overline{P}_R$. As in the case of benchmark, the first-order condition of profit maximization for the fringe can be represented as

$$\overline{P}_R - a = C'_F(q_F). \tag{4.27}$$

The monopolist determines its output taking the prices for renewable electricity $\overline{P}_R$ and network access a as given,

$$\max_{q_M} \pi_M, \quad \pi_M := PQ - C_M(q_M) - aq_M - \overline{P}_R q_F. \tag{4.28}$$

The first-order condition of profit maximization for the monopolist is

$$\frac{d\pi_M}{dq_M} = P + P'Q - C'_M - a = 0. \tag{4.29}$$

The conditions (4.27) and (4.29) determine the equilibrium output of the fringe and the monopolist as functions of a and $\overline{P}_R$. Then, the total output can be represented as $q_M(a, \overline{P}_R) + q_F(a, \overline{P}_R) = Q(a, \overline{P}_R)$. The independent operator in the network sector sets a price for access $a^*(\overline{P}_R)$:

$$a^*(\overline{P}_R) = \arg\max_a \pi_T, \quad \pi_T := (a - \theta)Q(a, \overline{P}_R) - F_T. \tag{4.30}$$

The first-order condition of profit maximization for the independent operator is

$$\frac{d\pi_T}{da} = (a - \theta)\frac{dQ}{da} + Q(a, \overline{P}_R) = 0. \tag{4.31}$$

4.2.3.3 Premium-Price FIT

The fringe determines its output taking the price for renewable electricity $P_R = P + t$ as given,

$$\max_{q_F} (P + t)q_F - C_F(q_F) - aq_F, \tag{4.32}$$

where t is a premium set by the policymaker. The first-order condition

$$P + t - a = C_F'(q_F) \tag{4.33}$$

generates renewable electricity supply $q_F(a, q_M; t)$. Note that output of the monopolist q_M is included as an argument because it influences the fringe output via the retail market price. Differentiating both sides of (4.33) with respect to q_M and rearranging terms yields the following:

$$\frac{\partial q_F}{\partial q_M} = \frac{P'}{C_F'' - P'}. \tag{4.34}$$

The monopolist is obliged to purchase renewable electricity from the fringe at price $P + t$ and incurs access price a per unit of electricity it produces,

$$\max_{q_M} PQ - C_M(q_M) - (P + t)q_F - aq_M. \tag{4.35}$$

The first-order condition of profit maximization is

$$\frac{d\pi_M}{dq_M} = P + P'q_M\left(1 + \frac{\partial q_F}{\partial q_M}\right) - C_M' - t\frac{\partial q_F}{\partial q_M} - a = 0. \tag{4.36}$$

4.2.3.4 RPS

The monopolist is obliged to satisfy purchase obligation $q_F \geq \beta q_M$. We assume that this constraint is satisfied by $q_F = \beta q_M$.

$$\max_{q_M} \pi_M, \quad \pi_M := PQ - C_M(q_M) - aq_M - P_R(q_F, a)q_F. \tag{4.37}$$

The first-order condition of the profit maximization is

$$\frac{d\pi_M}{dq_M} = (P + P'Q)\left(1 + \frac{dq_F}{dq_M}\right) - C_M'(q_M) - \left(P_R + \frac{dP_R}{dq_F}q_F\right)\frac{dq_F}{dq_M} - a = 0. \tag{4.38}$$

Substituting $q_F = \beta q_M$ and $P_R = a + C_F'(q_F)$ and rearranging the terms yields the following expression:

$$(P + P'Q)(1 + \beta) = C_M' + (C_F' + C_F''\beta q_M)\beta + (1 + \beta)a. \tag{4.39}$$

The left-hand side of (4.39) is the marginal revenue of increasing production. The first term of the right-hand side is the marginal production cost, the second term is the marginal procurement cost of renewable electricity, and the third term is the marginal price of transmission. The sum of these terms is the total marginal cost of increasing production. Recall that under vertical separation, the independent operator in the network sector sets a price for network access higher than marginal transmission cost; $a > \theta$. Comparing (4.20) and (4.39) indicates that the total marginal cost of increasing production is higher under vertical separation than vertical integration for any β. Then, we obtain the following proposition.

Proposition 4 *Under RPS, vertical integration generates higher production from renewable energy sources than vertical separation for any β. Therefore, vertical integration enables the policymaker to generate higher production from renewable energy sources than vertical separation.*

4.3 Comparison of Vertical Integration and Separation

In this section, we compare outcomes between vertical integration and separation using quadratic cost functions and a linear inverse demand function in the retail market; $C_M(q_M) := c_M(q_M)^2/2$, $C_F(q_F) := c_F(q_F)^2/2$, $P(Q) := A - BQ$.

We can calculate the output of the fringe in the case of benchmark under vertical integration as follows:

$$q_F^0 = \frac{(A - \theta)c_M}{2Bc_M + 4Bc_F + 2c_Mc_F}. \tag{4.40}$$

As Proposition 1 states, fixed-price and premium-price FITs yield the same fringe-production level as the above.

First, we investigate the case of fixed-price FIT. From (4.27), (4.29) and (4.31), the access price set by the independent operator in the case of fixed-price FIT under vertical separation is calculated as

$$a = \frac{\theta}{2} + \frac{Ac_F + c_M \overline{P}_R}{2(c_M + c_F)}, \tag{4.41}$$

and the corresponding output of the fringe is derived as follows:

$$q_F = \frac{(c_M + 2c_F)\overline{P}_R - (c_M + c_F)\theta - Ac_F}{2c_F(c_M + c_F)}. \tag{4.42}$$

Differentiating the above expression with respect $\overline{P}_R$, we obtain

$$\frac{dq_F}{d\overline{P}_R} = \frac{c_M + 2c_F}{2c_F(c_M + c_F)} > 0. \tag{4.43}$$

This finding indicates that increasing the fixed price under vertical separation increases the production of the fringe, although the effect of a higher fixed price is partially offset by the higher access price, as we can see from (4.41) that $da/d\overline{P}_R = c_M/2(c_M + c_F) < 1$. Because the fringe output in the case of fixed-price FIT is constant under vertical integration (as Proposition 1 states), a sufficiently higher fixed-price level under vertical separation generates higher fringe production than vertical integration. The following proposition summarizes the discussion above:

Proposition 5 *In the case of fixed-price FIT, a higher fixed price for renewable electricity increases the output of the fringe only if the network sector is vertically separated. Correspondingly, vertical separation with a sufficiently high fixed price enables the policymaker to achieve higher production from renewable energy sources than vertical integration.*

Next, we turn to the case of premium-price FIT. In the case of premium-price FIT, the price for network access set by the independent operator under vertical separation is given by

$$a = \frac{A + \theta}{2} + \left[\frac{2Bc_F + Bc_M + c_M c_F}{2Bc_F + Bc_M + c_M c_F + (c_F)^2} \right] t, \tag{4.44}$$

and the corresponding output of the fringe can be derived as

$$q_F = \frac{1}{B + c_F} \left[\left(\frac{N - Bc_F}{2N} \right) (A - \theta) + \left(\frac{(c_F)^2 + Bc_F}{N + (c_F)^2} \right) t \right], \tag{4.45}$$

where $N \equiv 2Bc_F + Bc_M + c_M c_F$. From (4.45), the effect of increasing the premium price on the fringe output is

$$\frac{dq_F}{dt} = \left(\frac{1}{B + c_F} \right) \frac{(c_F)^2 + Bc_F}{N + (c_F)^2} > 0. \tag{4.46}$$

This indicates that a higher premium has a positive effect on the fringe output with a constant magnitude.[3] Because the fringe output in the case of premium-price FIT is constant under vertical integration (as Proposition 1 states), a sufficiently higher premium under vertical separation generates a higher production of the fringe than vertical integration. The following proposition summarizes the discussion above:

Proposition 6 *In the case of premium-price FIT, a higher premium increases the output of the fringe only if the network sector is vertically separated. Correspondingly, vertical separation with a sufficiently high premium enables the policymaker to achieve higher production from renewable energy sources than vertical integration.*

Recall that Proposition 4 indicates that in the case of RPS, vertical integration generates higher fringe production than vertical separation does. The effect of vertical separation under RPS contrasts the cases of fixed-price and premium-price FIT policies, in which vertical separation enables the policymaker to increase the output of the fringe.

Finally, we compare the effectiveness of increasing the fringe production between policies under vertical integration. The output of the fringe in the case of RPS under vertical integration is calculated as

$$q_F = \frac{(\beta + \beta^2)(A - \theta)}{2B(1 + \beta)^2 + c_M + 2c_F\beta^2}. \tag{4.47}$$

Taking the limit of (4.47) with respect to β, we obtain

$$\lim_{\beta \to \infty} q_F = \frac{(A - \theta)}{2(B + c_F)}. \tag{4.48}$$

Note that the fringe outputs realized in the cases of fixed-price and premium-price FIT policies are same as those in the case of the benchmark, which is represented by (4.40). Comparing (4.40) and (4.48) derives the following proposition:

Proposition 7 *Under vertical integration, RPS is potentially more effective for promoting generation from renewable energy sources than fixed-price and premium-price FIT policies. In other words, RPS enables the policymaker to generate higher fringe production than fixed-price and premium-tariff FIT policies by setting the level of proportional obligation properly.*

4.4 Conclusion

It was shown that under vertical integration, the market equilibria of the fixed-price and premium-price feed-in tariff policy schemes are same as those of the benchmark case. This indicates that the effectiveness of both FIT policy schemes is fully offset

---

[3]Note that this linearity depends on the specified functional forms assumed here.

by strategic pricing in network access under vertical integration. It also indicates that under these policy schemes, the monopolist practically faces the same decision making as the benchmark case; it is able to set a price for renewable electricity as a monopsony through manipulating the access price. In contrast to the FIT policy schemes, RPS does not create an incentive for manipulation. This is because a higher access price shifts the inverse supply function upward, which in turn induces a higher cost for the vertically integrated monopolist to meet purchase obligations. Consequently, RPS is potentially more effective than FIT policy schemes under vertical integration.

It was also shown that vertical separation improves the effectiveness of both FIT policies but adversely reduces that of RPS. RPS under vertical separation gives the independent network operator room to increase its profit by raising the access price, whereas RPS under vertical integration does not create an incentive for access-price manipulation. In the case of fixed-price and premium-price feed-in tariff policies, the unbundling increases renewable electricity production because it makes it impossible for the monopolist to set a renewable electricity price as a monopsony.

Acknowledgements This work was supported by JSPS KAKENHI Grant Number 15K17058.

References

1. Amundsen ES, Bergmen L (2012) Green certificates and market power on the Nordic power market. Energy J 33(2):101–117
2. Amundsen ES, Mortensen JB (2001) The Danish green certificate system: some simple analytical results. Energy Econ 23(5):489–509
3. Couture TD, Cory K, Kreycik C, Williams E (2010) A policymaker's guide to feed-in tariff policy design. National Renewable Energy Laboratory, technical report NREL/TP-6A2-44849
4. Fischer C (2010) Renewable portfolio standards: When do they lower energy prices? Energy J 31(1):101–119
5. Hibiki A, Kurakawa Y (2013) Which is a better second best policy, the feed-in tariff scheme or the renewable portfolio standard scheme? RIETI discussion paper series 13-J-070 (in Japanese)
6. Jensen SG, Skytte K (2002) Interactions between the power and green certificate markets. Energy Policy 30(5):425–435
7. Ropenus S, Jensen SG (2009) Support schemes and vertical integration—Who skims the cream? Energy Policy 37(3):1104–1115
8. Siddiqui AS, Tanaka M, Chen Y (2016) Are targets for renewable portfolio standards too low? The impact of market structure on energy policy. Eur J Oper Res 250(1):328–341
9. Tamás MM, Shrestha SOB, Zhou H (2010) Feed-in tariff and tradable green certificate in oligopoly. Energy Policy 38(8):4040–4047
10. Tanaka M, Chen Y (2013) Market power in renewable portfolio standards. Energy Econ 39:187–196
11. Zhou H, Tamás MM (2010) Impacts of integration of production of black and green energy. Energy Econ 32(1):220–226

Chapter 5
The Welfare Effects of Environmental Taxation and Subsidization on Renewable Energy Sources in an Oligopolistic Electricity Market

Isamu Matsukawa

Abstract This chapter investigates the impact of environmental taxation and subsidization on non-polluting renewable power plants. Their dispatch is prioritized in a wholesale electricity market in which oligopolistic firms produce electricity both from polluting fossil-fuel inputs and non-polluting renewable energy sources, and competitive fringe firms produce electricity only from non-polluting renewable energy sources. To examine the welfare impacts of levying a tax on emissions of pollutants from fossil-fuel power plants and granting subsidies to renewable power plants, this chapter develops a three-stage model with the endogenous capacity of renewable power plants, forward contracts on electricity, and the operation of fossil-fuel power plants. An emission tax imposed on fossil-fuel use is expected to discourage firms from operating their fossil-fuel power plants, thereby promoting substitution of renewable energy sources for fossil-fuel inputs. A subsidy provided to firms building renewable power plants is expected to reduce the setup costs, thereby promoting the use of renewable energy sources. Using simulations, this chapter makes a welfare comparison between an emission tax on fossil-fuel use and a subsidy for renewable power plants and analyzes how these environmental policies affect a wholesale electricity market.

5.1 Introduction

Non-polluting renewable energy sources, such as solar and wind power, have drawn substantial attention as they are expected to replace conventional generation that relies on fossil fuels, such as coal and oil, thus helping in reducing pollution from electricity production. To promote the use of renewable energy sources in electricity production, many countries have provided firms with subsidies that compensate for a portion of the setup costs of renewable power plants [27, 31]. Examples of compensation schemes for renewable energy sources include the feed-in tariff (FIT), which enables firms to sell electricity at a fixed price that is high enough for them to shorten

I. Matsukawa (✉)
Faculty of Economics, Musashi University, Tokyo 176-8534, Japan
e-mail: matukawa@cc.musashi.ac.jp

© Springer Nature Singapore Pte Ltd. 2020
T. Hatanaka et al. (eds.), *Economically Enabled Energy Management*,
https://doi.org/10.1007/978-981-15-3576-5_5

the payback period of investment costs of renewable power plants. Alternatively, taxation on the use of fossil-fuel in electricity production is expected to discourage firms from operating fossil-fuel power plants, thereby promoting substitution of renewable energy sources for fossil-fuel in electricity production [34]. Examples of this kind of taxation include carbon pricing, which has drawn worldwide attention as a promising policy option that mitigates global warming. Under carbon pricing, the more carbon dioxide a firm emits, the more tax it must pay. Thus, carbon pricing provides electricity suppliers with pecuniary incentive to switch from fossil fuels to renewable energy sources.

This chapter investigates the impact of environmental taxation and subsidization on non-polluting renewable power plants whose dispatch is prioritized in a wholesale electricity market in which oligopolistic firms produce electricity both from polluting fossil-fuel inputs and non-polluting renewable energy sources and a competitive fringe produces electricity only from non-polluting renewable energy sources. Due to negligible marginal costs, the dispatch of renewable power plants is assumed to be prioritized in a wholesale electricity market. The priority dispatch of renewable power plants is expected to reduce electricity prices and damage costs associated with emissions of pollutants, thereby raising social welfare. However, in an oligopolistic market, incumbents' exercise of market power that reduces electricity production would offset the favorable impact of renewable energy sources on social welfare. Thus, an important policy question is the effectiveness of taxation and subsidization in an oligopolistic wholesale market from the viewpoint of social welfare.

To examine the welfare impacts of taxation on fossil-fuel power plants and subsidization for renewable power plants, this chapter develops a three-stage model with the endogenous capacity of renewable power plants, forward contracts on electricity, and the operation of fossil-fuel power plants. This model is developed by extending the two-stage model of electricity production and forward contracts in Acemoglu et al. [2] to account for investment in renewable power plants. Previous studies investigate the welfare impact of compensation policies for renewable energy sources in an oligopolistic electricity market [10, 25, 29]. While these studies focus on FIT and/or feed-in premium (FIP), which provides subsidy on top of the market price of electricity, this chapter investigates carbon pricing together with FIT and FIP. Furthermore, the welfare impact of investment tax credit (ITC), which is a subsidy on the investment in renewable power plants, is also examined. To make a welfare comparison among these policies, this chapter conducts a numerical simulation on the basis of parameters that approximate characteristics of the wholesale electricity market in Japan.

The rest of this chapter is organized as follows. Section 5.2 briefly reviews the literature on the merit-order effects of renewable energy sources. Section 5.3 presents a conceptual framework outlining how alternative policies affect a wholesale electricity market. Section 5.4 presents an illustrative example of welfare comparison of alternative policies for promoting renewable energy sources. Section 5.5 concludes this chapter. Derivation of key variables in equilibrium is presented in the appendices.

5.2   Merit Order Effects of Renewable Energy Sources

Due to negligible costs of operation, power plants using renewable energy sources such as solar and wind would receive priority in their dispatch in a wholesale electricity market that has been dominated by conventional generation using fossil fuels such as coal and oil. An increase in renewable energy sources with negligible costs of operation is expected to lower electricity prices in the market as well as to reduce emissions of pollutants such as greenhouse gases.

The impact of renewable energy sources on the market price of electricity is often referred to as the "merit-order effect," which has been investigated in the literature [5, 11, 13, 14, 26, 37–39]. The literature measures the magnitude of the merit-order effect using data on electricity markets in various countries or regions. For example, Cludius et al. [14] reveal that from 2008 to 2012 in Germany, an increase in the installed capacity of wind or solar power plants by 1 million kilowatts, which approximately corresponds to 20% of the annual average capacity of wind power or 50% of the annual average capacity of solar power, lowered the spot price of electricity by between 1.8 and 4.4%. Woo et al. [38] demonstrate that a 10% increase in the installed capacity of wind generation in Texas would reduce the spot price of electricity by between 2.1 and 8.9%.

Because of the presence of dominant generators who can affect electricity prices, wholesale electricity markets are often described as oligopolies. In an oligopolistic market, the merit-order effect of renewable energy sources would be offset by the market power of dominant generators, who can raise electricity prices in their favor. However, higher prices enable electricity suppliers to shorten the payback period of investment in renewable power plants, thereby promoting the use of renewable energy sources.

The literature on the merit-order effect has focused on the interaction of the effect with the market power of incumbents operating conventional generation in a wholesale electricity market. Green and Vasilakos [18] and Twomey and Neuhoff [35] argue that although the merit-order effect lowers the price of wholesale electricity, generators using wind power gain less than those using thermal power plants in the presence of market power of thermal generators. Ben-Moshe and Rubin [8], Acemoglu et al. [2], and Genc and Reynolds [17] highlight the possibility that the market power of incumbent generators could offset the merit-order effect when the incumbent generators own most of the renewable power plants in the market. Ritz [30] argues that the adverse effect of market power on the merit-order effect still remains even if forward contracts mitigate market power of incumbents. Milstein and Tishler [24] suggest that intermittency of photovoltaics strengthens the market power of incumbent generators, thereby causing frequent price hikes in California's electricity market.

To date, many countries have adopted economic policies that aim at promoting the use of renewable energy sources. Among others, a subsidy for renewable energy sources is often applied to provide sufficient economic incentives to those who install and operate power plants that use renewable energy sources [31]. Examples of subsidization include FIT, FIP, and ITC. The literature examines the effects of various forms of subsidies for renewable energy sources on an electricity market [1, 3, 12, 15, 16, 21, 28, 32, 36].

In the literature on subsidization for renewable energy sources, there is little focus on the interaction between the merit-order effect and market power in an electricity market. Reichenbach and Requate [29] examine the second-best FIT in the presence of learning effects on renewable power plants. Oliveira [25] compares the effects of socially optimal investment on renewable energy sources between FIT and FIP. Brown and Eckert [10] compare the effects between FIT and FIP on the outcome of a wholesale electricity market in the presence of a renewable procurement auction.

This chapter differs from previous studies on the effectiveness of subsidization for renewable energy sources in some important ways. First, in the analysis of the interaction between the merit-order effect and market power, this chapter examines a firm's determination of its capacity of renewable power plants, while previous studies assume that such capacity is fixed [2, 8, 10, 17, 18, 30, 35]. The endogenous capacity of renewable power plants enables us to evaluate the impact of renewable energy policies on external costs whose reduction is the primary objective for promoting the use of renewable energy sources.

Second, in the analysis of the welfare impacts of renewable energy policies, this chapter examines how carbon pricing of conventional generation as well as compensation schemes for renewable energy sources affects the merit-order effect in an oligopolistic electricity market, while the previous studies on the merit-order effects in oligopoly only focus on FIT and FIP [10, 25, 29]. Carbon pricing is a promising policy option that favors renewable energy sources [9]. The welfare impact of carbon pricing would be as important in practice as that of compensation schemes for renewable energy sources. Furthermore, this chapter examines the welfare impact of ITC, which has been used to encourage investment directly in renewable power plants such as photovoltaics.

5.3 The Model

Consider an oligopolistic wholesale electricity market where $n \geq 2$ firms produce electricity by operating both renewable energy sources and conventional generation technology that uses fossil fuels such as coal and oil. In addition to these oligopolists, competitive fringe firms using only renewable energy sources provide electricity to the wholesale market. The inverse demand function for wholesale electricity is assumed to take a linear form: $p = \alpha - \beta(Q + R)$, where p, Q, and R denote the wholesale price of electricity, total output of conventional generation, and total output of renewable energy sources, respectively. The parameters α and β are assumed to be strictly positive throughout the analysis.

The capacity of conventional generation is assumed to be fixed, and the total costs of conventional generation for firm i are assumed to be a linear function of its output, that is, $C_{q_i} = \gamma q_i$ where q_i denotes firm i's output of conventional generation and γ denotes a parameter which is strictly positive and less than α (i.e., $0 < \gamma < \alpha$). The fixed capacity of conventional generation implies practical restrictions on the expansion of oil- and coal-fired power plants. In fact, the International Energy Agency has required developed countries not to increase the capacity of oil-fired power plants for base load in the wake of the oil crisis in 1979. In addition, a growing concern about global warming discourages investors from buying stocks and bonds of electricity suppliers that will build coal-fired power plants.

The assumption about a linear cost function of conventional generation that is identical across oligopolists is in line with Ritz [30] and Acemoglu et al. [2]. The literature on wholesale electricity markets indicates that the quantitative relationship between marginal generation costs and electricity production is well approximated by a "hockey stick curve," which is essentially flat over a wide range of electricity production, but is close to a vertical line when electricity production comes close to the maximum generation capability [33]. This chapter focuses on a wholesale electricity market with a flat marginal cost curve of conventional generation. The assumption that a cost function of conventional generation is symmetric among firms facilitates the computation of the equilibrium outcome of an oligopolistic electricity market.

Total costs of renewable energy sources for firm i are assumed to be a quadratic function of their capacity, that is, $C_{R_i} = 0.5\omega R_i^2$, where R_i denotes firm i's renewable energy sources capacity and ω denotes a parameter that is strictly positive. In contrast with conventional generation, whose capacity is assumed to be fixed, renewable power plants are assumed to be newly built by both oligopolists and competitive fringe firms, and the cost function of renewable energy sources represents the investment costs of renewable power plants. The assumption of a quadratic cost function of renewable energy sources enables oligopolists to choose the capacity of renewable power plants that satisfies a second-order condition for maximizing their profits. Power plants using renewable energy sources are assumed to receive priority dispatch in the wholesale market and the output of renewable energy sources for any firm in the market is assumed to be equal to the capacity. The former assumption implies negligible costs of operation of renewable power plants while the latter implies that no intermittency of renewable energy sources is considered in this chapter.

Prior to the production of electricity, oligopolists can create forward contracts associated with their output of conventional generation. Each oligopolist is risk-neutral and their forward position is known in public. The market for forward contracts is assumed to be competitive so that each firm cannot affect the equilibrium price of forward contracts. Competitive fringe firms are assumed not to be engaged with any forward contract. Before forward contracts are made, both oligopolists and competitive fringe firms install facilities that use renewable energy sources.

To promote the use of renewable energy sources, the government is assumed to apply either subsidization or taxation to firms in the wholesale electricity market. The amount of subsidy depends on either output or total costs of renewable energy sources of oligopolists and competitive fringe firms. Firm i selling output of renewable energy sources receives either sR_i or φC_{R_i}, where s and φ denote the rate of subsidy. FIP is a typical form of subsidy sR_i to firm i that sells electricity produced by renewable power plants to the wholesale market while ITC is a practical way of providing subsidy φC_{R_i} to that firm. Instead of paying sR_i or φC_{R_i} to firm i, the government requires transmission utilities to purchase R_i from firm i at the fixed price s and forces consumers to pay the sum of sR_i over all firms operating renewable power plants. FIT corresponds to this form of subsidy to renewable energy sources. Regarding taxation, an emission tax is assumed to be imposed on oligopolists using conventional generation. They must pay τq_i to the government. Carbon pricing is a practical form of emission tax.

Given taxation or subsidization, oligopolists are assumed to determine the output of conventional generation, quantity of forward contracts, and capacity of renewable energy sources in a multistage game, while competitive fringe firms are assumed to determine capacity of renewable energy sources. In the subsequent sections, oligopolists' decision-making is assumed to be described by a one-shot, pure strategy, subgame perfect Cournot–Nash equilibrium, while the capacity of renewable energy sources of competitive fringe firms is assumed to be determined by a break-even condition. Backward induction is used to investigate oligopolists' decision-making.

5.3.1 Conventional Generation

In the third stage of competition in a wholesale electricity market, oligopolist i determines the output of conventional generation, q_i, to maximize the profits, π_i, given other oligopolists' output through conventional generation, the capacity of renewable energy sources for all oligopolists and competitive fringe firms, and forward contracts on conventional generation: For all $j \neq i$, $i, j = 1, \ldots n$, given $R_F, p_i^f, p_j^f, q_i^f, q_j^f, R_i, R_j, q_j,$

$$
\begin{aligned}
\max_{q_i} \pi_i = {} & p\left(q_i - q_i^f\right) + (\delta p + s)R_i + p_i^f q_i^f \\
& - C_q(q_i) - (1 - \varphi)C_R(R_i) - \tau q_i, \\
& i = 1, \ldots n,
\end{aligned}
\tag{5.1}
$$

where p_i^f and q_i^f denote the price and quantity of forward contracts for firm i, respectively. R_F denotes the capacity of all renewable power plants installed by competitive fringe firms. The binary variable δ indicates whether FIT is applied to the wholesale electricity market (i.e., whether the government purchases the output of renewable energy sources at the fixed price s); $\delta = 0$ if FIT is applied, and $\delta = 1$ otherwise. Note that the government is assumed to choose only one policy instrument among carbon pricing, FIT, FIP, and ITC. This implies that if $\delta = 0$, then $s > 0$, $\tau = 0$, and $\varphi = 0$. If $\delta = 1$, either FIP with $s > 0$, ITC with $\varphi > 0$, or carbon pricing with $\tau > 0$ is applied and firm i using renewable energy sources obtains revenue by selling its output (R_i or R_F) to the wholesale electricity market at price p.

With a linear inverse demand function and a linear cost function of conventional generation, the mutual best response that leads to Cournot–Nash equilibrium outcome in the third stage of the multistage game, q_i^*, is derived from the first-order condition for (5.1) and the market clearing condition:

$$
q_i^* = \frac{1}{\beta(n+1)} \left[\alpha - \tau - \gamma - \beta R_F + \beta n q_i^f - \beta \sum_{j \neq i}^{n} q_j^f \right.
$$

$$
\left. -(\beta + \delta\beta n) R_i - \beta(1-\delta) \sum_{j \neq i}^{n} R_j \right],
$$

$$
i = 1, \ldots, n. \tag{5.2}
$$

where the conjectural variation for firm i, $\partial Q/\partial q_i$, is assumed to be unity. See Appendix A.1 for the derivation of (5.2).

Equation (5.2) implies that holding other things equal, an increase in firm i's quantity of forward contracts raises its production of conventional generation but lowers rivals' production of conventional generation. The positive impact of each firm's forward contracts on its production is consistent with Allaz and Vila [4], who indicate that the presence of forward markets mitigates the adverse impact of market power on efficiency.

Equation (5.2) also implies that holding other things equal, renewable power plants of competitive fringe firms displace conventional generation according to the merit-order effect [8, 25, 30, 35]. The substitution of renewable energy sources of each oligopolist for their conventional generation is also present, but this depends on renewable energy policy. If FIT is applied (i.e., $\delta = 0$), the effect of a rise in renewable energy sources of each oligopolist on their conventional generation is the same as the effect of a rise in fringe firms' renewable energy sources on the conventional generation of that oligopolist. In contrast, if either FIP, ITC, or carbon pricing is applied (i.e., $\delta = 1$), the effect of a rise in renewable energy sources of each oligopolist on their conventional generation exceeds the effect of a rise in fringe firms' renewable energy sources on the conventional generation of that oligopolist.

Renewable energy policy also affects the impact of a rise in renewable energy sources of oligopolist j on the conventional generation of oligopolist i ($i \neq j$). Holding other things equal, the increased capacity of oligopolist j's renewable power plants lowers oligopolist i's conventional generation under FIT (i.e., $\delta = 0$), but this adverse impact is absent under FIP, ITC, and carbon pricing (i.e., $\delta = 1$).

5.3.2 Forward Contracts on Electricity

In the second stage of the game, oligopolist i determines the quantity of forward contracts associated with conventional generation to maximize the profits given other oligopolists' quantity of forward contracts and the capacity of renewable energy sources for all oligopolists and competitive fringe firms. Assuming that $p_i^f = p$ in equilibrium (i.e., the condition for no arbitrage), the first-order conditions for profit maximization of firm i in both the first and second stages of the game, together with the market clearing condition, yield the mutual best response that leads to the Nash equilibrium outcome in the second stage, q_i^{f*}, which is given by

$$q_i^{f*} = \left[\frac{n-1}{\beta(n^2+1)} \right] \left[\alpha - \tau - \gamma - \beta R_F - \beta(1-\delta) \sum_{j=1}^{n} R_j \right],$$

$$i = 1, \ldots, n. \tag{5.3}$$

See Appendix A.2 for the derivation of (5.3).

Equation (5.3) indicates that holding other things equal, competitive fringe firms' capacity of renewable power plants has an adverse impact on the quantity of forward contracts. Due to the merit-order effect, which makes the wholesale electricity market less attractive for oligopolists, a rise in competitive fringe firms' renewable generation reduces oligopolists' incentive to make forward commitments [30]. The increase in each oligopolist's capacity of renewable power plants also reduces their forward contracts, as well as those of other oligopolists, if FIT is applied [10]. However, if FIP, ITC, or carbon pricing is applied, each oligopolist's capacity of renewable power plants has no effect on their forward contracts or those of other oligopolists. This is because, as shown later in this section, the merit-order effect is completely offset by market power of each oligopolist if FIP, ITC, or carbon pricing is applied. Note that in equilibrium, the quantity of forward contracts in (5.3) is identical across all oligopolists, implying symmetry.

Given the capacity of renewable power plants of each firm, the merit-order effect in the second stage of the game is derived from (5.2) to (5.3). First, inserting (5.3) into (5.2) yields the equilibrium output of conventional generation in the second stage of the game:

$$q_i^* = \frac{n}{\beta(n^2+1)} \left[\alpha - \tau - \gamma - \beta R_F - \beta \left(1 + \delta n - \delta + \frac{\delta}{n} \right) R_i \right.$$

$$\left. - \beta(1-\delta) \sum_{j \neq i}^{n} R_j \right],$$

$$i = 1, \ldots, n. \tag{5.4}$$

Note that given the capacity of renewable power plants, the equilibrium quantity of conventional generation in (5.4) is identical among oligopolists if FIT is applied. The symmetry of conventional generation also holds in the second stage of the game if FIP, ITC, or carbon pricing is applied to a market in which renewable capacity is identical among oligopolists. Second, inserting (5.4) into the inverse demand function yields the equilibrium price of wholesale electricity:

$$p^* = \frac{1}{(n^2 + 1)} \left[\alpha + n^2(\tau + \gamma) - \beta R_F - \beta(1 - \delta) \sum_{j=1}^{n} R_j \right]. \quad (5.5)$$

Finally, differentiating the equilibrium price in (5.5) with respect to R_i yields the merit order effect of renewable energy sources of oligopolist i:

$$\frac{dp^*}{dR_i} = \begin{cases} \frac{-\beta}{n^2+1}, & \text{if } \delta = 0 \\ 0, & \text{if } \delta = 1 \end{cases}, \quad i = 1, \ldots, n. \quad (5.6)$$

Equation (5.6) implies that under FIP, ITC, and carbon pricing (i.e., $\delta = 1$), the merit-order effect is completely offset by oligopolists' market power, because oligopolists reduce conventional generation to offset the price fall so that they avoid the loss of profits from their renewable energy sources. If $\delta = 1$, every unit increase in oligopolist's renewable energy sources leads to a unit decrease in conventional generation, as shown by (5.4). Acemoglu et al. [2] also reveal the absence of merit-order effects for the case in which conventional generation of oligopolists exhibits a constant marginal cost and neither taxation nor subsidization is applied to firms.

In contrast, if FIT is applied (i.e., $\delta = 0$), the merit-order effect of oligopolists' renewable generation is present, because the fixed price of renewable energy sources under FIT prevents them from losing profits from their renewable energy sources. Under FIT, the merit-order effect depends on the slope of the inverse demand function as well as the number of oligopolists. The greater the consumers' response to the price (i.e., smaller β), the smaller the merit-order effect becomes. Similarly, the larger the number of oligopolists, the smaller the merit-order effect becomes.

For competitive fringe firms, differentiating the equilibrium price in (5.5) with respect to R_F yields the merit-order effect:

$$\frac{dp^*}{dR_F} = \frac{-\beta}{n^2 + 1}, \quad (5.7)$$

which holds for both $\delta = 0$ and $\delta = 1$. Ritz [30] also indicates that the merit-order effect is present for the case in which intermittency of renewable generation is absent for competitive fringe firms. Equations (5.6) and (5.7) imply that under FIT (i.e., $\delta = 0$), the merit-order effect of renewable generation of competitive fringe firms coincides with that of oligopolists.

5.3.3 Investment in Renewable Energy Sources and Effects of Taxation/Subsidization

In the first stage of competition in a wholesale electricity market, competitive fringe firms determine the capacity of renewable energy sources so that their revenue is equal to their total cost, while each oligopolist determines the capacity of renewable energy sources to maximize their profits given other firms' capacity of renewable energy sources. The break-even condition for competitive fringe firms indicates that in equilibrium, the following condition holds:

$$\delta p + s = 0.5(1 - \varphi)\omega R_F. \tag{5.8}$$

The first-order conditions for profit maximization of oligopolist i in all stages of the game together, with the break-even condition for competitive fringe firms in (5.8), yield the mutual best response in the Cournot–Nash equilibrium in the first stage of the game, R_i^*, and equilibrium capacity of renewable energy sources of competitive fringe firms, R_F^*. Under FIT (i.e., $\delta = 0$), the equilibrium capacity for each oligopolist and that for competitive fringe firms are given by

$$R_i^* = \frac{s\left[4\beta n + \omega(n^2 + 1)^2\right] - 2\omega n(\alpha - \gamma)}{\omega\left[\omega(n^2 + 1)^2 - 2\beta n^2\right]}, \quad i = 1, \ldots, n, \tag{5.9}$$

$$R_F^* = 2s/\omega. \tag{5.10}$$

Under FIP, ITC, and carbon pricing (i.e., $\delta = 1$), these capacities are given by

$$R_i^* = (s + \tau + \gamma)/[(1 - \varphi)\omega], \quad i = 1, \ldots, n, \tag{5.11}$$

$$R_F^* = \frac{2s(n^2 + 1) + 2\left[\alpha + n^2(\tau + \gamma)\right]}{(1 - \varphi)\omega(n^2 + 1) + 2\beta}. \tag{5.12}$$

See Appendix A.3 for the derivation of (5.9)–(5.12). Note that the equilibrium capacity of renewable power plants is identical among all oligopolists, as shown by (5.9) and (5.11).

Also note that for R_i^* to be strictly positive in the case of $\delta = 0$, either one of the following two conditions must hold:

$$s > \frac{2\omega n(\alpha - \gamma)}{4\beta n + \omega(n^2 + 1)^2}, \quad \text{if } \omega > \frac{2\beta n^2}{(n^2 + 1)^2}, \tag{5.13}$$

$$s < \frac{2\omega n(\alpha - \gamma)}{4\beta n + \omega(n^2 + 1)^2}, \quad \text{if } \omega < \frac{2\beta n^2}{(n^2 + 1)^2}, \tag{5.14}$$

Table 5.1 Policy effects on the capacity of renewable energy sources

y	dR_i^*/dy	dR_F^*/dy
s: $\delta = 0$ (FIT)	$\dfrac{4\beta n+\omega(n^2+1)^2}{\omega[\omega(n^2+1)^2-2\beta n^2]}$	$\dfrac{2}{\omega}$
s: $\delta = 1$ (FIP)	$\dfrac{1}{\omega}$	$\dfrac{2(n^2+1)}{\omega(n^2+1)+2\beta}$
φ (ITC)	$\dfrac{\gamma}{\omega(1-\varphi)^2}$	$\dfrac{2\omega(n^2+1)(\alpha+n^2\gamma)}{[\omega(1-\varphi)(n^2+1)+2\beta]^2}$
τ (carbon pricing)	$\dfrac{1}{\omega}$	$\dfrac{2n^2}{\omega(n^2+1)+2\beta}$

Note The variable "y" in the table denotes s, φ, or τ

These conditions imply that under FIT, the rate of subsidy, s, should be large (small) if the cost parameter of renewable energy sources, ω, is large (small). Since FIT is usually applied to renewable power plants whose investment costs are sufficiently high, this chapter only focuses on the case of a large ω in (5.13), which satisfies the second-order condition for maximizing the profit of each oligopolist in the first stage of the game [i.e., ω must exceed $2\beta n/(n^2+1)^2$].

Differentiating the equilibrium capacity of renewable power plants in (5.9)–(5.12) with respect to s, φ, or τ yields the effect of each policy on the capacity of renewable energy sources. Table 5.1 summarizes the policy effects on renewable energy sources. Under all policies, an increase in the rate of subsidy or tax raises the capacity of renewable power plants of oligopolists and competitive fringe firms, as shown by $dR_i^*/dy > 0$ and $dR_F^*/dy > 0$ for any y, which denotes s, φ, or τ in Table 5.1.

5.3.4 Second-Best Taxation and Subsidization

The government can raise social welfare by properly applying taxation or subsidization to the electricity market. Social welfare, denoted by W, is defined as the difference between gross surplus from electricity consumption and the sum of production costs and external costs associated with pollutants emitted from conventional generation:

$$W = \int_0^{Q+R} p(x)dx - \sum_{i=1}^{n}\left(C_{q_i}+C_{R_i}\right) - C_{R_F} - D, \tag{5.15}$$

where D denotes external costs. Specifically, external costs are assumed to be a quadratic function of total conventional generation, $D = 0.5\upsilon Q^2$ where υ denotes a positive parameter.

Differentiating (5.15) with respect to each policy variable, denoted by y, yields the first-order condition for achieving the second-best allocation in the electricity market:

$$p^* \left(n \frac{\mathrm{d}q_o^*}{\mathrm{d}y} + n \frac{\mathrm{d}R_o^*}{\mathrm{d}y} + \frac{\mathrm{d}R_F^*}{\mathrm{d}y} \right) - n \left(\gamma + \upsilon Q^* \right) \frac{\mathrm{d}q_o^*}{\mathrm{d}y} - \omega R_F^* \frac{\mathrm{d}R_F^*}{\mathrm{d}y} - \omega n R_o^* \frac{\mathrm{d}R_o^*}{\mathrm{d}y} = 0,$$

$$(5.16)$$

where $q_i^* = q_i^* = q_o^*$ and $R_i^* = R_i^* = R_o^*$ for any $i, j = 1, \ldots, n (i \neq j)$, due to symmetry in equilibrium. A policy variable that satisfies (5.16) leads to a second-best allocation, which deviates from the welfare-optimal, first-best allocation that satisfies the equality of the social marginal benefit from electricity consumption with social marginal costs of electricity production. The second-best allocation under FIT, FIP, and carbon pricing is analytically computed by solving a linear Eq. (5.16) associated with s or τ, while the second-best allocation under ITC is numerically computed by solving a quartic Eq. (5.16) associated with φ. Section 5.4 presents an illustrative example of the second-best allocation under each policy for a counterfactual case of a Japanese wholesale electricity market.

5.4   Welfare Comparison: Numerical Example of Duopoly

Using a numerical example of duopoly, this section compares the effects of second-best taxation/subsidization policies on market outcomes and social welfare. The following example of welfare comparison illustrates how taxation and subsidization affect social welfare in duopoly. The assumption of duopoly reduces complications that make it difficult to tractably investigate policy effects on the market with more than two oligopolists. Although the numerical example does not represent a simulation of actual wholesale electricity markets in Japan, where three generators in the metropolitan areas have been dominant in the wholesale electricity market [23], parameters that are assumed in the example approximate characteristics of the markets such as marginal costs of conventional generation and renewable power plants, price elasticities of electricity demand, and external costs of conventional generation.

FIT has been the main compensation policy for renewable energy sources in Japan. The Japanese government first applied FIT only to photovoltaics in 2009. Since 2012, FIT has been extended to wind, geothermal, and hydro power, and biomass energy. The purchase price per kilowatt hour and contract period of FIT depend on renewable energy sources, the commencement year of operation, and the amount of electricity production. FIP has not yet been applied to the electricity market in Japan.

Since April 2018, the Japanese government has also applied ITC that raises the depreciation rate of power plants to biomass energy, geothermal power, and hydro power, but the application of this ITC is scheduled to end in March 2020 [20]. Although the use of coal, oil, and gas for generation has been subject to fuel tax [22, p. 8], carbon pricing, which imposes a uniform tax rate on unit emission of carbon dioxide from any economic activity, has not yet been implemented in Japan.

Table 5.2 Assumptions on the parameters of the model under the basic scenario

Parameters	n	α	β	γ	ω	υ
Value	2	100	2	3.9 U.S. cents per kWh	6.5 U.S. cents per kWh	2.4 U.S. cents per kWh

Note 1 U.S. dollar is assumed to be 100 yen

5.4.1   Assumptions

Table 5.2 summarizes the key parameters of the model in the basic scenario in which two incumbents with market power and competitive fringe firms supply electricity to a wholesale market. The duopolists are assumed to operate coal-fired power plants whose capital costs have been fully recovered. In terms of fossil fuels for generation, coal has been mainly used for base load power plants in Japan. The duopolists and competitive fringe firms are assumed to determine the capacity of their wind power plants. Although electricity production by wind power has been smaller than that by photovoltaics and biomass energy in Japan, there is a large potential for developing wind power plants, particularly offshore ones. The operating costs of wind power plants are assumed to be zero and such plants are assumed to be dispatched prior to the dispatch of coal-fired power plants. The inverse demand function is assumed so that the absolute value of the elasticity of electricity demand with respect to the price is approximately 0.26 for the case in which no policy is applied. The price elasticity of 0.26 in absolute terms corresponds to the short-run price elasticity of electricity demand in Japan [22, pp. 41–42].

The parameters associated with costs approximate the wholesale electricity market in Japan. The parameter associated with the capital cost of a wind power plant, ω, is assumed to be 6.5 U.S. cents per kWh (1 U.S. dollar = 100 yen), which corresponds to the lowest unit capital cost for typical onshore wind power plants in Japan. A fuel cost of a coal-fired power plant, γ, is assumed to be 3.9 cents per kWh, which corresponds to the lowest fuel cost for typical coal-fired power plants in Japan. These data on electricity costs are obtained from Japan Ministry of Economy, Trade and Industry (JMETI) [19]. The parameter associated with external costs, υ, is assumed to be 2.4 cents per kWh, which is close to the marginal abatement cost of carbon dioxide in Japan [7].

5.4.2   Results of the Basic Scenario

Table 5.3 summarizes the computational results of the capacity and output of wind power plants, output of coal-fired power plants, amounts of forward contracts, and wholesale price of electricity under alternative second-best policies for renewable energy sources. For comparison, the market outcomes without any policy are also presented in Table 5.3. The "No policy" column assumes that wind power plants receive

Table 5.3 Equilibrium outcome under second-best taxation/subsidization: the basic scenario

	No policy: $\delta = 1$	FIT: $s^* = 37.174, \delta = 0$	FIP: $s^* = 21.382, \delta = 1$	ITC: $\varphi^* = 0.501, \delta = 1$	Carbon pricing: $\tau^* = 19.697, \delta = 1$
R_o^*	0.600	4.344	3.890	1.203	3.630
R_F^*	6.334	11.438	12.192	11.437	10.651
R^*	7.534	20.127	19.971	13.842	17.912
Q^*	32.173	22.338	20.907	26.885	14.779
Q_{f^*}	16.686	11.169	14.343	14.645	11.020
p^*	20.586	15.069	18.243	18.545	34.617

Note Policy variables s^*, φ^*, and τ^* denote those for the second-best allocation in an oligopolistic market of wholesale electricity

priority dispatch without taxation/subsidization. In Table 5.3, the binary variable δ is assumed to be one for FIP, ITC, and carbon pricing, and zero for FIT. The second-best subsidies and tax rate are obtained from the solution to (5.16). The second-best rate of subsidy under ITC, φ^*, is obtained by numerically solving a quartic equation, which is derived from (5.16) with respect to φ. Among the four solutions, $\varphi^* = 0.501$ is the only one that satisfies a constraint that $0 \leq \varphi^* \leq 1$. The other second-best policy variables (i.e., s^* and τ^*) are computed by solving a linear equation, which is derived from (5.16) with respect to these variables.

In comparison to the "No policy" case, all policies promote substitution of electricity produced by wind power plants, R^*, for electricity produced by coal-fired power plants, Q^*. Among these policies, the second-best subsidy under FIT results in the largest amount of electricity produced by wind power plants while the second-best carbon pricing results in the smallest amount of electricity produced by coal-fired power plants. Because of the merit-order effects, the second-best FIT, FIP, and ITC lower the market price of electricity. In contrast, the second-best carbon pricing raises the price of electricity because the tax rate, which is close to the market price under the "No policy" case, leads to a substantial upward shift of firms' supply function, thereby more than offsetting the merit-order effect of wind power plants.

Note that under FIT, forward contracts cover exactly half of the electricity produced by coal-fired power plants, as indicated by (5.3) and (5.4) with $n = 2$. For FIP, ITC, and carbon pricing, the coverage of forward contracts exceeds half of the electricity produced by those plants.

Although the substitution of wind power for coal in electricity production contributes to a reduction in external costs, additional loss of efficiency by taxation and subsidization may offset the environmental benefit. Figure 5.1 illustrates welfare comparison among alternative policies for the basic scenario. For comparison, the figure also presents the outcome under the social planner's problem (i.e., the first-best allocation) that maximizes social welfare in (5.15). Welfare and market outcomes

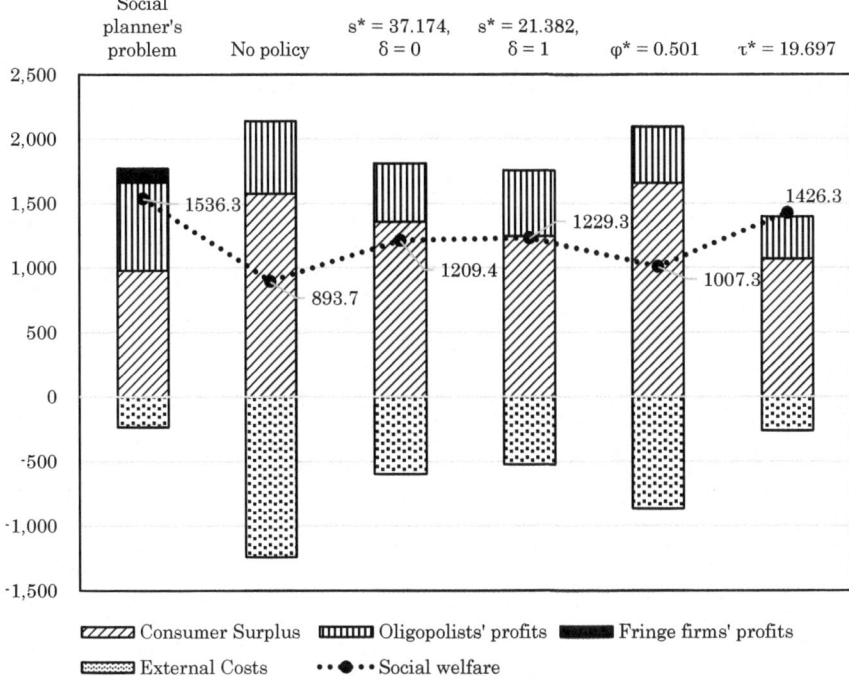

Fig. 5.1 Welfare comparison of renewable energy policies under the basic scenario. *Notes* External costs are in negative values. Except for the case of the social planner's problem, fringe firms' profits are zero in equilibrium because of a break-even constraint. The total amount of subsidy payment is subtracted from consumer surplus under FIT and FIP. Social welfare excludes the total amount of subsidy payment under ITC. Social welfare includes the total amount of tax revenue under carbon pricing

for the social planner's problem are computed using first-order conditions for welfare maximization, which indicate the equality of the market price of electricity with the social marginal costs (i.e., marginal production costs plus external cost). Note that except for the first-best allocation, competitive fringe firms receive zero profits because of the break-even constraint on them at equilibrium.

Among alternative policies for wind power, carbon pricing exhibits the largest welfare while ITC exhibits the lowest. Although carbon pricing raises the market price of electricity and reduces consumer surplus, it significantly contributes to the reduction in external costs that overwhelms the reduced surplus, thereby achieving a larger social welfare than FIT, FIP, and ITC.

5.4.3 Sensitivity Analysis

To illustrate how demand structure and capital costs of wind power plants affect market outcomes and social welfare, alternative scenarios regarding the values of β and ω are assumed in the numerical example. Specifically, it is assumed that $\beta = 4$ so that market demand curve shifts downward. Under this alternative scenario, the response of electricity demand to price changes is slightly smaller than the basic scenario; the price elasticity in absolute terms is approximately 0.23 for the "No policy" case. Alternatively, the parameter of investment costs of wind power plants, ω, is assumed to be larger than the basic scenario. It is assumed to be 12.8 cents per kWh, which corresponds to the largest estimate for a typical onshore wind power plant in Japan [19].

Tables 5.4 and 5.5 present the market outcomes under each policy for the alternative scenarios. With a market demand function that shifts downward, substitution

Table 5.4 Equilibrium outcome under second-best taxation/subsidization: an alternative scenario for demand structure ($\beta = 4$)

	No policy: $\delta = 1$	FIT: $s^* = 22.071, \delta = 0$	FIP: $s^* = 6.663, \delta = 1$	ITC: $\varphi^* = 0.091, \delta = 1$	Carbon pricing: $\tau^* = 7.282, \delta = 1$
R_o^*	0.600	2.115	1.625	0.660	1.720
R_F^*	5.709	6.791	7.354	6.160	7.147
R^*	6.909	11.022	10.604	7.480	10.588
Q^*	13.453	10.403	10.087	12.972	8.605
Q^{f*}	7.327	5.201	6.669	7.146	6.023
p^*	18.553	14.303	17.237	18.192	23.228

Note Policy variables s^*, φ^*, and τ^* denote those for the second-best allocation in an oligopolistic market of wholesale electricity

Table 5.5 Equilibrium outcome under second-best taxation/subsidization: an alternative scenario for capital costs of wind power plants ($\omega = 12.8$)

	No policy: $\delta = 1$	FIT: $s^* = 45.655, \delta = 0$	FIP: $s^* = 28.916, \delta = 1$	ITC: $\varphi^* = 0.513, \delta = 1$	Carbon pricing: $\tau^* = 26.604, \delta = 1$
R_o^*	0.305	2.678	2.564	0.625	2.383
R_F^*	3.400	7.134	7.652	6.571	6.530
R^*	4.009	12.489	12.780	7.821	11.296
Q^*	35.111	28.449	27.191	31.933	17.808
Q^{f*}	17.860	14.224	16.159	16.592	11.287
p^*	21.760	18.124	20.059	20.492	41.791

Note Policy variables s^*, φ^*, and τ^* denote those for the second-best allocation in an oligopolistic market of wholesale electricity

of electricity produced by wind power plants, R^*, for electricity produced by coal-fired power plants, Q^*, becomes modest under all policies, as shown in Table 5.4. A rise in investment cost parameter ω from the basic scenario also leads to modest substitution of wind power for coal in electricity production under all policies, as shown in Table 5.5. While the second-best FIT records the largest investment of wind power under the scenario with a downward shift of the market demand curve, the second-best FIP records the largest capacity of wind power under the scenario with a rise in the investment cost parameter.

Figures 5.2 and 5.3 illustrate welfare comparison among renewable energy policies for the alternative scenarios. Even if the investment cost parameter rises from 6.5 to 12.8, carbon pricing still exhibits the largest welfare among four alternative policies, as shown by Fig. 5.3. However, with a market demand function that shifts downward, FIT exhibits the largest welfare, as shown by Fig. 5.2. These results imply that the optimal choice of the renewable energy policy from the viewpoint of social welfare depends on the structure of market demand for electricity and investment costs of renewable energy sources.

5.5 Conclusion

Using a three-stage model with the endogenous capacity of renewable power plants, forward contracts on electricity, and the operation of fossil-fuel power plants, this chapter investigates how taxation and subsidization affect an oligopolistic wholesale electricity market. In the analysis, oligopolistic firms are assumed to produce electricity both from polluting fossil-fuel inputs and non-polluting renewable energy sources while competitive fringe firms are assumed to produce electricity only from non-polluting renewable energy sources. The dispatch of non-polluting renewable power plants that exhibit negligible operating costs is assumed to be prioritized in a wholesale electricity market. With regard to taxation, carbon pricing is assumed to be applied to conventional generation. Regarding subsidization, either FIT, FIP, or ITC is assumed to be applied to firms operating renewable power plants.

A theoretical investigation of the model implies that merit-order effects, which indicate that an increase in renewable energy sources with negligible costs of operation is expected to lower electricity prices in the market, are present for renewable power plants operated by competitive fringe firms under FIT, FIP, ITC, and carbon pricing. However, for renewable power plants operated by oligopolists, merit-order effects are completely offset by oligopolists' market power that raises the electricity price under FIP, ITC, and carbon pricing. The merit-order effects of oligopolists' renewable power plants exist only under FIT.

Numerical examples of duopoly illustrate how alternative policies affect market outcomes and social welfare. In the example, parameters approximate the characteristics of the wholesale electricity markets in Japan such as marginal costs of conventional generation and renewable power plants, price elasticities of electricity demand, and external costs of conventional generation. Both a tax on emissions from

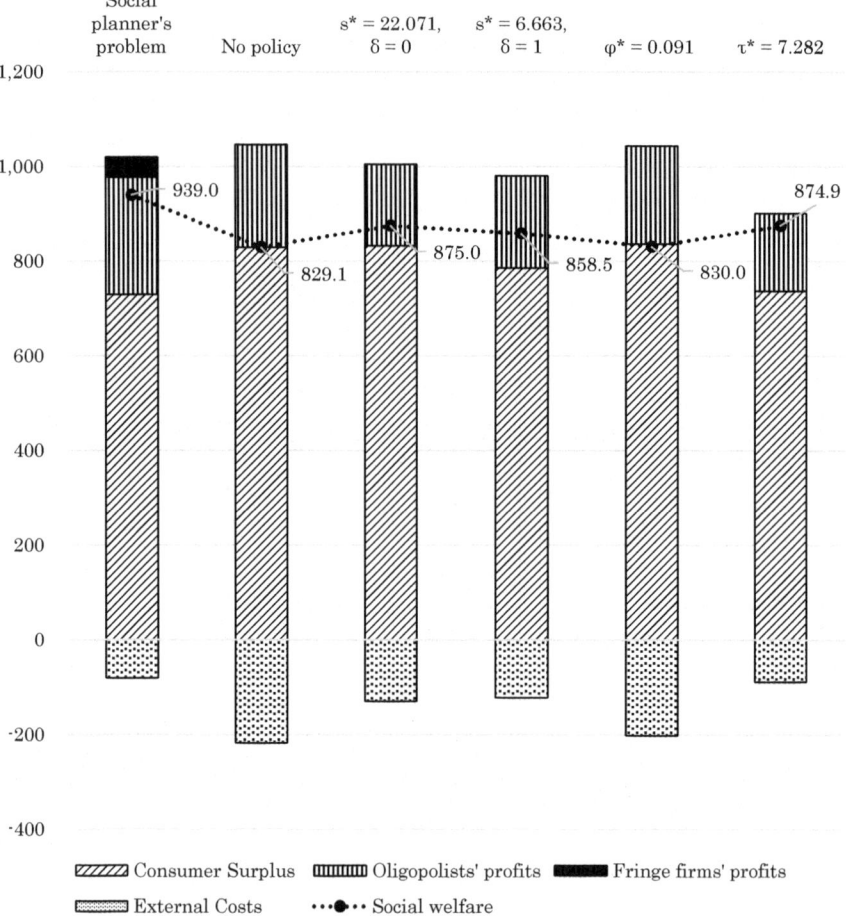

Fig. 5.2 Welfare comparison of renewable energy policies under an alternative scenario for demand structure ($\beta = 4$). *Notes* External costs are in negative values. Except for the case of the social planner's problem, fringe firms' profits are zero in equilibrium because of a break-even constraint. The total amount of subsidy payment is subtracted from consumer surplus under FIT and FIP. Social welfare excludes the total amount of subsidy payment under ITC. Social welfare includes the total amount of tax revenue under carbon pricing

coal-fired power plants and a subsidy for wind power plants promote substitution of wind power for coal in electricity production, thereby contributing to a reduction in external costs that arise from the use of coal for generation. However, additional loss of efficiency arises under both taxation and subsidization, and this efficiency loss may offset the reduction in external costs. Although carbon pricing raises the market price of electricity and reduces consumer surplus, it may greatly contribute to the reduction in external costs that could exceed the reduced consumer surplus, thereby achieving a larger social welfare than subsidization policies. However, from a social

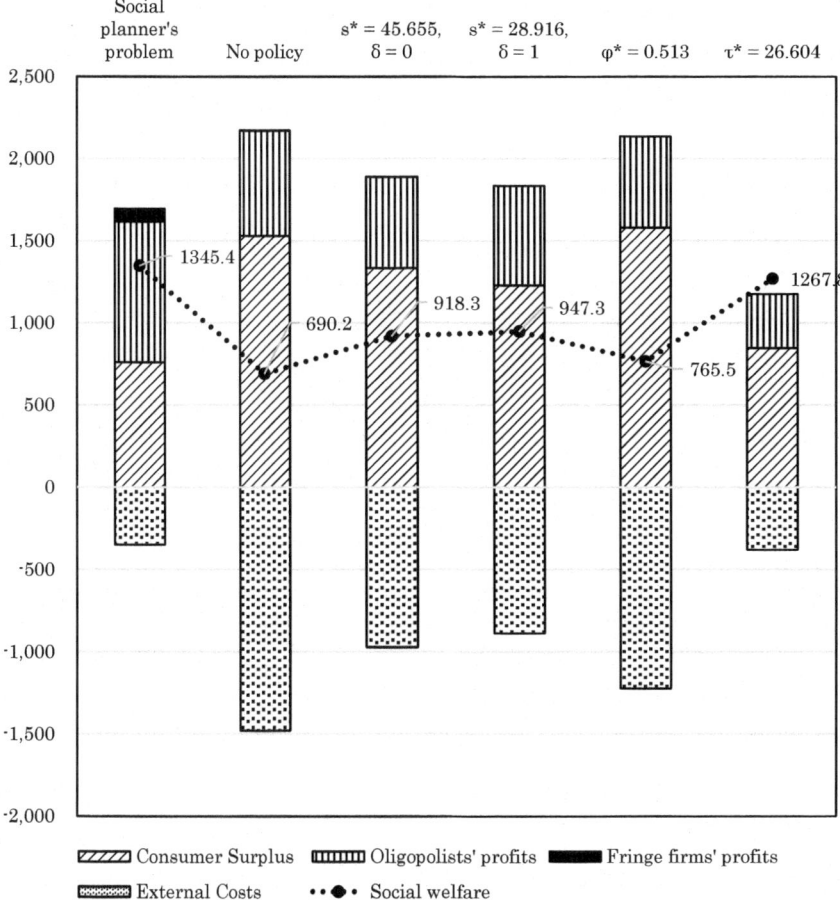

Fig. 5.3 Welfare comparison of renewable energy policies under an alternative scenario for investment costs of renewable power plants ($\omega = 12.8$). *Notes* External costs are in negative values. Except for the case of the social planner's problem, fringe firms' profits are zero in equilibrium because of a break-even constraint. The total amount of subsidy payment is subtracted from consumer surplus under FIT and FIP. Social welfare excludes the total amount of subsidy payment under ITC. Social welfare includes the total amount of tax revenue under carbon pricing

welfare viewpoint, whether carbon pricing performs better than compensation policies such as FIT, FIP, and ITC depends on the demand structure and investment costs of renewable energy sources, as implied by a sensitivity analysis of the numerical example.

Future research may address the adverse impacts of an increase in renewable energy sources on transmission capacity constraints, voltage and frequency control, and the stability of the entire power network. See Ardian et al. [6] for the adverse effects of renewable energy sources on network congestion. Furthermore, volatility of market prices of electricity, which may be caused by intermittency of renewable energy sources such as solar and wind power, is an important issue for renewable energy policies.

Acknowledgements This work was supported by JSPS KAKENHI Grant Number 18K01626.

Appendix: Derivation of Key Variables in Equilibrium

A.1. Derivation of the Equilibrium Production by Conventional Generation in (5.2)

The first-order condition for (5.1) with respect to q_i yields

$$\frac{\partial p}{\partial q_i}\left(q_i - q_i^f\right) + p^* + \delta R_i \frac{\partial p}{\partial q_i} - C_q' - \tau = 0, \quad i = 1, \ldots, n, \tag{A.1}$$

where p^* denotes the market price in equilibrium. Due to the assumption about the inverse demand function, $p = \alpha - \beta(Q + R)$, and the cost function of conventional generation, $C_q = \gamma q_i$, (A.1) can be rewritten as

$$(-\beta)\left(q_i - q_i^f\right) + \alpha - \beta Q - \beta R - \delta\beta R_i - \gamma - \tau = 0, \quad i = 1, \ldots, n. \tag{A.2}$$

Taking a sum over all i from (A.2) yields the aggregate production by conventional generation in equilibrium at the third stage of the game, denoted by Q^*:

$$Q^* = \frac{1}{\beta(n+1)}\left[n(\alpha - \gamma - \tau) + \beta \sum_{i=1}^{n} q_i^f - \beta n R - \delta\beta \sum_{i=1}^{n} R_i\right]. \tag{A.3}$$

Substituting Q^* for Q in (A.2) yields firm i's optimal production by conventional generation, q_i^*, in (5.2).

A.2. Derivation of the Equilibrium Forward Contracts in (5.3)

With firm i's optimal production by conventional generation, q_i^*, in (5.2), the first-order condition for profit maximization with respect to q_i^f in the second stage of the game yields

$$(-\beta)\frac{\partial Q^*}{\partial q_i^f}(q_i^* + \delta R_i) + \frac{\partial q_i^*}{\partial q_i^f}(p^* - \gamma - \tau) = 0, \quad i = 1, \ldots, n. \tag{A.4}$$

Inserting (A.2) into (A.4) yields

$$(-\beta)\left[q_i^{f*}\frac{\partial q_i^*}{\partial q_i^f} + (q_i^* + \delta R_i)\sum_{j\neq i}^{n}\frac{\partial q_j^*}{\partial q_i^f}\right] = 0, \quad i = 1, \ldots, n, \tag{A.5}$$

where q_i^{f*} denotes firm i's optimal forward contracts on electricity in the second stage of the game. From (5.2), we obtain

$$\frac{\partial q_i^*}{\partial q_i^f} = \frac{n}{n+1}, \quad \frac{\partial q_j^*}{\partial q_i^f} = -\frac{1}{n+1}, \quad i, j = 1, \ldots, n; i \neq j. \tag{A.6}$$

Inserting (A.6) into (A.5) yields

$$nq_i^{f*} = (n-1)(q_i^* + \delta R_i), \quad i = 1, \ldots, n. \tag{A.7}$$

From (A.7) and (5.2), we obtain firm i's optimal forward contracts on electricity in the second stage of the game, q_i^{f*}, in (5.3).

A.3. Derivation of the Equilibrium Capacity of Renewable Power Plants in (5.9)–(5.12)

With firm i's optimal forward contracts, q_i^{f*}, in (5.3) and production by conventional generation, q_i^*, in (5.4), the first-order condition for profit maximization with respect to R_i in the first stage of the game yields

$$(-\beta)\left(1 + \frac{\partial Q^*}{\partial R_i}\right)(q_i^* + \delta R_i^*) + \frac{\partial q_i^*}{\partial R_i}(p^* - \gamma - \tau) + \delta p^* + s - \omega(1-\varphi)R_i^* = 0,$$
$$i = 1, \ldots, n, \tag{A.8}$$

where R_i^* denotes oligopolist i's optimal capacity of renewable power plants. Using (A.7), equation (A.2) in equilibrium can be rewritten as

$$p^* - \gamma - \tau = \beta\left(q_i^* - q_i^{f*} + \delta R_i\right) = \left(\frac{\beta}{n-1}\right)q_i^{f*}, \quad i = 1, \ldots, n. \tag{A.9}$$

From (5.4), we obtain

$$\frac{\partial Q^*}{\partial R_i} = \frac{-(n^2 + \delta)}{n^2 + 1}, \quad \frac{\partial q_i^*}{\partial R_i} = -\frac{n + \delta(n^2 - n + 1)}{n^2 + 1}, \quad i = 1, \ldots, n. \tag{A.10}$$

Inserting (A.7), (A.9), and (A.10) into (A.8) yields

$$q_i^{f*} \left\{ \frac{-\beta[2n + \delta(n-1)^2]}{(n-1)(n^2+1)} \right\} + \delta p^* + s - \omega(1-\varphi)R_i^* = 0, \quad i = 1, \ldots, n.$$

(A.11)

By replacing q_i^{f*} and p^* in (A.11) with q_i^{f*} in (5.4) and p^* in (5.5), equation (A.11) for $\delta = 0$ can be rewritten as

$$2\beta n R_F^* + 2\beta n \sum_{j=1}^{n} R_j^* + (n^2+1)^2 (s - \omega R_i^*) - 2n(\alpha - \gamma) = 0, \quad i = 1, \ldots, n,$$

(A.12)

and (A.11) for $\delta = 1$ can be rewritten as

$$s + \gamma + \tau - \omega(1-\varphi)R_i^* = 0, \quad i = 1, \ldots, n. \qquad (A.13)$$

Notice that if $\delta = 0$, only FIT is applied so that $\tau = 0$ and $\varphi = 0$.

In the case that $\delta = 0$, rearranging (5.8) yields the equilibrium capacity of renewable power plants of competitive fringe, R_F^*, which is equal to $2s/\omega$ under FIT, as indicated by (5.10). Assuming symmetry $R_i^* = R_j^*$ for $i \neq j$ and inserting $R_F^* = 2s/\omega$ into (A.12) yields oligopolist i's optimal capacity under FIT, as indicated by (5.9). For $\delta = 1$, (A.13) implies that oligopolist i's optimal capacity under FIP, ITC, and carbon pricing is given by (5.11). Using (5.5), (5.8), and (5.11), we obtain the equilibrium capacity of renewable power plants of competitive fringe firms in (5.12).

References

1. Abrella J, Koscha M, Rausch S (2019) Carbon abatement with renewables: evaluating wind and solar subsidies in Germany and Spain. J Public Econ 169:172–202
2. Acemoglu A, Kakhbod A, Ozdaglar A (2017) Competition in electricity markets with renewable energy sources. Energy J 38(SI):137–155
3. Aldy J, Gerarden T, Sweeney R (2018) Investment versus output subsidies: implications of alternative incentives for wind energy. In: NBER working paper 24378
4. Allaz B, Vila J (1993) Cournot competition, forward markets, and efficiency. J Econ Theor 59(1):1–16
5. Annan-Phan S, Roques F (2018) Market integration and wind generation: an empirical analysis of the impact of wind generation on cross-border power prices. Energy J 39(3):1–23
6. Ardian F, Concettini S, Creti A (2018) Renewable generation and network congestion: an empirical analysis of the Italian power market. Energy J 39(SI2):3–39
7. Arimura T, Iwata K (2015) An evaluation of Japanese environmental regulations. Springer
8. Ben-Moshe O, Rubin O (2015) Does wind energy mitigate market power in deregulated electricity markets? Energy 85:511–521
9. Bhandari V, Giacomoni AM, Wollenberg B, Wilson EJ (2017) Interacting policies in power systems: renewable subsidies and a carbon tax. Electr J 30(6):80–84
10. Brown D, Eckert A (2018) Imperfect competition in electricity markets with renewable generation: the role of renewable compensation policies. In: Working paper 2018-12. University of Alberta, Department of Economics

11. Bushnell J, Novan K (2018) Setting with the sun: the impacts of renewable energy on wholesale power markets. In: Energy Institute Working Paper 292, Energy Institute at Haas
12. Butler L, Neuhoff K (2008) Comparison of feed-in tariff, quota and auction mechanisms to support wind power development. Renew Energy 33(8):1854–1867
13. Ciarreta A, Espinosa MP, Pizarro-Irizar C (2017) Has renewable energy induced competitive behavior in the Spanish electricity market? Energy Policy 104:171–182
14. Cludius J, Hermann H, Matthes FC, Graichen V (2014) The merit order effect of wind and photovoltaic electricity generation in Germany 2008–2016: estimation and distributional implications. Energy Econ 44:302–313
15. Couture T, Gagnon Y (2010) An analysis of feed-in tariff remuneration models: implications for renewable energy investment. Energy Policy 38(2):955–965
16. Garcia A, Alzate JM, Barrega J (2012) Regulatory design and incentives for renewable energy. J Regul Econ 41(3):315–336
17. Genc TS, Reynolds SS (2019) Who should own a renewable technology? Ownership theory and an application. Int J Ind Organ 63:213–238
18. Green R, Vasilakos N (2010) Market behaviour with large amounts of intermittent generation. Energy Policy 38(7):3211–3220
19. Japan Ministry of Economy, Trade and Industry (2012) Report on costs of power plants, energy and environmental council (in Japanese)
20. Japan Ministry of Economy, Trade and Industry (2018) Taxation that aims at promoting investment for renewable energy (in Japanese). https://www.enecho.meti.go.jp/category/saving_and_new/new/information/180404a/summary.html. Accessed 17 July 2019
21. Lesser JA, Su X (2008) Design of an economically efficient feed-in tariff structure for renewable energy development. Energy Policy 36(3):981–990
22. Matsukawa I (2016) Consumer energy conservation behavior after Fukushima: evidence from field experiments. Springer
23. Matsukawa I (2019) Detecting collusion in retail electricity markets: Results from Japan for 2005 to 2010. Utilities Policy 57:16–23
24. Milstein I, Tishler A (2015) Can price volatility enhance market power? The case of renewable technologies in competitive electricity markets. Resour Energy Econ 41:70–90
25. Oliveira T (2015) Market signals and investment in variable renewables. Available at SSRN: https://ssrn.com/abstract=2695596
26. Percebois J, Pommeret S (2018) Cross-subsidies tied to the introduction of intermittent renewable electricity: an analysis based on a model of the French day-ahead market. Energy J 39(3):245–267
27. Pollitt M, Anaya K (2016) Can current electricity markets cope with high shares of renewables? A comparison of approaches in Germany, the UK and the State of New York. Energy J 37(SI2):69–88
28. Reguant M (2018) The efficiency and sectoral distributional implications of large-scale renewable policies. In: NBER working paper 24398
29. Reichenbach J, Requate T (2012) Subsidies for renewable energies in the presence of learning effects and market power. Resour Energy Econ 34(2):236–254
30. Ritz R (2016) How does renewable competition affect forward contracting in electricity markets? Econ Lett 146:135–139
31. Schmalensee R (2012) Evaluating policies to increase electricity generation from renewable energy. Rev Environ Econ Policy 6(1):45–64
32. Schneider I, Roozbehani M (2017) Wind capacity investments: inefficient drivers and long-term impacts. In: MIT Center for energy and environmental policy research working paper CEEPR WP 2017-002
33. Stoft S (2002) Power system economics: designing markets for electricity. Wiley-IEEE Press
34. Tietenberg TH (2013) Reflections—carbon pricing in practice. Rev Environ Econ Policy 7(2):313–329
35. Twomey P, Neuhoff K (2010) Wind power and market power in competitive markets. Energy Policy 38(7):3198–3210

36. von der Fehr NHM, Ropenus S (2017) Renewable energy policy instruments and market power. Scand J Econ 119(2):312–345

37. Woo CK, Horowitz I, Moore J, Pacheco A (2011) The impact of wind generation on the electricity spot-market price level and variance: the Texas experience. Energy Policy 39(7):3939–3944

38. Woo CK, Moore J, Schneiderman B, Ho T, Olson A, Alagappan L, Chawla K, Toyama N, Zarnikau J (2016) Merit-order effects of renewable energy and price divergence in California's day-ahead and real-time electricity markets. Energy Policy 92:299–312

39. Wurzburg K, Labandeira X, Linares P (2013) Renewable generation and electricity prices: taking stock and new evidence for Germany and Austria. Energy Econ 40(S1):S159–S171

Chapter 6
Behavioral Study of Demand Response: Web-Based Survey, Field Experiment, and Laboratory Experiment

Takanori Ida, Yoshiaki Ushifusa, Kenta Tanaka, Kayo Murakami, and Takunori Ishihara

Abstract This chapter analyzes customers' electricity conservation behavior when demand response is called for. Behavioral economics provides very useful insights to account for human behavioral anomalies such as *status quo bias*, *loss aversion*, *overconfidence*, *moral cost*, and *default bias*. The chapter is composed of five sections. The second section investigates a Web-based survey of residential electricity plan choice. The third section investigates a field experiment on residential electricity plan choice. The fourth section investigates a laboratory experiment on residential energy conservation. The fifth section investigates a field experiment on building electricity conservation.

6.1 Introduction

The Great East Japan Earthquake of 2011 resulted in the closure of several nuclear power plants, placing considerable pressure on the supply and demand of electric power. This in turn increased focus on demand response (DR). electric power suppliers implemented power usage reduction programs and increased electricity rates

T. Ida (✉) · T. Ishihara
Graduate School of Economics, Kyoto University, Kyoto 606-8501, Japan
e-mail: ida@econ.kyoto-u.ac.jp

T. Ishihara
e-mail: takunori.ishihara@gmail.com

Y. Ushifusa
Faculty of Economics and Business Administration, The University of Kitakyushu, Fukuoka 802-8577, Japan
e-mail: ushifusa@kitakyu-u.ac.jp

K. Tanaka
Faculty of Economics, Musashi University, Tokyo 176-0011, Japan
e-mail: k-tanaka@cc.musashi.ac.jp

K. Murakami
Graduate School of Agriculture, Kyoto University, Kyoto 606-8502, Japan
e-mail: murakamikayo@gmail.com

© Springer Nature Singapore Pte Ltd. 2020   117
T. Hatanaka et al. (eds.), *Economically Enabled Energy Management*,
https://doi.org/10.1007/978-981-15-3576-5_6

where electrical grids were destabilized or supply could not meet demand. This section presents an analysis of how consumers—both residential customers with home energy management systems (HEMS) and business customers with building energy management systems (BEMS)—responded to DR requests. To that end, *behavioral economics* provides useful insights regarding complicated anomalies in human behavior, such as *status quo bias* (Sect. 6.2), *loss aversion* (Sect. 6.3), *overconfidence* (Sect. 6.3), *moral cost* (Sect. 6.4), and *default bias* (Sect. 6.5).

The second section in this chapter presents an online survey of residential electricity plans and analyzes status quo bias against alternatives that became newly available following deregulation of Japan's residential electricity market. A choice experiment conducted via an online survey before deregulation revealed that the average Japanese consumer experiences status quo bias in electricity plan choice. Specifically, consumers preferred to remain with their current provider despite a clear 5% savings that would result from switching providers. Through a simulation of the potential share of new providers in the newly liberalized market, we found a larger potential market share for a 50% renewable energy plan compared with a plan offering a 7% bill reduction under price competition.

The third section considers a field experiment for investigating residential electricity plan choice. Specifically, this section examines differences in attitudes toward various plans according to whether information about participants' past electricity consumption is presented on electricity bills. We conducted a randomized control trial with a stated preference (SP) design on electricity rate choice before liberalization. The SP experiment suggested that informing participants of benefit or loss from switching corrected tendencies toward overconfidence, and that there is a decrease in evaluation values attached to the potential benefits of switching. Separate analyses of benefits and losses revealed that this decrease reflects a tendency toward loss aversion tendency. The evaluation value greatly drops when information of receiving negative result is provided.

The fourth section describes a laboratory experiment on residential energy conservation to analyze effects of social comparison information. This experiment provided some important findings. First, the results show that information provision based on social comparison encourages electricity conservation. Of course, such effects have been revealed in previous studies, but those studies focused on average behavioral trends and better behavior in society. We reveal that other types of social comparisons are also able to contribute to electricity conservation. Second, the experimental results imply that social comparison schemes have differing effects depending on the initial electricity demand of each individual. When individuals know trends in others' behavior for social betterment, they tend to align their own behavior to the social trends in order to maximize utility or minimize loss of utility (moral cost). When they realize they are not perceived as better behaved, however, or are unable to eliminate some undesirable social behavior without significant effort, such efforts are often abandoned. These findings imply that the effects of social comparison schemes depend on the initial demand situation of the distribution of targeted goods or services.

The fifth section investigates a field experiment on building electricity conservation in a working place with personalized lighting control, wherein a quasi-randomized controlled trial (QRCT) is presented, and the impact of automated DR on employees is analyzed. In particular, we reveal under what conditions the employees agree to participate in DR based on the experimental results. The results showed conditions that encourage employees' participation in DR. To this end, we prepared the following treatment groups: an opt-in performance incentive, an opt-in fixed incentive, an opt-out performance incentive, and an opt-out fixed incentive. The opt-out performance incentive group is then experimentally shown to achieve the highest participation rate among these four. We then estimated two average treatment effects, namely intent-to-treat and treatment-on-the-treated for DR. Electricity consumption during peak hours was found to be significantly reduced for all four treatment groups. We observed that although the DR participation rate was high in the opt-out group, average energy saving effects in that group were not significantly different from those in the opt-in group, and average energy savings of opt-in consumers exceeded those of opt-out consumers. Participation rates and power-saving effects showed a similar trade-off effect among employees who participated in DR.

This chapter reviews a series of the authors' publications [21, 22, 34, 41], which correspond to Sects. 6.2, 6.3, 6.4, and 6.5, respectively, extracting only key results from these papers. Readers who are interested in the details of each section are encouraged to refer to these original publications.

6.2   Web-Based Survey of Residential Electricity Plan Choice

6.2.1   Motivation

Since electricity deregulation in 2000, Japanese electricity market has been experiencing the largest deregulation in its national history, as in the USA and European countries in past decades. The Japanese market was completely liberalized on April 1, 2016. Japan newly opened the market of about 70 billion dollars that had been monopolistically operated by 10 electricity power companies (EPCOs). The readers are recommended to refer [31] for a brief summary of the history of Japanese electricity deregulation.

Deregulation of the electricity retail market will provide two remarkable benefits. First, Japanese consumers can expand their options for electricity services and providers, which means that, to some extent, consumers may make a impact on the future energy mix and lead to its restructuring. Another important benefit from deregulation is that we can achieve expected peak-time energy savings in the demand-side management.

Despite the solid benefits for consumers expected from the consumer-oriented future energy mix and the peak-time energy savings, default flat-rate tariffs are preferred by most of the Japanese residential electricity customers. One reason for this tendency is the fact that retail deregulation has just started so that many of the new electricity plans tend to give benefits to only those customers who consume daily large amounts of electricity, instead of numerous low consumption customers. As another potential reason, we should mention "status quo bias." Changing electricity providers or consumption plans may probably accompany some psychological burdens for Japanese customers who have so far bought electricity from local monopolistic electricity companies. In light of this, we can expect that consumers remain with their default providers or plans despite obvious savings gained by switching from the current provider or plan.

In many market and non-market situations [1, 5, 9, 15, 16, 32], including the electricity market [17, 24, 40], status quo bias in plan choice has been observed; for example, in a field experiment of electricity customers in the Sacramento Municipal Utility District (SMUD), we observed opt-in enrollment rates of less than 20% in dynamic pricing tariffs in sharp contrast to more than 90% in the case of opt-out enrollment [24, 40]. This example shows that a small percentage of customers select to opt-out when they are enrolled automatically to a new plan by default, while only about 20% of customers enroll when they face to take opt-in options. Using a choice experiment, we investigate status quo bias against new alternatives after the deregulation in the residential electricity market in Japan.

The objective of this section is to characterize the status quo bias of Japanese residential electricity customers when choosing optional plans, and to evaluate the extent of the bias. In addition, we will discuss effects of the consumer choice in the electricity retail market after deregulation on the future energy mix plan, based on understanding Japanese consumers' recent preferences for renewable and nuclear sources in the fuel energy mix.

6.2.2 Method

6.2.2.1 Web-Based Survey

Our Web-based survey was conducted in January 2016, three months before the complete deregulation of the electricity market on April 1. We randomly extracted 11,000 households in Japan from respondents registered with MyVoice Communications, Inc., a Japanese research agency, while considering the geographic characteristics, gender, and age, so as to represent the average Japanese population. We asked the respondents to answer our 17 queries that include eight hypothetical electricity plan choice situations. We paid a small remuneration to respondents in exchange for the questionnaire.

In order to grasp respondents' current circumstances associated with electricity consumption, the questionnaire is designed so as to enquire about a recent monthly

Table 6.1 Attributes and levels used in the choice experiment [reprinted from Murakami and Ida [34])

Attributes	Levels
Electricity provider	The current provider you traditionally subscribe to (default) A new provider (new)
Dynamic pricing	Flat-rate plan (default) Time-of-use plan (new)
Renewable power generation (%)	0, 20, 40
Nuclear power generation (%)	0, 20, 40
Monthly electricity bill	No reduction, 10% lower, 20% lower

electricity bill in the summer, the current electricity tariff type they subscribed to, ownership of private power generation systems typified by a rooftop photovoltaic (PV) system, and their awareness of electricity deregulation in the retail market. We also added information about the total electricity deregulation to the questionnaire in order to promote the respondents' understanding such that: they would be able to choose better options among the available plans, e.g., discount plans with other family bills such as gas (other energy), cell phone, or Internet (telecom), dynamic pricing rate plans, and electricity from renewable sources. Consequently, we surveyed respondents' intention to switch from the current provider to another provider. We also surveyed key factors that determined the provider choice after the retail deregulation. Finally, we posed eight hypothetical electricity choice situations, as described below.

In Table 6.1, the attributes of the choice experiment are summarized. We established 16 profiles through the orthogonal planning method, classified them into alternatives 1 and 2, and posed them with the status quo option (alternative 3). In Table 6.2, we show an example of one of the choice sets provided in the questionnaire. In the questionnaire, we asked the respondents to select their preferred option from three alternatives. Each respondent answered eight hypothetical plan choice questions.

6.2.2.2 Econometric Model

We analyzed statistically the response data collected from the surveys by using a random parameter logit (RPL) model. The RPL model is built based on the random utility theory and allows random taste variation [30], under the assumption that utilities vary at random. A utility function, consisting of a defined term V and a random term ε, is given by

$$U_i = V_i(x_i, m_i) + \varepsilon_i, \tag{6.1}$$

Table 6.2 Example of one of the choice sets provided in the questionnaire (Reprinted from Murakami and Ida [34])

	Plan 1 New alternative by a new provider	Plan 2 New alternative by the current provider	Plan 3 as a default
Electricity provider	A new provider	The current provider	The current provider
Dynamic pricing	Flat-rate plan	Time-of-use plan	Flat-rate plan
Renewable power generation (%)	40	20	10
Nuclear power generation (%)	20	40	10
Monthly bill	10% lower	10% lower	No reduction

Note Plan 1 was always provided by a new provider, while Plans 2 and 3 were provided by the current provider. Each alternative was presented under the same provider label across the eight choice situations

where x_i denotes an attribute vector of an alternative i, which specifies a new provider, TOU rate, and dependency on renewable and nuclear power generations, and m_i denotes a monetary attribute, which is the bill reduction rate (%).

The utility function has a linear-in-parameter form and can be written as follows:

$$V_{nit} = \gamma' m_{it} + \beta'_n x_{it}, \qquad (6.2)$$

where m_{it} and x_{it} are observable variables, γ is a fixed parameter vector indicating a numeraire, and β_n is a random parameter vector. Subscript n shows distinctive parameters for each individual, and subscript t shows the choice situation. Thus, V_{nit} is the conditional utility for respondent n who chooses energy service i in choice situation t.

Assuming that parameter β_n is distributed with a density function $f(\beta_n)$, we specify the model in such a way that each respondent is allowed to choose the constant coefficients of the model repeatedly by varying them in each respondent's choice situation. We express the logit probability of respondent n taking alternative energy service i in choice situation t as follows:

$$L_{nit}(\beta_n) = \prod_{t=1}^{T} \left[\exp(V_{nit}(\beta_n)) / \sum_{j=1}^{J} \exp\left(V_{njt}(\beta_n)\right) \right], \qquad (6.3)$$

which is the product of normal logit formulas given parameter β_n and the observable portion of utility function V_{nit} for all alternatives $j = 1, ..., J$ ($J = 3$ in this study) and all situations $t = 1, ..., T$ ($T = 8$ in this study). Then, choice probability is given as a weighted average of logit probability $L_{nit}(\beta_n)$ evaluated at parameter β_n with the density function $f(\beta_n)$, which is given by

$$P_{nit} = \int L_{nit}(\beta_n) f(\beta_n) d\beta_n. \tag{6.4}$$

On the basis of the above formulas, we demonstrate the parameters' variations at the individual level using the maximum simulated likelihood (MSL) method for estimation, with a set of 200 Halton draws. Accordingly, carrying out calculations based on Formula (6.2), we derive Willingness to Pay (WTP) values for each attribute.

Since γ indicates the marginal utility of the electricity bill (%), the marginal WTP (MWTP) for the lth attribute can be calculated, with the ratio of β_l to γ, as follows.

$$\mathrm{MWTP}_l = -\frac{\beta_l}{\gamma} \times \frac{\text{average monthly bill}}{100}. \tag{6.5}$$

6.2.3 Main Results

The estimation results using the utility function (6.2) are shown in Table 6.3 for all respondents. The parameters were assumed to be distributed normally. All random parameters are statistically significant at the 1% level. ASC1 (−), ASC2 (−), TOU (+), renewable (+), and nuclear (−) are represented by the statistical mean estimates, where the symbols in the parentheses are the signs for each estimate. Table 6.3 also displays the average WTP values, which are calculated with Eq. (6.5) in both Japanese Yen (JPY) and US Dollar (USD). Moreover, we find in Table 6.3 that the respondents had a negative WTP of −$4.55 on a monthly basis for a new electricity alternative led by a new entrant. Similarly, the respondents had a negative WTP of −$1.30 for a new electricity alternative from the current company. From these results, we see that respondents had a strong status quo bias against new alternatives, especially focusing on electricity led by a new entrant. Summarizing the above observations, we could say that consumers would not consider switching to a new provider unless a 5% bill reduction is expected, while a 2% bill reduction sufficed for them to select switching to a new plan presented by the current company.

In Table 6.3, we also see that the respondents were willing to pay $1.16 for the TOU tariffs, which implies that they expected larger savings than the average by switching to a TOU tariff. As for the WTP for different energy sources, Japanese consumers' WTP for a 1% increase in dependency rate on renewable power and nuclear power resulted in $0.18 and −$0.35 per month, respectively, together with some decrease in three years since the previous study.[1]

Focusing on the electricity bill and the dependency on renewables, we analyzed the potential market share of new entrants with respect to several feasible competitive strategies, based on the estimated results shown in Table 6.3. More specifically, using

---

[1] According to the results of [33], the average marginal WTPs for a 1% dependency on renewable and nuclear sources were $0.28 (31 JPY if $1=110 JPY) and −$0.65 (−72 JPY if $1=110 JPY), respectively.

Table 6.3 Estimated utility function (Reprinted from Murakami and Ida [34])

	Coefficient	
	Mean	s.d.
Fixed parameter		
Monthly bill (Current status = 1)	−17.7834*** (0.2432)	
Random parameters		
ASC1 [Plan 1] (New plan by a new provider)	−0.9497*** (0.0362)	2.1396*** (0.0378)
ASC2 [Plan 2] (New plan by a current provider)	−0.2704*** (0.0346)	2.0077*** (0.0366)
TOU (0,1)	0.2411*** (0.0276)	1.5982*** (0.0345)
Renewable (%)	0.0375*** (0.0011)	−0.0592*** (0.0013)
Nuclear (%)	−0.0720*** (0.0014)	0.1024*** (0.0015)
Average monthly bill (JPY)	9375	
Marginal utility for 1 JPY	−0.002	
WTPs	WTPs (JPY)	WTPs (USD)
ASC1 [Plan 1 by a new provider]	−501	−4.55
ASC2 [Plan 2 by a current provider]	−143	−1.30
TOU (0,1)	127	1.16
Renewable (%)	20	0.18
Nuclear (%)	−38	−0.35
LR chi2 (5)	23,284	
Log likelihood	−48,625.006	

Note Halton = 200; 1 USD = 110 JPY; N. of observation = 194,088 (3 × 8 × 8087); *** indicates statistical significance at 1% level

the estimated parameters in the left column of Table 6.3, we calculated the relative shares of new entrants [Plan 1] compared to current providers' new alternatives [Plan 2] under certain given conditions of Plan 2 and Plan 3, where the conditions are defined by the current provider, flat-rate plan, 15% renewables, 2% nuclear power generation, and 5% lower monthly bill for Plan 2 and by the current provider, flat-rate plan, 15% renewable power generation, 2% nuclear power generation, and current level of monthly bill for Plan 3.

Table 6.4 Potential shares of new providers (Reprinted from Murakami and Ida [34])

		Renewable energy		
Bill (%)		RE15 (%) (current)	RE50 (%)	RE100 (%)
+10		2	8	39
+5		5	17	61
±0	(current)	**12**	33	79
−5		25	55	90
−10		44	75	96

Note We calculated the relative shares of new entrants [Plan 1] to current providers' new alternatives [Plan 2] under the given conditions of Plan 2 (the current provider, flat-rate plan, renewable power generation 15%, nuclear power generation 2%, monthly bill 5% lower), and Plan 3 (the current provider, flat-rate plan, renewable power generation 15%, nuclear power generation 2%, monthly bill 0% lower (current level)) by using estimated parameters in Table 6.3

From the simulated results shown in Table 6.4, we see that new alternatives that enhance renewables pose a large potential ability to increase the shares of new entrants. Specifically, under the condition that a current provider carries out a 5% bill reduction plan as a new alternative (Plan 2), the current share of new providers is 12%. Despite this drawback, the strategic plan of 50% dependency on renewable energy (RE50) brings a 34% potential share, which is larger than the expected share when they execute a 7% bill reduction plan. Moreover, when new providers choose a 100% renewable energy plan (RE100), the share of new providers has the potential to be 39%, even though they would incur a 10% additional cost. This target is equivalent to that they present a 9% bill reduction plan for price competition.

6.3   Field Experiment on Residential Electricity Plan Choice

6.3.1   Motivation

After the power retail liberalization of the Japanese electricity market in April 2016, the switching rate by March 2017 was about 8.8% (or 5.53 million households), of which the rate of switching to a new power company was about 4.7%. In addition, the conversion rate from a regulated tariff plan to a free tariff plan within the large power companies was only approximately 4.1%. The reason is inertia [18]. In other words, consumers do not seek alternative tariff plans or utilities [19], and they do not desire to exert energy to look for another plan or power company. Hence, even if the current plan is more expensive than that of the others, consumers may select to retain their current plan. In such situations, it is possible that some suitable information induces consumers to change their choice [4, 13].

Moreover, the converted consumers may believe that they would be able to reduce their monthly electricity bills by the Time-of-Use (TOU) tariff and may have made the switch independent of their actual consumption patterns. This tendency is in general termed "overconfidence," and it has been revealed in the attitude toward investment [3] and public expectations about their own abilities [6, 8, 10]. Hence, even if the consumers attempt to switch the power plan and the company, there remains a possibility that they may not choose suitable ones due to insufficient understanding of their own consumption patterns.

The Read, Evaluate, and Compare Alternative Prices (RECAP) approach to giving information is known for its effectiveness [42]. Under the situation of information asymmetry whereby the enterprise has more information on the consumers than themselves, providing the consumers' information on their own consumption and payments will make it possible that they understand which plan will benefit them and improve consumer welfare [25]. Kling et al. [26] empirically inspected changes in consumer choice by giving such RECAP-type information. They also revealed that about 11% of the consumers will change their decision adequately if provided with information on the current plan and the cheapest plan in the context of health plans for the elderly in the insurance market.

In this chapter, the authors propose an experiment to investigate and examine if consumers would select a proper rate plan in the presence of the RECAP-type information. Especially, stated preference (SP) regarding the choice of power plan before liberalization is considered. In order to verify the effect of information provision, a randomized control trial (RCT) is performed. In RCT, subjects are randomly assigned to control groups and treatment groups. The control groups do not receive the intervention, whereas treatment groups do it. Due to the random assignment, the covariates in the treatment and control groups are homogeneous on average and the difference in average outcome can be interpreted as that caused by the intervention. Hence, the internal validity of the average treatment effect is secured by conducting RCT. The intervention performed in the experiment regards the detailed electric power usage data per 30 min for each subject in the summer of 2015 as the RECAP-type information provision. Each subject in the treatment group receives monthly electricity rates calculated by each tariff plan: a flat plan that does not change the tariff rate, and a TOU plan in which the daytime rate and the night rate are different. The price differences are also provided. Then, it is observed that the overconfidence is avoided and the overall evaluation value decreases when consumers receive information on gaining or losing by changing their electricity rate plan in the SP experiment. It is particularly found that, when consumers know that money by switching will be lost, loss aversion is obvious, and the evaluation value is greatly reduced. Moreover, there was no response to the information based on a potential gain.

6.3.2 Method

6.3.2.1 Experimental Design

The authors performed a SP experiment on electricity tariff plans before the power retail liberalization in April 2016. 1063 households that are along the Tokyu Railway Line in Aoba Ward, Yokohama City, Kanagawa Prefecture, Japan participated in this experiment. The target households installed HEMSs that provide power consumption data at every 30 min from June 2015 to September 2015. Note that, due to the protection requirements of personal information, power consumption could not be acquired outside the survey period.

The schematic of the experiment is illustrated in Fig. 6.1. We randomly divided the 1063 participants into two groups: the control group without intervention ($n = 531$) and the treatment group ($n = 532$) with intervention. The SP survey experiment was performed from February to March 2016, just before the liberalization in Japan on April 2016. As a result, 983 participants were contacted in the experiment, but the remaining participants did not respond.

The information provision is provided as follows. For the 531 participants in the treatment group, we enclosed the questionnaire shown in Fig. 6.2. In this intervention text in Fig. 6.2, we offer the following three information. (1) We provided the monthly electricity bill payment calculated with the standard electricity tariff plan (25 yen per 1 kWh; FLAT) based on the data on electricity consumption in summer 2015. (2) We presented the monthly electricity bill payment per month calculated based on the time-of-use (TOU) tariff plan. (3) We presented information on how much

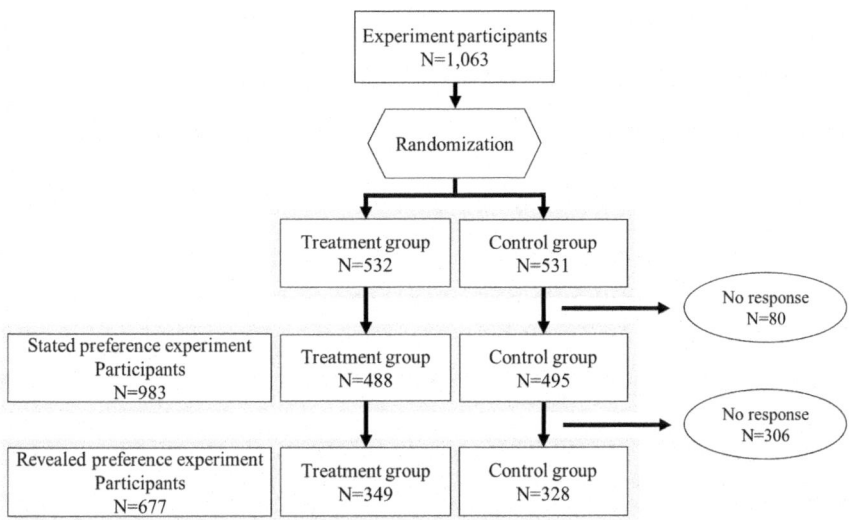

Fig. 6.1 Flow of the experiments (Reprinted from Ishihara and Ida [22])

Mr. ○○ ○○

Questionnaire response procedure

Kyoto University
Graduate School of Economics

Please follow the steps below to answer the questionnaire.

(1) First of all, please read the article " Which rate plan is better for you?" in the box.

(2) Based on this article, please continue to answer the "Electricity Questionnaire Survey". (Please fill in the answer postcard.)

Which rate plan is better for you?

We use your household's summer 2015 (July to September) electricity usage data (measured every 30 minutes by HEMS) to calculate electricity rate with two virtual electricity rate plans and compare them.

① **Per-use lighting rate plan (24-hour flat rate)**

Generally, it is said to be a recommended menu for those who want to use electricity without worrying about time of day or day of the week, or those who with a high rate of usage in the daytime.

[Virtual rate plan]   24 hours uniform   Approximately 25 yen per 1 kWh

Summer electricity rate calculation result of your home   **15,000** yen per month

② **Time-of-use rate plan**

Under this plan, usage at night is more advantageous. It is recommended for those households where the family is out during the daytime and there are many people at home at night.

[Virtual rate plan]   Daytime (9 a.m. to 9 p.m.)   Approximately 40 yen per 1kWh
                      Nighttime (9 p.m. to 9 a.m.)   Approximately 8 yen per 1kWh

Summer electricity rate calculation result of your home   **10,000** yen per month

③ **Comparison of ① Per-use lighting rate plan and ② Time-of-use rate plan**

Comparing the two virtual electricity rate plans,

When changing from ① **Per-use lighting rate plan** to ② **Time-of-use rate plan**,

5,000 yen cheaper per month

Fig. 6.2 Intervention flyer (Reprinted from Ishihara and Ida [22])

lower/higher the cost per month would be brought by the switch from the general electricity tariff plan to the TOU.

The power unit price for daytime from 9:00 a.m. to 9:00 p.m. of the TOU plan is 40 yen per kWh, and the power unit price for night from 9:00 p.m. to 9:00 a.m. is 8 yen per kWh. Revenue neutrality for the unit price of this TOU plan is assumed. Specifically, we initially set the unit price of electricity for daytime to be 40 yen per kWh, and then set that for night so that the monthly electricity rate payment among the participants is equivalent on average regardless of the selected plan.

It is shown in Table 6.1 and Fig. 6.3 that revenue neutrality holds. In Table 6.1, for each participant, the HEMS data for the summer of 2015 are used in order to compute the payments for the FLAT plan and the TOU plan, and the average price per month. Here, assuming that price elasticity is zero, the electricity rate and the average difference per month between FLAT and TOU are calculated. Consequently, the monthly average electricity rate was 9428.82 yen for the FLAT plan and 9428.80 yen for the TOU plan, respectively. The difference is about 0.02 yen, and we see that the monthly electricity rate is ±0 as a whole, on average, when to switch from FLAT plan to TOU (Table 6.5).

The gap between the FLAT and the TOU plan is distributed as shown in Fig. 6.3, wherein the gap on the vertical axis is arranged for customers who switch on the horizontal axis. It is immediate to see that the half of people gain benefits from the switch from FLAT to TOU while another half do not.

In order to validate the randomization for the aforementioned control and treatment groups, the balance check was performed through t test for the data on the average power consumption, the data during the daytime and the data at night (Table 6.6), which is based on power consumption from July to September 2015. We see from the table that the power consumptions for the control group and treatment group

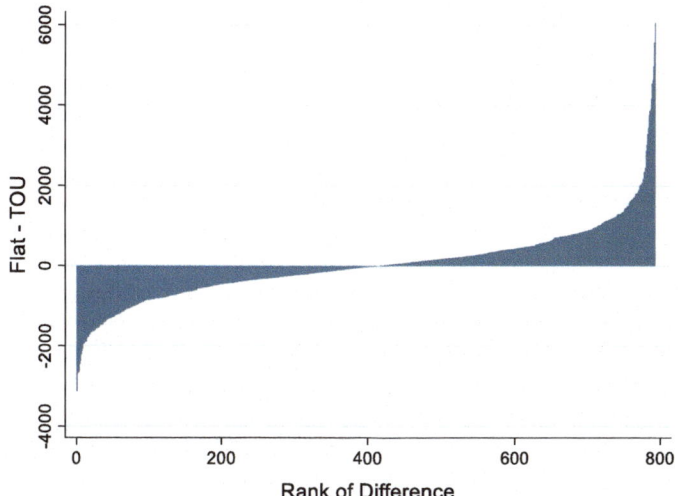

Fig. 6.3 Distribution of monthly payment difference (Reprinted from Ishihara and Ida [22])

Table 6.5 Monthly rate difference according to the tariff plan (Reprinted from Ishihara and Ida [22])

	FLAT plan (JPY)	TOU plan (JPY)	Difference (JPY)
Mean	9428.819	9428.802	0.017
Standard deviation	4433.424	4515.220	883.082
Min	0	0	−3534
Max	39,007	37,627	6052
N	1063	1063	1063

Table 6.6 Balance check of average power consumption (Reprinted from Ishihara and Ida [22])

	Control group	Treatment group	Difference
Daily (kWh) (s.d.)	12.9222 (5.6715)	13.0235 (5.4440)	−0.1013 (s.e. = 0.3464)
Daytime only (kWh) (s.d.)	6.8686 (3.1844)	6.9714 (3.0190)	−0.1028 (s.e. = 0.1933)
Nighttime only (kWh) (s.d.)	6.0536 (2.7949)	6.0521 (2.7258)	0.0015 (s.e. = 0.1721)
N	531	532	

were about 12.92 kWh and 13.02 kWh, respectively. The gap, say −0.10, is not large enough (t value $= -0.29$). The difference between these groups in the daytime consumption is −0.10, which is also not large (t value $= -0.53$). The difference in the nighttime, about 0.001 kWh, is also not statistically significant (t value $= 0.0008$). Consequently, we do not find any significant difference in the averages for daily, or daytime and nighttime only, power consumption. Hence, the two groups are concluded to be well balanced.

6.3.2.2 Stated Preference Experiment

We performed experiments from February 2016 to March 2016 before the retail liberalization. A 1000 yen Quo voucher is provided to the participants in the experiment. Nine hundred eighty-three 983 (C: $n = 495$, T: $n = 488$) participated in the experiment, whereas the remaining potential participants did not respond.

In this experiment, we asked people, in both of the control and treatment group, eight questions using the stated preference method questionnaire to choose the most preferable plan among three virtual power plans (Fig. 6.4). The stated preference method is a type of conjoint analysis whereby we assume that good service consists of several attributes. The experimenter presents a plurality of virtual options to the

	Plan 1	Plan 2	Plan 3
Type of power company	New power company	Power company currently under contract	Power company currently under contract
Type of tariff plan	Time-of-use or Flat rate plan	Time-of-use or Flat rate plan	Flat rate plan
The ratio of renewable energy	0%, 20%, 40%	0%, 20%, 40%	10%
The ration of nuclear energy	0%, 20%, 40%	0%, 20%, 40%	10%
Monthly electricity charge	Same as the present, 10% decrease, 20 decrease	Same as the present, 10% decrease, 20 decrease	Same as the present

Fig. 6.4 Stated preference method question attribute values (Reprinted from Ishihara and Ida [22])

subject for which the property levels of the options are gradually changed, and the subjects choose an option. Consequently, one can analyze what is the dominant attribute to what extent based on the option that the subject chose.

As shown in Fig. 6.4, we presented the three options, namely "plan by new power company" (Plan 1), "new plan by currently under contract" (Plan 2), and "existing charge plan by company currently under contract" (Plan 3). The attribute related to the power company in Plan 1 is fixed to "new power company," while that in Plan 2 is to "the company currently under contract." In both Plan 1 and 2, the attributes on the rate plan are set to the alternatives "time-of-use plan" or "Flat rate plan." Moreover, we prepare three options, 0, 20, and 40%, for "renewable energy ratio" and "nuclear power ratio," respectively. Lastly, we also set three options "the same as the present," 10% lower, and 20% lower for "Monthly electricity charges." All the attribute values in Plan 3 are fixed to those in Fig. 6.4.

On the basis of the data acquired by the above experiment, we hypothesize a random utility model formulated as below and then estimate the evaluation for each attribute by the maximum likelihood method.

$$U_{ni} = \beta_n x_{nit} + \gamma m_{nit} + \delta I_{nit} + \varepsilon_{ni} \qquad (6.6)$$

The subscript n describes the number of the individual, i denotes the plan ($i \in (1, 2, 3)$) in each question, and t represents the number of the question ($t \in (1, 2, \ldots, 8)$). In other words, U_{ni} represents the utility of individual n when he/she chooses rate plan i. The symbol x_{it} denotes the variable indicating each attribute level of the plan i presented in query t. The symbol m_{it} indicates the attribute level of monthly electricity rates, and I_{nit} is the indicator function that takes 1 when the individual n chooses Plan 3, and 0 otherwise. Moreover, β_n is a random parameter that obeys the normal distribution, which indicates the marginal utility for the change of one attribute level of each attribute of the respondent, γ is the marginal utility associated with the decrease in monthly electricity charges, and δ is the basic utility for the current plan. The symbol ε_{ni} is an error term and is assumed to obey the type 1 extreme value distribution.

Let us now assume that an individual n chooses option i that maximizes (6.6). The probability that individual n chooses option i is then given as below for any $j(\neq i)$:

$$P_{ni} = \Pr\left(U_{ni} > U_{nj}\right) \tag{6.7}$$

In addition, the log-likelihood function of this probability is expressed as follows.

$$L(\beta, \gamma, \delta) = \sum_n \sum_j I_{nj} \ln P_{ni}, \tag{6.8}$$

where $I_{ni} = I\left(U_{ni} > U_{nj}\right)$ is an indicator function that takes 1 if individual n chooses option i, and 0 otherwise.

Assuming a mixed logit model for individual selection probabilities, Eq. (6.7) can be rewritten as follows.

$$P_{ni} = \int \frac{\exp(V_{ni})}{\sum_j \exp\left(V_{nj}\right)} f(\beta) d\beta, \tag{6.9}$$

where $V_{ni} = \beta_n x_{nit} + \gamma m_{nit} + \delta I_{nit}$. Note that the selection probability (6.9) cannot be solved algebraically. It is thus required to approximately acquire the parameter through simulation such as in the Halton sequence method. Moreover, as the parameters acquired from the simulation hardly present intuitive interpretations, marginal willingness to pay (MWTP) is calculated from each random parameter [43]. As the parameters indicate the marginal utilities of the attributes, MWTP is acquired by dividing them on the basis of the marginal utility of money per yen. Meanwhile, the marginal utility of money cannot be estimated directly from our question attributes. Hence, the marginal utility per yen is computed by dividing γ, by the monthly electricity bill computed for the individuals' FLAT payments.[2]

Additionally, although the above parameter is an average parameter, we can compute its conditional distribution according to the respondent by the Bayesian theorem in the mixed logit model. Accordingly, MWTP for each individual is calculated by dividing the random parameters by the marginal utility per yen.

---

[2]In estimating the selection probability, the attribute value of "Monthly electricity charges" is replaced with 1 if it is "the same as the present", 0.9 if it is "10% decrease", and 0.8 if it is "20% decrease".

6.3.3 Main Results

6.3.3.1 Estimation Result of Marginal Willingness to Pay

The stated preference method-based analyses for both treatment and control group are summarized in Table 6.7. Let us first focus on the constant parameters. The coefficient on the electricity bill payment ratio for the control group was about −23.91, while that for the treatment group was about −23.15, both of which were significant at the 1% level. Moreover, we see from Table 6.7 that the parameter on

Table 6.7 Estimated result by stated preference method (Reprinted from Ishihara and Ida [22])

	Control group	Treatment group
Fixed parameters		
Electricity payment ratio	−23.9120***	−23.1459***
(s.e.)	(1.2888)	(1.1785)
Status quo	0.0136***	0.3555***
(s.e.)	(0.1104)	(0.1100)
Random parameters (mean)		
New company	−0.5961***	−0.5771***
(s.e.)	(0.0981)	(0.1060)
TOU	0.2516***	−0.1379
(s.e.)	(0.1217)	(0.1216)
Renewable energy ratio (%)	0.0527***	0.0514***
(s.e.)	(0.0056)	(0.0059)
Nuclear energy ratio (%)	−0.1245***	−0.1348***
(s.e.)	(0.0077)	(0.0079)
Random parameters (standard deviation)		
New company	1.6129***	1.7954***
(s.e.)	(0.1115)	(0.1210)
TOU	1.7964***	1.8356***
(s.e.)	(0.1318)	(0.1496)
Renewable energy ratio (%)	0.0785***	0.0853***
(s.e.)	(0.0067)	(0.0058)
Nuclear energy ratio (%)	0.1280***	0.1378***
(s.e.)	(0.0078)	(0.0077)
R^2	0.4121	0.3892
LRI	−2557.5447	−2619.7779
obs	3960	3904

*** indicates statistical significance at 1% level, ** indicates statistical significance at 5% level, and * indicates statistical significance at 10% level

the status quo was 0.014 in the control group and was significant at the 5% level. In the treatment group, it was 0.355, which was significant at the 1% level.

We next focus on the random parameters. The parameter on the new company was −0.60 in the control group, while −0.58 in the treatment group, which were significant at the 1% level. The TOU coefficient for the control group was 0.25 which is significant at the 1% level. On the other hand, the coefficient for the treatment group was −0.14, and there was no significant result. The parameters on the renewable energy ratio for control and treatment group were 0.053 and 0.051, respectively, both of which were significant at the 1% level. With respect to the parameters on the nuclear power ratio, the control group had −0.12, while the treatment group had −0.13; significant differences were found at 1% level for both. Table 6.7 also presents the verifications on the random parameters based on the standard deviation. The authors omit presenting the details in this chapter, but it was confirmed that all the parameters were statistically significant at the 1% level, indicating that the parameters vary within each group. The readers are recommended to refer to [22] for more detailed analysis.

In summary, we conclude that all parameters are significant, except the TOU parameter for the treatment group.

Table 6.8 summarizes the marginal utility of money and estimates of the WTP for the attributes, where the former is calculated by dividing the estimation result of the electricity bill payment ratio parameter by the monthly electricity bill. The marginal utility of money averaged −0.0034 in the control group and was statistically significant at the 1% level (t value $= -12.27$). The mean the treatment group was −0.0032 and also significant at the 1% level (t value $= -11.48$). We cannot find any significant difference in these differences (t value $= -0.42$).

Table 6.8 Marginal willingness to pay (Reprinted from Ishihara and Ida [22])

	Control group	Treatment group
Marginal utility of money (s.e.)	−0.0034*** (0.00028)	−0.0032*** (0.00028)
Marginal willingness to pay (JPY)		
New company (s.e.)	−238.0432*** (26.9045)	−252.9766*** (28.8248)
TOU (s.e.)	125.9255*** (26.5207)	−9.0380 (26.6762)
Renewable energy ratio (%) (s.e.)	22.4005*** (1.2210)	23.7203*** (1.4834)
Nuclear energy ratio (%) (s.e.)	−49.0696*** (2.2122)	−56.4047*** (2.5634)
N	464	488

*** indicates statistical significance at 1% level, ** indicates statistical significance at 5% level, and * indicates statistical significance at 10% level

Moreover, Table 6.8 also presents the MWTP for each attribute calculated from the above marginal utility of money and the estimates of the random parameters. The estimated random parameters are divided by the marginal utility of money and then multiplied by -1. In terms of the new companies, MWTP was -238.04 yen (t value $= -8.85$) in the control group, while -252.98 yen (t value $= -8.78$) in the treatment group. Note that both have statistical significance at the 1% level. There was no significant difference in the difference in MWTP between the control and treatment groups (t value $= 0.34$). With respect to TOU, the MWTP was 125.93 yen (t value $= 4.75$) in the control group. The MWTP was significant at the 1% level. Meanwhile, in the treatment group, the MWTP was -9.04 yen (t value $= -0.34$). This result was itself not significant, but there was a statistically significant difference at the 1% level (t value $= 3.59$). Regarding the renewable energy ratio, the control group had 22.40 yen (t value $= 18.35$), while the treatment group had 23.72 yen (t value $= 15.59$), which were both statistically significant at the 1% level. We found no significant difference in these differences (t value $= -0.69$). Last, the nuclear energy ratio was 49.07 yen for the control group (t value$= -22.18$), and it was 56.40 yen for the treatment group (t value$= -22.00$). They were statistically significant at the 1% level. The difference between the two groups was also statistically significant at the 5% level (t value $= 2.17$).

6.3.3.2 Estimation Results of Average Treatment Effect

Let us finally estimate the treatment effects on the basis of the following equation.

$$\text{MWTP}_n = \alpha + \beta \cdot \text{Recap}_n + \varepsilon_n, \qquad (6.10)$$

where MWTP_n, which is the dependent variable, is the MWTP for TOU of individual n. The explanatory variable Recap_n is a dummy variable that takes 1 if the individual n belongs to the treatment group and 0 otherwise. Due to the random assignment for both the groups, the coefficient β is interpreted as the average treatment effect (ATE) for receiving information. The constant parameter α is regarded as the average MWTP for the control group, and ε_n is an error term that obeys the zero mean normal distribution.

We then estimate the effect of information treatment on MWTP for TOU, and the results are summarized in Table 6.9, wherein we focus only on the presence/absence of the effect of the intervention. The effect and constant were about -135 yen and 126 yen, respectively, which were both statistically significant at the 1% level.

It is observed that the estimated value of MWTP for TOU is higher than 0 yen in the absence of the RECAP information. The RECAP information on the plan changes would eliminate this tendency, and the evaluation value would get close to 0. Indeed, when testing the sum of the constant term and the coefficient of the RECAP dummy, we found no statistical significance (t value $= 0.12$).

In summary, we revealed differences in the evaluation of each plan depending on the situations, loss or gain, and with/without additional information. If people

Table 6.9 OLS estimation results—overall treatment effect (SP) (Reprinted from Ishihara and Ida [22])

	(1)
Cons	125.9255***
(s.e.)	(26.9529)
RECAP	−134.9636***
(s.e.)	(37.6456)
R^2	0.0133
adj R^2	0.0123
N	952

*** indicates statistical significance at 1% level, ** indicates statistical significance at 5% level, and * indicates statistical significance at 10% level

are in the gain area, we found no remarkable difference between with and without information. On the other hand, people in the loss area are rather sensitive for the label that they would lose by switching and tend to greatly underestimate the value of the option.

6.4 Laboratory Experiment on Residential Energy Conservation

6.4.1 Motivation

Energy conservation programs without any price control have gained increasing interest in recent years. One of the most effective ways to enhance voluntary electricity conservation is by providing information based on the so-called social comparison. The social comparison involves considering comparative information with other people. Recently, economists are devoted to constructing a utility model that incorporates psychological factors. Levitt and List [28] summarized such ideas as "moral cost." Allcott [2] implies that people can receive utility through electricity conservation. He mentioned that the term moral utility depends on beliefs about the social norm. It seems likely that untreated households believe that they are closer to the social norm than they actually are, implying that the treatment causes low (high) usage households to update their beliefs about the social norm upward (downward). Therefore, an information provision scheme that focuses on social comparison is expected to encourage the voluntary conservation of electricity. Allcott [2] revealed that the social comparison trial by OPWER can decrease electricity consumption from 1.4 to 3.3%.

However, recent studies found that the effects of social comparison schemes have a cross effect with the subject's characteristics. For example, Costa and Kahn [7] revealed that the effectiveness of an energy conservation scheme based on social comparison depends on an individual's ideology. In addition, information provision by OPWER's home energy report (HER) includes several kinds of information that may affect the consumer's behavior. For example, each consumer is not only informed of the average electricity use of their neighbor by HER. Additionally, they are informed the average electricity use of their neighbors that are the efficient user of electricity. In such a case, it is difficult to know which information is more effective for enhancing electricity conservation. In principle, "Moral cost" includes several kinds of factors, such as the need for admiration, preference for equality, a feeling of guilt, and so on. Previous studies have already found several types of factors regarded as "moral cost." Therefore, each piece of information has a different type of effect that may be classified as "moral cost."

In this study, we reveal the social comparison effect for electricity conservation through a laboratory experiment based on a hypothetical decision-making situation of electricity use. It should be emphasized that our study is far from a simple laboratory experiment. Actually, the experiment invites actual residential people as subjects. Moreover, the experimental setting is determined based on the situation of actual electricity usage by each subject. The subjects in this experiment have already participated in the field experiment in the Yokohama area. In this field experiment, the electricity company distributed HEMS to each subject which allows us to use real data for the subjects. Based on hourly electricity use data, we set up the initial setting (electricity used in summer season) for each subject in this experiment. This is why our experiments are expected to capture more realistic behavior than a typical laboratory experiment.

6.4.2 Method

In the present experiment, each subject hypothetically determines the temperature set point of an air conditioner. Each session consists of 10 periods. In each period, the subject makes a decision based on a hypothetical situation. The experiment is conducted in summer, the hottest season in Japan. Each subject determines the set point during the peak hours (from 13:00 to 16:00). The electricity price is set to 25 Japanese yen per kWh that is determined based on the general pricing in and around Tokyo region (the consumer electricity price of Tokyo Electric Power Company Holdings, Incorporated). However, the price of electricity may fluctuate, as critical peak pricing (CPP) is implemented when the demand gets close to the capacity limit. In this experiment, the total demand for electricity is decided exogenously and accordingly each subject's decision-making does not affect the occurrence of CPP). Each subject can choose temperatures from 25 °C (comfortable temperature) to 29 °C (discrete choice). If the subject makes a decision of not using the air conditioner, the room's temperature becomes 30 °C. If he/she increases the set point by 1 °C from the base temperature of 25 °C, the total electricity usage of each period (day) is reduced by 10%. In other words, the subject can decrease the electricity fee by choosing a higher temperature.

Although the subject can decrease their own electricity consumption by opting for a higher temperature set point, he/she must be willing to accept a possibly uncomfortable situation due to a higher temperature. It is necessary to add such preferences about a comfortable temperature to live in. In this experiment, we measure the willingness to accept (WTA) higher temperatures through a questionnaire done before the initial period, where we ask each subject the following question: "How much money do you require to sustain the uncomfortable situation with an increase in temperature?" The same question is asked for all five possible temperatures from 25 to 30 °C to reflect the non-constant increment of WTA with respect to the 1 °C rise of the temperature. Based on the results of the questionnaire, the payoff is calculated as follows:

The payoff

   = Initial endowment−Electricity price × Electricity usage

   −WTA for higher temperature

   = Initial endowment−Electricity price

   × (Initial amount of electricity usage − Conservation amount of electricity)

   −WTA for higher temperature

In this formula, the initial amount of electricity usage is fixed based on the actual consumption data of each subject at the same time as the situation of the experiment. We calculate the actual average electricity consumption of each subject at the peak time in summer season. Then, we apply the data to initial setting of the experiment.

In each period, subjects can monitor the information about electricity usage, namely their own electricity usage in each period, before each period, and the electricity price in the current period. Besides, we provide information about another subject's electricity usage to a group of the subjects, termed treatment group, and do not provide such information to another group, termed control group. The treatment group receives this information at the beginning of the next period.

We further divide the treatment group into subgroups and provide different information among these subgroups. In treatment 1, each subject receives information about the electricity usage of a "Non-efficient" person, where the non-efficient subject is selected from the top 10% of subjects in the same session. In treatment 2, subjects receive the average consumption amount of all subjects in the same session. In treatment 3, each subject receives information about the electricity usage of "Efficient" subjects, where the efficient subjects belong to the bottom 10% of subjects in the same session. Allcott [2] and other related work on OPWER include the effects of our treatment 2 and treatment 3. If the direction of social comparison effect is symmetrically the same, the results of our experiments are expected to show the same amount of reduction of electricity use. These information updates in the next period based on the result of the decision in the previous period.

In this study, we implemented seven sessions to reveal the effects of the information supplied to each treatment group. Table 6.10 describes the details about the

Table 6.10 Details about the experiment sessions (Reprinted from Tanaka et al. [41])

Session no.	Date and time	The number of subjects	Treatment
1	January, 20, 2017 (10:00)	26	Treatment 2
2	January, 21, 2017 (10:00)	25	Treatment 1
3	January, 21, 2017 (14:00)	27	Treatment 3
4	January, 27, 2017 (10:00)	25	Treatment 2
5	January, 27, 2017 (14:00)	23	Treatment 1
6	January, 28, 2017 (10:00)	39	Control
7	January, 28, 2017 (14:00)	37	Treatment 3

experiment session including the date, the number of the subjects, and the treatment. The total number of subjects in this experiment was 202. These experiments were implemented by z-Tree (see details in [11]). We gathered the voluntary participants around Yokohama area in Japan. The sessions were conducted at the laboratory experimental room at the Faculty of Political Science and Economics and PC room of the School of Social Sciences, Waseda University.

The subjects first took a value orientation test, developed in the field of psychology (see, e.g., [14, 29]) that examines each subject's preference for cooperation. The results of the test provide the weights on his/her own welfare and welfare of others that the individual has [35]. Our test is designed based on [36]. We measured the level of altruism based on the vector measured by the choices of the value orientation test. Subjects' characteristics are then classified into five categories, namely "Altruistic," "Cooperative," "Individualistic," "Competitive," and "Aggressive."

6.4.3 Main Results

Table 6.11 and Fig. 6.5 both describe the fraction of the temperature set points selected in each treatment group. In the control group, the probability of "No use of air conditioner" is lowest in all groups. Furthermore, the choice probability of 26 and 27 °C is highest in all treatment groups. Meanwhile, treatment 3 records the highest probability of choosing "No use of air conditioner" and 29 °C. These results indicate that each information provision scheme can encourage electricity conservation behavior. Moreover, we see that the impact of the information differs depending on what kind of information is supplied to the individuals. In particular,

Table 6.11 Choice probability of the temperature of air conditioner (Reprinted from Tanaka et al. [41])

Temperature (°C)	Control (%)	Treatment 1 (%)	Treatment 2 (%)	Treatment 3 (%)
25	5.90	9.58	9.62	6.41
26	24.36	14.79	10.38	8.59
27	24.62	22.29	18.27	14.06
28	20.51	21.88	30.77	25.47
29	19.23	17.92	18.65	24.53
No use	5.38	13.54	12.31	20.94

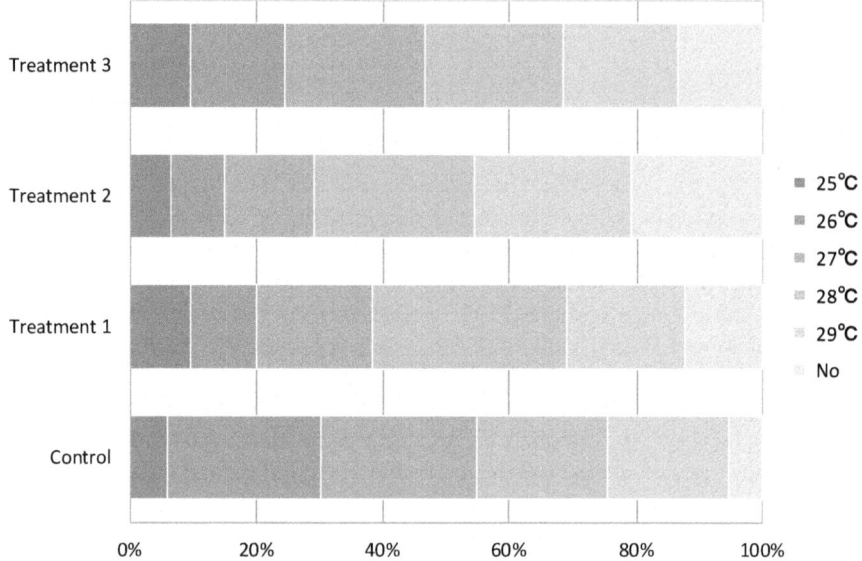

Fig. 6.5 Choice probability of temperatures of air conditioner in each treatment (Reprinted from Tanaka et al. [41])

the experiment results show that information about efficient persons is the most favorable information in the sense of the electricity usage reduction.

We have to notice that the above results may be affected by several factors, although our experiments control the situation of decision-making. To address the issue, we analyzed the factors that affect the choice of the set points, using ordered logit estimation. The estimation model is given as follows:

$$\text{Choice}_{i,t} = \text{Tr}_1 + \text{Tr}_2 + \text{Tr}_3 + \text{price}_{i,t} + \text{val}_{i,t} + \text{period}_{i,t} + \text{initial}_{i,t}$$
$$+ \text{initial}_{i,t} \times \text{Tr}_1 + \text{initial}_{i,t} \times \text{Tr}_2 + \text{initial}_{i,t} \times \text{Tr}_3 \quad (6.11)$$

The dependent variable (Choice) is an ordinal variable dependent on the choice of a particular temperature that is equal to 0 if 25 °C is selected. Thus $Choice_{i,t}$ is i's temprature choice in each period t. As the temperature increases, the variable also increases, and it scores 5 when the subject chooses "No use." Equation (6.11) means that the variables with positive coefficients have positive effects for electricity conservation.

The model includes dummy variables Tr_1, Tr_2, and Tr_3 that are introduced to analyze the effect of each treatment. Tr_1 is the dummy variable of treatment group 1. Moreover, Tr_2 and Tr_3 are the dummy variables of treatment group 2 and group 3. Some previous studies point out the correlation between electricity conservation and altruism. In order to analyze the effects of such preferences, we use the aforementioned level of altruism (Val) measured by the value orientation test. Roughly speaking, if the subject chooses an option that increases the partner's payoff, it gets larger. The readers are recommended to refer [41] for more details on the estimation of Val.

In order to analyze the effects of the initial electricity demand on the electricity conservation, we add the variable Initial to the model. In general, a person having a larger initial demand tends to be a "Non-efficient" person. In each treatment, such persons may exhibit different behavior than other persons. To reflect the factor, we add the initial and cross term of initial and dummy variables in Eq. (6.11).

Table 6.12 describes the estimation result, where Model 1 means the base model, while Models 2 and 3 are models reflecting the effects of the initial electricity demand. It is seen for all models that dummy variables have a positive correlation in each treatment. This implies that information provision based on social comparison enhances electricity conservation. In particular, treatment 3 has the most significant effect on conservation behavior.

Meanwhile, the cross-terms of initial and dummy variables in treatments 1 and 3 show a negative correlation with the dependent variable. In other words, persons who have a greater initial demand are discouraged from electricity conservation behavior under the social comparison schemes. One of the reasons for these results is that persons who have a larger initial demand give up attempting to imitate the behavior of efficient persons. In addition, if the "Non-efficient" persons fail to follow "Efficient" persons in a certain period, they surrender their efforts to conserve electricity. Our results may capture such behavioral effects for the information provision.

6.5   Field Experiment on Building Electricity Conservation

6.5.1   Motivation

In contrast to many previous field experiments treating manual household DR, we designed and executed a field experiment to investigate automated DR (ADR) in commercial spaces, particularly office buildings. ADR systems automatically notify users

		Model 1	Model 2	Model 3
Table 6.12 Factor analysis of temperature choice (ordered logit) (Reprinted from Tanaka et al. [41])	Tr_1	0.350*** (2.90)	0.364*** (3.01)	1.261*** (4.35)
	Tr_2	0.425*** (3.58)	0.431*** (3.61)	0.746*** (2.87)
	Tr_3	0.991*** (8.51)	0.961*** (8.22)	1.909*** (7.87)
	Val	−0.005*** (−4.41)	−0.006*** (−4.94)	−0.006*** (−4.96)
	Price	0.044*** (13.06)	0.045*** (13.10)	0.045*** (13.15)
	Period	0.008 (0.53)	0.008 (0.55)	0.007 (0.49)
	Initial		−0.240*** (−4.42)	−0.067 (0.73)
	Initial×Tr_1			−0.556*** (−3.43)
	Initial×Tr_2			−0.199 (−1.39)
	Initial×Tr_3			−0.617*** (−4.45)
	Log likelihood	−3323.3517	−3313.4524	−3301.1918

of DR requests or changes in electricity rates and can control household appliances, such as changing air conditioner settings, turning devices on and off, or reducing light intensity [27, 37]. A previous study [44] implemented a residential DR pilot program for Belgian households, but did not apply RCT. Many technical trials on DR field tests for business facilities have been reported in the literature, but past studies have faced difficulty in ensuring an adequate number of participants [23, 39, 44]. To our knowledge, any field experiment that examines individuals' energy conservation and electricity-saving actions using RCT or quasi-RCT (QRCT) has not been conducted.

Household DR field experiments in [20, 38] have revealed the conditions and environments under which consumers participate in DR to conserve electricity. For example, DR participation rates vary significantly according to whether participation is opt-in or opt-out. Opt-in means that people do not participate by default and report only when they wish to join, while opt-out means that people do participate by default and report only when they wish to drop out. Specifically, it was reported in [20, 38] that the opt-out condition achieves a higher participation rate than the opt-in. On the other hand, the household DR field experiments have shown a trade-off between participation rates and power-saving effects; although DR participation rates are high under opt-out conditions, energy savings increase under opt-in conditions [12]. These results suggest that the default condition is a key factor in determining the human actions.

In this section, we divided employees in an office building into four treatment groups, namely opt-in performance incentive, opt-in fixed incentive, opt-out performance incentive, and opt-out fixed incentive, and a control group in order to quantify the electricity-saving effects for each group (i.e., the amount of electric power consumed by each employee's task light). The term "performance incentive group" implies that employees in the groups are incentivized for each DR event if they participated in DR and actually saved power during the event. On the other hand, employees in the fixed incentive groups receive a fixed incentive regardless of participation.

6.5.2 Method

6.5.2.1 Outline of the Experiment

Let us first describe the outline of the experiment summarized in Table 6.13. The project was initiated in 2013 by Takenaka Corporation, which is a large-scale general contractor, and Kyoto University. The experiment was conducted in the fifth and seventh floors in Takenaka Corporation's Tokyo headquarters for about two weeks from July 14 to August 1, 2014, excluding Saturdays, Sundays, and holidays. Requests for electricity savings (DR requests) were issued between 13:30 and 15:30 on six midsummer days, generally days with high power demand.

The characteristics of ADR are explained next. We conducted a DR event in workspaces with personal lighting control illustrated in Fig. 6.6. Task lights were connected to smart outlets allowing remote shutoff and power consumption measurements. On each day that a DR event occurred, a BEMS sent employees DR requests via e-mail. Each worker was asked to reply with "participate" or "not participate" in an e-mailed response. The BEMS turned off the work lights of only those employees replying "participate" at 13:30, then turned them back on at 15:30 (Fig. 6.6). In order to purely analyze the electricity saving and eliminate the effects of environmental factor, say sunlight, window blinds were closed during the period.

Table 6.13 Experiment outline (Reprinted from Ida et al. [41] Copyright 2019 Elsevier Ltd.)

Field building	Office (Koto-ku, Tokyo)
Area	About 1000 m^2 (excluding common space and toilets)
Experiment period	From July 14, 2014 (Monday) to August 1, 2014 (Friday)
DR event day*	July 16, 18, 23, 24, 30 and August 1
DR notice time	8:30 current day
DR event time	13:30–15:30 (2 h)

*The ADR system does not inform workers of the DR day in advance, but on the morning of the DR day

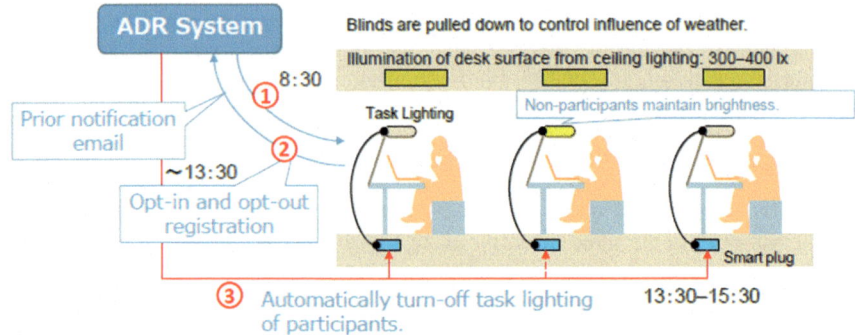

Fig. 6.6 ADR experiment (Reprinted from Ida et al. [41] Copyright 2019 Elsevier Ltd.). *Notes* [1] During the morning of each DR event day, the ADR system sends a DR request via e-mail to each employee (treatment groups). [2] Opt-in group: If employees in the opt-in group participate in the DR, they reply "participation." Opt-out group: If employees in the opt-out group do not participate in the DR, they reply "non-participation." [3] At the DR event time (13:30), the task lights of the employees who chose to participate are automatically turned off and then back on when the event ends (15:30). Because this study is an electricity-saving field experiment for task lights, window blinds are closed during the period to remove the effects of brightness dependent on the weather

We applied QRCT to divide the employees into a control group that did not receive DR requests (66 fifth-floor employees) and a treatment group that received DR requests (211 seventh-floor employees). The treatment group was further divided into the aforementioned four subgroups based on conditions of DR participation (Table 6.14). Among 211 employees, 48 were assigned to the opt-in performance incentive group (T1), 51 were to opt-in fixed incentive group (T2), 58 were to opt-out performance incentive group (T3), and 54 were to opt-out fixed incentive group (T4). Control group and treatment group members were selected for the experiment because they primarily did deskwork and worked on PCs.

We selected the fifth floor as the control group and the seventh floor as the treatment group because work and attendance times of employees on those floors were

Table 6.14 Outline of the groups (Reprinted from Ida et al. [41] Copyright 2019 Elsevier Ltd.)

Group	Control group	Treatment group			
		T1	T2	T3	T4
Participation type	–	Opt-in	Opt-in	Opt-out	Opt-out
Incentive	–	Performance	Fixed	Performance	Fixed
No. of participants	66	48	51	58	54

Note In the case of a performance incentive, employees were given a coffee coupon worth 500 yen (1 cup) each time they participated in a DR. If employees participated on all six days, they received coffee coupons worth 3000 yen. In the case of a fixed incentive, employees were given coffee coupons worth 1000 yen (2 cups) irrespective of their participation through the six days. The two types of incentives (reward)—the opt-in and opt-out types—were combined to design four treatment groups

similar. However, this separation of floors could result in differences in observed and unobserved attributes. We therefore controlled for differences in attributes by using the difference-in-differences (DID) technique. ATE can be precisely identified as long as the "common trend" assumption is satisfied.

As stated, two incentive types were offered to the treatment group employees. Performance incentives were coffee coupons worth 500 yen (1 cup) each time they participated in a DR, so employees participating in all six days received coupons worth 3000 yen. Fixed incentives were coffee coupons worth 1000 yen (2 cups) irrespective of participation throughout the six days. The two incentive (reward) types, opt-in and opt-out performance types, were combined to design the four treatment groups.

The ceiling lights in the building operate in blocks, so the seventh floor was divided into eight blocks, and the four treatment groups were randomly distributed among them as illustrated in Fig. 6.7. In this experiment, therefore, employees were quasi-randomly ("as if" randomly) divided among the block units. Internal validity was verified using data on pre-experiment power consumption for task lighting in each group. Across-group comparison indicates statistical balance in electricity use due to the quasi-random assignment $[F(3, 4868) = 2.64, p < 0.05]$.

7F Floor Plan (Treatment Group)

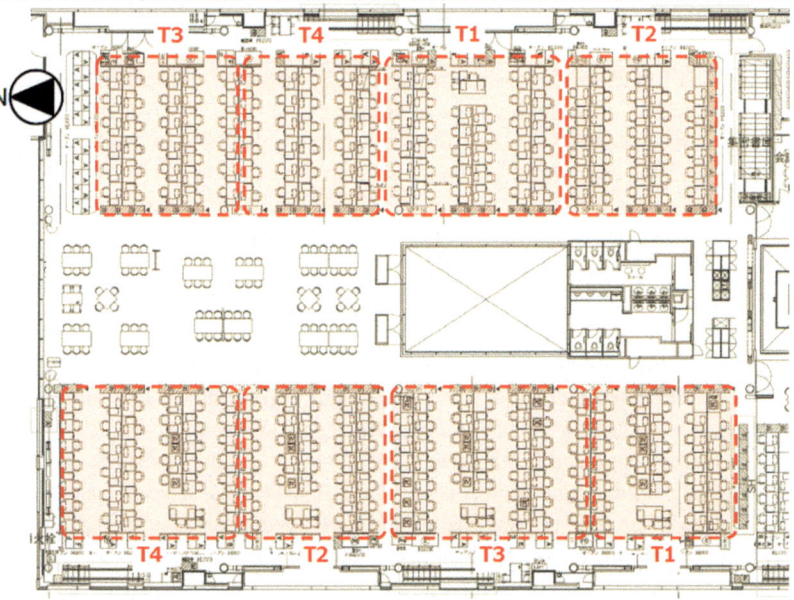

Fig. 6.7 Floor plan of treatment groups (Reprinted from Ida et al. [41] Copyright 2019 Elsevier Ltd.). *Note* Employees on the seventh floor were divided into eight blocks, which were then randomly assigned to the four treatment groups

Using the above verification design, we quantitatively analyzed the following research question in terms of employee responses to the DR events: "Do the average treatment effects in each treatment group (intention to treat [ITT]) vary according to DR participation?"

This experiment used data on task light power consumption and DR participation rates of employees. Specifically, these data are from 9:00–11:00 and 13:30–15:30 on July 16, 18, 23, 24, and 30 and August 1, 2014. Employees in each group primarily did deskwork, with negligible changes in attendance between morning and afternoon. For each employee, power consumption data for task lighting were collected at five-minute intervals. DR participation rates for each treatment group were calculated by converting each employee's DR participation or non-participation into a dummy variable. We estimated the electricity-saving effects based on these data. Morning power consumption for each event day before a DR request was treated as baseline data. To measure treatment effects, we estimated differences in the electricity consumption between this baseline and data during the DR request period for the treatment and control groups, as well the difference-in-differences to measure the treatment effects.

6.5.2.2   Estimation Model

The quantity of electricity consumption in the four treatment groups varied with conditions for DR request participation, with the control group treated as the baseline. We measured ITT to estimate the treatment effect on quantity of electricity consumption. To estimate ITT, namely the gross treatment effect in each treatment group, we conducted a panel data analysis controlling for individual effects and time effects for each employee. This estimation method is described as

$$y_{it} = \sum_{g \in (T1,T2,T3,T4)} \beta_{ITT}^g \cdot D_{it}^g + \theta_i + \lambda_t + \varepsilon_{it}, \qquad (6.12)$$

where y_{it} is the amount of power consumption for task lighting over five minutes in time period t by employee I, β_{ITT}^g is the average treatment effect of treatment group g, and D_{it}^g is a treatment index that is 1 if employee i belongs to the treatment groups and a time period including the DR request in time period t (in this case, 13:30–15:30 for each DR event day) and 0 otherwise. The symbols θ_i and λ_t denote individual fixed effects of employee and time fixed effects, respectively. They are included to control for worker-specific characteristics and time-specific shocks such as weather. ε_{it} is the error term. The baseline used to estimate the average treatment effects of ITT is 9:00–11:00 on each DR event day.

6.5.3 Main Results

Table 6.15 summarizes the average treatment effects by ITT (reduction in electric power consumption) that is illustrated in Fig. 6.8.

We found that ITTs for each treatment group were significant at the 1% level. It was also confirmed that power consumption decreased significantly in every treatment group. We also found that T3 showed the largest reduction, followed by T1, T4, and T2. This result indicated that the performance incentive groups (T1 and T3) achieved greater electricity-saving effects than for fixed incentive groups (T2 and T4), implying more active engagement in energy savings under a performance incentive that rewards saving electricity. The p values in the ITT column in Table 6.15 are the results of pairwise testing of ITT equality. For ITT, all null hypotheses were rejected at the 1% significance level, indicating differences in ATE between all treatment groups.

Table 6.15 ITT estimation results (Reprinted from Ida et al. [41] Copyright 2019 Elsevier Ltd.)

		Unit: Wh
Treatment group		ITT
T1		−19.18*** (0.93)
T2		−8.72*** (0.95)
T3		−22.42*** (0.88)
T4		−12.51*** (0.90)
p-value	T1 = T2	0.00***
	T1 = T3	0.00***
	T1 = T4	0.00***
	T2 = T3	0.00***
	T2 = T4	0.00***
	T3 = T4	0.00***
Observations		76,934

Standard errors are in parentheses

Note The results of estimating the treatment effects for ITT and the statistical significance of the differences between the treatment effects of each group were verified. The average treatment effects of each group were significant at the 1% level, confirming that the quantity of electric power consumed by task lights in each treatment group was significantly reduced. The quantity of reduction was highest for the opt-out performance incentive group (T3), followed by the opt-in performance incentive group (T1), opt-out fixed incentive group (T4), and opt-in fixed incentive group (T2); *** indicates statistical significance at 1% level

Fig. 6.8 Average treatment effects for each group (Reprinted from Ida et al. [41] Copyright 2019 Elsevier Ltd.). *Note* The vertical bars are the estimated values of average treatment effects and the vertical lines the standard errors of average treatment effects

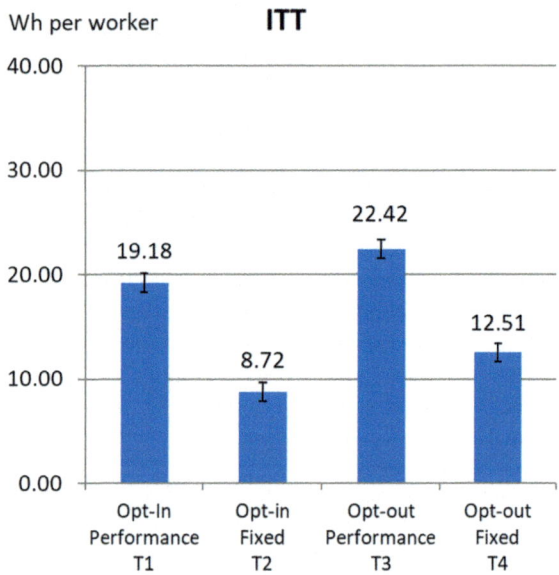

We therefore conclude the following regarding ITT: First, the effects of performance incentives are larger than those of fixed incentives. Second, opt-out achieves greater effects than opt-in. In other words, comparing the power-saving effects of opt-in and opt-out defaults for each incentive, an opt-out approach contributes to saving electricity.

6.6 Conclusion

This chapter analyzed customers' electricity conservation behavior when DR is called for. Behavioral economics provides very useful insights to account for human behavioral anomalies such as status quo bias, loss aversion, overconfidence, moral cost, and default bias.

The second section investigated a Web-based survey of residential electricity plan choice. A choice experiment is performed using online survey three months before deregulation. The results indicated that the average Japanese consumer experiences status quo bias in electricity plan choice; consumers are willing to keep their default provider even in the presence of the obvious 5% bill savings that could be gained from switching to a new provider. By simulating the potential share of new providers in the liberalized market, we revealed that a 50% renewable energy plan has a larger potential market share than a plan offering a 7% bill reduction under price competition.

The third section investigated a field experiment on residential electricity plan choice. It was seen from the results of the SP experiment that the evaluation amount (the limit payment intention amount) was about 126 yen in the absence of information on the gain or loss resulting from switching, and the evaluation amount decreased to nearly 0 yen in the presence of the information. Additionally, participants receiving the information that they would lose money due to switching changed their evaluation value significantly. By contrast, in the case of gains, the evaluation value did not change with/without information.

The fourth section investigated a laboratory experiment of residential energy conservation. The experiment results include some important findings. First, our results showed that an information provision scheme based on social comparison encourages electricity conservation behavior. Second, our results implied that the effects of social comparison schemes differ depending on the initial electricity demand of each individual. When individuals know the trend of other people's behavior for the betterment of society, they try to match other people's behavior.

The fifth section investigated a field experiment on building electricity conservation, wherein we treated four treatment groups: opt-in performance incentive, opt-in fixed incentive, opt-out performance incentive, and opt-out fixed incentive. The experimental results showed that the opt-out performance incentive group achieved the highest participation rate. We also observed a significant reduction in electricity consumption for all of the treatment groups. In addition, despite the high DR participation rate for the opt-out group, we could not find significant difference in the average energy saving as compared with the opt-in group, and actually the opt-in consumers showed larger energy-saving effects than the opt-out consumers.

References

1. Abildtrup J, Garcia S, Olsen SB, Stenger A (2013) Spatial preference heterogeneity in forest recreation. Ecol Econ 92:67–77
2. Allcott H (2011) Social norms and energy conservation. J Publ Econ 95:1082–1095
3. Barber BM, Odean T (2001) Boys will be boys: gender, overconfidence, and common stock investment. Q J Econ 116:261–292
4. Bertrand M, Morse A (2011) RECAP disclosure, cognitive biases, and payday borrowing. J Finance 66:1865–1893
5. Birol E, Karousakis K, Koundouri P (2006) Using a choice experiment to account for preference heterogeneity in wetland attributes: the case of Cheimaditida wetland in Greece. Ecol Econ 60:145–156
6. Clark J, Friesen L (2009) Overconfidence in forecasts of own performance: an experimental study. Econ J 119(534):229–251
7. Costa DL, Kahn ME (2013) Energy conservation "nudges" and environmentalist ideology: evidence from a randomized residential electricity field experiment. J Eur Econ Assoc 11:680–702
8. Eil D, Rao JM (2011) The good news-bad news effect: asymmetric processing of objective information about yourself. Am Econ J Microecon 3:114–138
9. Einav L, Finkelstein A, Ryan S, Schrimpf P, Cullen MR (2013) Selection on moral hazard in health insurance. Am Econ Rev 103(1):178
10. Ertac S (2011) Does self-relevance affect information processing? Experimental evidence on the response to performance and non-performance feedback. J Econ Behav Organ 80:532–545

11. Fischbacher U (2007) z-Tree: Zurich toolbox for ready-made economic experiments. Exp Econ 10:171–178
12. Fowlie M, Wolfram C, Spurlock CA, Todd A, Baylis P, Cappers P (2017) Default effects and follow-on behavior: evidence from an electricity pricing program. Energy Institute at Haas, Working Papers WP-280
13. Giné X, Mazer RK (2016) Financial (Dis-)RECAP: evidence from a multi-country audit study. Policy research working papers
14. Griesinger DW, Livingston JW (1973) Toward a model of interpersonal motivation in experimental games. Behav Sci 18:173–188
15. Handel BR (2013) Adverse selection and inertia in health insurance markets: when nudging hurts. Am Econ Rev 103(7):2643–2682
16. Handel BR, Kolstad JT (2015) Health insurance for "Humans": information frictions, plan choice, and consumer welfare. Am Econ Rev 105(8):2449–2500
17. Herter K (2007) Residential implementation of critical-peak pricing of electricity. Energy Policy 35(4):2121–2130
18. Hartman RS, Doane MJ, Woo C-K (1991) Consumer rationality and the status quo. Quart J Econ 106:141–162
19. Hortaçsu A, Madanizadeh SA, Puller SL (2017) Power to choose? An analysis of consumer inertia in the residential electricity market. Am Econ J Econ Policy 9:192–226
20. Ida T, Wang W (2015) A field experiment on dynamic electricity pricing in Los Alamos: Opt-in versus opt-out. Kyoto University, Discussion Paper No. E-14-010
21. Ida T, Motegi N, Ushifusa Y (2019) Behavioral study of personalized automated demand response in the workplace. Energy Policy 132:1009–1016
22. Ishihara T, Ida T (2019) Effects of information provision on stated preferences and revealed preferences: field experiment of electricity plan choice before and after deregulation. Kyoto University, Discussion Paper No. E-19
23. Jain RK, Taylor JE, Peschiera G (2012) Assessing eco-feedback interface usage and design to drive energy efficiency in buildings. Energy Build 48:8–17
24. Jimenez LR, Potter JM, George SS (2013) Smart pricing option interim evaluation. Sacramento Municipal Utility District
25. Kamenica E, Mullainathan S, Thaler R (2011) Helping consumers know themselves. Am Econ Rev 101:417–422
26. Kling JR, Mullainathan S, Shafir E, Vermeulen LC, Wrobel MV (2012) Comparison friction: Experimental evidence from medicare drug plans. Quart J Econ 127:199–235
27. Krioukov A, Dawson-Haggerty S, Lee L, Rehmane O, Culler D (2011) A living laboratory study in personalized automated lighting controls. 3rd ACM Workshop on Embedded Sensing Systems for Energy-Efficiency in Buildings, Seattle, WA
28. Levitt SD, List JA (2007) What do laboratory experiments measuring social preferences reveal about the real world? J Econ Perspect 21:153–174
29. Liebrand WBG (1984) The effect of social motives, communications and group sizes on behavior in an n-person multi-stage mixed motive game. Eur J Soc Psychol 14:239–264
30. McFadden D, Train KE (2000) Mixed MNL models of discrete choice models of discrete response. J Appl Econom 15:447–470
31. METI (Japan's Ministry of Economy, Trade and Industry) (2018) Electricity system and market in Japan, presented from Tatsuya Shinkawa in Electricity and Gas Market Surveillance Commission
32. Miravete EJ (2003) Choosing the wrong calling plan? Ignorance and learning. Am Econ Rev 93(1):297–310
33. Murakami K, Ida T, Tanaka M, Friedman L (2015) Consumers' willingness to pay for renewable and nuclear energy: a comparative analysis between the US and Japan. Energy Econ 50:178–189
34. Murakami K, Ida T (2019) Deregulation and status quo bias: evidence from stated and revealed switching behaviors in the electricity market in Japan. Graduate School of Economics, Kyoto University, Discussion Paper E-19-01

35. Offerman T, Sonnemans J, Arthur A (1996) Value orientation, expectations and voluntary contributions in public goods. Econ J 106:817–845
36. Park E (2000) Warm-glow versus cold-prickle: a further experimental study of framing effects on free-riding. J Econ Behav Organ 43:405–421
37. Piette MA, Schetrit O, Killacotte S, Cheung I, Li BZ (2015) Costs to automate demand response—taxonomy and results from field studies and programs. Lawrence Berkeley National Laboratory Report, LBNL Report Number 1003924
38. Sacramento Municipal Utility District (2014) Smart pricing options final evaluation: the final report on pilot design, implementation, and evaluation of the Sacramento Municipal Utility District's Consumer Behavior Study
39. Siero FW, Bakker AB, Dekker GB, Van Den Burg MTC (1996) Changing organizational energy consumption behavior through comparative feedback. J Environ Psych 16:235–246
40. SMUD (2014) Smart pricing options final evaluation: the final report on pilot design, implementation, and evaluation of the Sacramento Municipal Utility District's Consumer Behavior Study
41. Tanaka K, Kurakawa H, Ishihara T, Ida T, Akao K (2020) Moral utility or moral tax? Experimental study of electricity conservation by social comparison. Kyoto University, Discussion Paper No. E-19-011
42. Thaler RH, Sunstein CR (2009) Nudge: improving decisions about health, wealth, and happiness. Penguin Books, London
43. Train KE (2009) Discrete choice methods with simulation, 2nd edn. Cambridge University Press, New York
44. Vanthournout K, Dupont B, Foubert W, Stuckens C, Claessens S (2015) An automated residential demand response pilot experiment, based on day-ahead dynamic pricing. Appl Energy 155:195–203

Chapter 7
Economic Impact and Market Power of Strategic Aggregators in Energy Demand Networks

Yusuke Okajima, Kenji Hirata and Vijay Gupta

Abstract This chapter investigates the strategic behavior of aggregators in the three-layered energy demand network optimization problem. Participants of the network are a utility company, who plays a role of energy supply source, multiple aggregators, and a large number of consumers. We suppose that the network will be optimized through price response-based or market-based optimization process. Under this assumption, the main interest of this chapter is in strategic behavior of aggregators. Although the aggregators are an intermediate entity and expected to solve scalability issues arisen in a large-scale optimization problem, our interest is in an economic impact of the aggregators in energy market. In order to formulate strategic behaviors of the aggregators, we focus on the two specific problem settings. In the first problem, the aggregators will try to pursue market power as well as its own benefit. The other considers strategic operations of battery storage by the aggregators. We use numerical case studies and show that the strategic decision making by the aggregators could provide some useful insights in qualitative analysis of large-scale energy demand network.

This chapter was developed and enhanced from earlier papers published as [15] © 2017 IEEE and [16] © 2018 IEEE.

Y. Okajima (✉)
Joetsu University of Education, 1 Yamayashiki Joetsu, Niigata 943 8512, Japan
e-mail: okajima@juen.ac.jp

K. Hirata
University of Toyama, 3190 Gofuku, Toyama 930 8555, Japan
e-mail: hirata@eng.u-toyama.ac.jp

V. Gupta
University of Notre Dame, Notre Dame, IN 46556, USA
e-mail: vgupta2@nd.edu

© Springer Nature Singapore Pte Ltd. 2020
T. Hatanaka et al. (eds.), *Economically Enabled Energy Management*,
https://doi.org/10.1007/978-981-15-3576-5_7

7.1 Introduction

Liberalization of the energy market and enhancing distributed energy resources such as renewable energy encourage researches on analysis and design of decentralized control architecture for power supply and demand networks. One promising approach is price response-based or market-based decentralized control [6, 7, 10, 11, 19, 22]. Efficient design of price response-based control may include significant challenges. The network involves a large number of consumers. Thus the utility, who plays a coordination role for energy transactions, needs to formulate and solve a large-scale optimization problem. Another issue from the economic point of view is that each consumer only has a negligible ability to affect the price. The difficulties might be redeemed by utilizing hierarchical control architecture, and this encourages introducing aggregators [3, 8, 18]. The aggregators are new entities in energy markets which act as mediators between energy supply sources and consumers. The aggregator is expected to solve a scalability issue arisen in an optimization problem of large-scale energy network because the utility does not need to face a large number of consumers. The aggregator is also expected to have enough negotiation power on behalf of associated consumers because it aggregates a significant amount of total demands.

This research focuses on clarifying the fundamental role the aggregator can play in the market from economic point of view or qualitatively evaluating the economic impact of the aggregator rather than solving a specific design problem of well-worked aggregator algorithm. In this chapter, we investigate the strategic behavior of aggregators and its economic impact in the two specific problem settings. We formulate the problems under the three-layered energy demand network optimization process. The participants of the network are a utility company, who will act as an energy supply source, multiple aggregators, and a large number of consumers. We suppose that the network will be optimized through price response-based optimization process. Under this assumption, in the first problem, we consider market power of the aggregator. Market power is defined as the ability to affect the market price profitably from competitive levels [13, 20]. If the market power of the utility company is too large compared to the aggregator's, most of the benefit may be brought to the utility company. On the other hand, if the market power of the aggregator is large, the benefit may be properly allocated to the associated consumers. The aggregator will try to pursue market power as well as its own benefit. By using numerical case studies, we discuss if this strategic behavior of the aggregator is actually beneficial to the consumers or not. The second problem considers the strategic operation of battery storage by the aggregator. We suppose that the aggregator is equipped with storage and can choose the total capacity of the storage to be installed as well as ratio of high to low performance storage. The aggregator will strategically decide the total capacity and ratio of the storage to increase its own benefit. By using numerical case studies, we evaluate that this strategic behavior of the aggregator also concludes a benefit to the society or not.

The strategic decision making by the aggregator could provide some useful insights in qualitative analysis of large energy demand network. Numerical case studies for market power optimization actually indicate that, for example, oligopoly by the aggregator may not be beneficial to the consumers. In case of strategic storage managements, a benefit maximization by the aggregator may not conclude the social welfare maximization, and this may encourage researches, for example, an incentive design problem for the future energy market. Some technical results in this chapter have been reported in [15, 16] and other numerical case studies can also be found in [15, 16].

The remainder of this chapter is organized as follows. Section 7.2 describes the three-layered energy demand network and the benefit-maximization problem of each participant. Section 7.3 reviews price response-based optimization process. Section 7.4 discusses optimization problems which try to evaluate strategic behavior of aggregators. Section 7.5 shows numerical case studies in which the aggregator will try to pursue market power as well as its own benefit. Numerical case studies in Sect. 7.6 consider strategic battery storage operations by the aggregator. Section 7.7 gives concluding remarks.

7.2 Three-Layered Optimization Model of Energy Demand Network

As a conceptual model of future energy market, we consider the three-layered energy demand network. The three-layered energy demand network is depicted in Fig. 7.1. The model includes several generation companies and aggregators, which are connected to each other, and many consumers who are associated with each aggregator. In this model, the utility company may be expected to play a coordination role for energy transactions between the multiple generation companies and aggregators. Although the energy demand network depicted in Fig. 7.1 may be realistic, a simplified setting is still enough for the main purpose of this chapter, namely clarifying an impact of strategic behavior of aggregators in future energy market.

Fig. 7.1 A realistic model of three-layered energy demand network. Reprinted from [15]

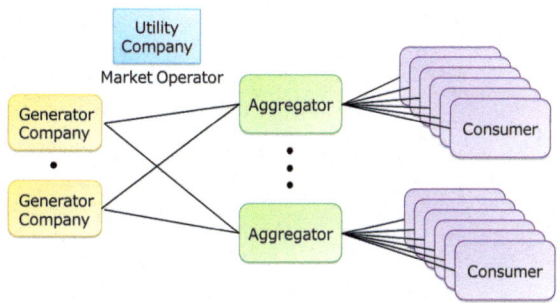

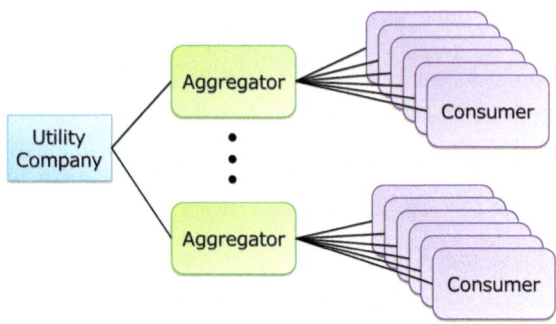

Fig. 7.2 A simplified model of three-layered energy demand network. Modified from [15]

A simplified model of the three-layered energy demand network is depicted in Fig. 7.2. In this setting, we consider a single utility company who plays a role of energy supply source as well as coordination for energy transactions. The utility company or energy supply source is connected to the wholesale market and purchases total energy for the network. The multiple aggregators are connected to the utility company. Consumers are connected to the utility company through the aggregators.

This chapter considers the three-layered energy demand network illustrated in Fig. 7.2. We investigate market-based optimization through pricing with the supply and demand balance constraints. Under this optimization process, our main interest is on strategic behavior of aggregators and its impact on the energy market.

7.2.1 Utility Company

The utility company purchases electricity from the wholesale market and sells it to the aggregators on price $p_0 \in \mathbb{R}^P$, where $P > 0$ denotes number of time slots. We suppose that the time horizon for optimization is divided into P time slots. If one considers a day-ahead market, the time horizon may be 24 h and P could be 48 and each time slot corresponds to 30 min. If one is interested in a minute-by-minute optimization horizon for a short-term energy scheduling, then each time slot should be a few seconds. The generation cost of electricity $u_0 \in \mathbb{R}^P$ purchased by the utility company generally follows a convex function [12]. We consider the quadratic cost function as

$$J_0^{\sharp}(u_0) = u_0^{\top} Q_0 u_0 + R_0 u_0 + C_0.$$

Since the utility company sells the purchased energy u_0 to the aggregators on price p_0, the benefit-maximization problem of the utility company is given by

$$\max_{u_0} \quad J_0^{\sharp}(u_0) + p_0^{\top} u_0. \tag{7.1}$$

7.2.2 Aggregators

Aggregators $A_i, i \in N = \{1, \ldots, n\}$ purchase electricity on price p_0 from the utility company and sell it to the consumers on price $p_i \in \mathbb{R}^P$. The benefit-maximization problem of aggregator A_i is given by

$$\max_{u_i} \quad J_i(r_i; u_i) + (p_i - p_0)^\top u_i, \tag{7.2}$$

where $J_i(r_i; u_i)$ represents the operational cost of the aggregator. We assume that the operational cost of the aggregator is a convex function with respect to the purchased amount of electricity u_i.

The cost function $J_i(r_i; u_i)$ includes the parameter r_i which should be adjusted by aggregator A_i according to its strategic behavior. The parameter r_i may represent its operational cost. If the aggregator is equipped with storage, it may represent amount and performance of the installed battery storage. Details on strategic behavior of aggregators will be described in Sect. 7.4.

7.2.3 Consumers

Let $A_{ij}, j \in N_i = \{1, \ldots, n_i\}$ denote the consumers belonging to aggregator A_i, $i \in N$. Consumer A_{ij} purchases electricity $u_{ij} \in \mathbb{R}^P$ from aggregator A_i on price p_i and consumes it by using appliances. The demand function of consumer A_{ij} generally follows a convex function and can be expressed in a quadratic form [2, 19]. We consider

$$J_{ij}^\sharp(u_{ij}) = u_{ij}^\top Q_{ij} u_{ij} + R_{ij} u_{ij} + C_{ij},$$

as the demand function of consumer A_{ij}. See [17] and Appendix A for detailed examples of the explicit representation of the demand function including both of the dynamic and static appliances. The benefit-maximization problem of consumer A_{ij} is given by

$$\max_{u_{ij}} \quad J_{ij}^\sharp(u_{ij}) - p_i^\top u_{ij}. \tag{7.3}$$

7.2.4 Optimization of Three-Layered Energy Demand Network

The three-layered energy demand network should be optimized so as to maximize the social welfare through pricing. We define the social welfare as the sum of the benefits of the utility company, aggregators, and consumers. We formulate the social welfare maximization problem as

$$\max_{\substack{u_0 \\ u_i \ i \in N \\ u_{ij} \ j \in N_i}} \quad J_0^{\sharp}(u_0) + \sum_{i \in N} J_i(r_i; u_i) + \sum_{\substack{i \in N \\ j \in N_i}} J_{ij}^{\sharp}(u_{ij}) \tag{7.4a}$$

$$\text{subject to} \quad u_0 = \sum_{i \in N} u_i \tag{7.4b}$$

$$u_i = \sum_{j \in N_i} u_{ij} \quad i \in N. \tag{7.4c}$$

We note that the cost function (7.4a) can be recognized as the sum of the cost functions of the benefit-maximization problems in (7.1), (7.2) and (7.3), since the terms depending on the prices p_0 and p_i will cancel out under the supply and demand balance constraints in (7.4b) and (7.4c). We denote by u_0^*, u_i^* and u_{ij}^* the unique optimal solution to (7.4).

The selfish optimization by each participant in (7.1), (7.2) and (7.3) may not align supply and demand balance. An appropriate pricing strategy should be designed to enforce supply and demand balance. Next Sect. 7.3 will consider pricing and optimization processes which align the individual benefit-maximization problem with the social welfare maximization.

7.3 Optimization Processes Through Pricing

This section briefly reviews the well-known optimization processes so called supply function bidding and tâtonnement processes. These two processes may be extremal models: the former requires rigorous information of the market participants and a large computational burden, while the optimal solution might be immediately obtained as it is required in a short-term energy scheduling; the latter may require an large number of iterations to obtain the optimal solution and it may be applicable to a day-ahead market-type problem, while it needs no rigorous model of the market participants.

The aggregators are expected to moderate difficulties arisen in a large-scale energy demand network optimization. We investigate an optimization process unitizing information exchange or aggregation by the aggregators. It can be recognized as an intermediate model between supply function bidding and tâtonnement processes.

7.3.1 Supply Function Bidding Process

We start with the dual problem of (7.4) given by

$$\min_{\substack{\lambda_0 \\ \lambda_i \ i \in N}} \max_{\substack{u_0 \\ u_i \ i \in N \\ u_{ij} \ j \in N_i}} \quad J_0^{\sharp}(u_0) + \sum_{i \in N} J_i(r_i; u_i) + \sum_{\substack{i \in N \\ j \in N_i}} J_{ij}^{\sharp}(u_{ij})$$

$$- \lambda_0^{\top} \left(-u_0 + \sum_{i \in N} u_i \right) - \sum_{i \in N} \lambda_i^{\top} \left(-u_i + \sum_{j \in N_i} u_{ij} \right), \quad (7.5)$$

where $\lambda_0 \in \mathbb{R}^P$ and $\lambda_i \in \mathbb{R}^P$ denote Lagrange multipliers.

Let λ_0^* and λ_i^* denote the dual optimal of (7.5), which can be interpreted as the shadow prices [1]. The social welfare maximization problem (7.4) can be decomposed into the subproblems of the utility company, aggregators, and consumers by using optimal prices $p_0 = \lambda_0^*$ and $p_i = \lambda_i^*$. Utility company A_0 can maximize the benefit by solving (7.1) with $p_0 = \lambda_0^*$ and obtain u_0^*. Aggregator A_i can maximize the benefit by solving (7.2) with $p_0 = \lambda_0^*$, $p_i = \lambda_i^*$ and obtain u_i^*. Similarly, consumer A_{ij} can maximize the benefit by solving (7.3) with $p_i = \lambda_i^*$ and obtain u_{ij}^*.

A centralized operator, who will determine the optimal prices $p_0 = \lambda_0^*$ and $p_i = \lambda_i^*$, needs to correct all cost functions of the utility company, aggregators and consumers to formulate the optimization problem (7.4) or (7.5). The centralized operator also needs to solve a large size optimization problem (7.5), which is essentially equivalent to solve (7.4).

In the supply function bidding process, the centralized operator needs to have large enough communication bandwidth and computing power, but the optimal prices can be immediately obtained and broadcasted to the aggregators and consumers. The supply function bidding process may be useful if one is interested in a short-term energy scheduling problem.

7.3.2 Tâtonnement Process

Let k be an integer and consider $\lambda_0^k, \lambda_i^k \in \mathbb{R}^P$. We consider the benefit-maximization problems (7.1), (7.2) and (7.3) with $p_0 = \lambda_0^k$ and $p_i = \lambda_i^k$ and denote the optimal solutions as u_0^k, u_i^k and u_{ij}^k.

We substitute u_0^k, u_i^k, and u_{ij}^k into (7.5) and consider the gradients at the point λ_0^k and λ_i^k, which are given by $u_0^k - \sum_{i \in N} u_i^k$ and $u_i^k - \sum_{j \in N_i} u_{ij}^k$, respectively. Thus, we have updating rules of the dual variables as

$$\lambda_0^{k+1} = \lambda_0^k - \epsilon \left(u_0^k - \sum_{i \in N} u_i^k \right), \quad (7.6a)$$

$$\lambda_i^{k+1} = \lambda_i^k - \epsilon \left(u_i^k - \sum_{j \in N_i} u_{ij}^k \right), \quad (7.6b)$$

where $\epsilon > 0$ is a step-size parameter.

The iteration process according to (7.1), (7.2), (7.3), and (7.6) with the pricing rules $p_0 = \lambda_0^k$ and $p_i = \lambda_i^k$ is known as tâtonnement process or gradient play. Under appropriate assumptions, the solution will converge to the optimal solutions λ_0^*, λ_i^*, u_0^*, u_i^*, and u_{ij}^* as k will increase, if the step-size parameter $\epsilon > 0$ is sufficiently small.

In contrast to the supply function biding process, the updating rule (7.6), which may be executed by the utility company in the problem settings of this chapter, does not require any specific information on the cost functions of the market participants. The updating rule (7.6) is also simple enough, and it may be easily implemented without large computing power. Although tâtonnement process allows private information such as the cost functions of the participants to be undisclosed, a large number of iterations may be required to obtain the optimal solution. Thus, it may be useful if one considers a day-ahead market-type problem, while preserving the private information of the market participants.

7.3.3 Information Exchange via Aggregators

Supply function bidding and tâtonnement processes may involve difficulties if it applies to a large-scale energy market optimization. The aggregators are expected to moderate the amount of information exchange and computational burden or number of iterations to obtain the optimal solutions. We propose an information exchange or aggregation procedure by the aggregators, where the aggregators will define the cost function of their subnetwork and submit it to the utility company.

Consumer A_{ij} associated to aggregator A_i submits the cost function $J_{ij}^{\sharp}$ to A_i. Aggregator A_i gathers the cost functions of all consumer A_{ij}, $j \in N_i$ and determines the cost function of the subnetwork. The cost function $J_i^{\sharp}(r_i; \cdot)$ of the subnetwork is given by

$$J_i^{\sharp}(r_i; u_i) = J_i(r_i; u_i) + \max_{u_{ij}\ j\in N_i} \sum_{j\in N_i} J_{ij}^{\sharp}(u_{ij}) \tag{7.7a}$$

$$\text{subject to} \quad u_i = \sum_{j\in N_i} u_{ij}, \tag{7.7b}$$

where u_i is a specified amount of energy that aggregator A_i purchases.

The function $J_i^{\sharp}(r_i; u_i)$ decides the optimal allocation of energy u_{ij}, $j \in N_i$ when the purchased amount u_i by aggregator A_i is specified. For general convex functions $J_i(r_i; \cdot)$ and $J_{ij}^{\sharp}$, it may not be easy to obtain an explicit representation of $J_i^{\sharp}(r_i, \cdot)$. If $J_i(r_i; \cdot)$ is in a quadratic form as similar to $J_{ij}^{\sharp}$, then the cost function of the subnetwork $J_i^{\sharp}(r_i, \cdot)$ also has a quadratic form and its coefficients will be determined by algebraic manipulations considering the Karush–Kuhn–Tucker (KKT) conditions for (7.7), see Appendix B for details.

Each aggregator A_i will submit the cost function of the subnetwork $J_i^\sharp(r_i, \cdot)$ to the utility company. Utility company A_0, in the next step, tries to determine the optimal price $p_0 = \lambda_0^*$ by using the submitted cost functions. Utility company A_0 rewrites the social welfare maximization problem (7.4) as

$$\max_{\substack{u_0 \\ u_i \ i \in N}} J_0^\sharp(u_0) + \sum_{i \in N} J_i^\sharp(r_i; u_i)$$

$$\text{subject to} \quad u_0 = \sum_{i \in N} u_i,$$

and considers its dual problem as

$$\min_{\lambda_0} \max_{\substack{u_0 \\ u_i \ i \in N}} J_0^\sharp(u_0) + \sum_{i \in N} J_i^\sharp(r_i; u_i) - \lambda_0^\top \left(-u_0 + \sum_{i \in N} u_i \right). \tag{7.8}$$

Utility company A_0 decides the optimal price $p_0 = \lambda_0^*$ by solving (7.8) and broadcast it to the aggregators. The utility company also maximizes the benefit by solving (7.1) with $p_0 = \lambda_0^*$ and obtain u_0^*.

By using the received $p_0 = \lambda_0^*$, aggregator A_i maximizes the benefit by solving

$$\max_{u_i} \quad J_i^\sharp(r_i; u_i) - p_0^\top u_i,$$

and obtain u_i^*. Another task of aggregator A_i is to determine the price p_i for the sub-network. Once u_i^* is obtained, aggregator A_i formulates

$$\max_{\substack{u_{ij} \ j \in N_i}} \sum_{j \in N_i} J_{ij}^\sharp(u_{ij})$$

$$\text{subject to} \quad u_i^* = \sum_{j \in N_i} u_{ij},$$

and considers its dual problem as

$$\min_{\lambda_i} \max_{\substack{u_{ij} \\ i \in N \ j \in N_i}} \sum_{j \in N_i} J_{ij}^\sharp(u_{ij}) - \lambda_i^\top \left(-u_i^* + \sum_{j \in N_i} u_{ij} \right). \tag{7.9}$$

By solving (7.9), aggregator A_i will determine the optimal price $p_i = \lambda_i^*$.

Finally, aggregator A_i broadcasts the optimal price $p_i = \lambda_i^*$ to the consumers, and consumer A_{ij} maximizes the benefit by solving (7.3) with $p_i = \lambda_i^*$ and obtain u_{ij}^*.

Figure 7.3 illustrates the information exchange or aggregation process by the aggregators according to the proposed optimization process. The aggregators can actually moderate the amount of information exchange and computational burden, since, by comparing (7.5) and (7.8), the number of cost functions which should be gathered to the utility company is reduced to $1 + n$ from $1 + n + \sum_{i \in N} n_i$, and the dimension of dual variable that should be determined by the utility company is also reduced to P from $(1 + n)P$. The remaining tasks are equitably shared by n aggregators and solved in (7.7) and (7.9) in a decentralized manner. The proposed optimization process also does not require any communications between the utility company and consumers due to iterative computations. It may be applicable to a short-term energy scheduling problem.

Other optimization processes in which the aggregators moderate the tasks of the utility company has been considered [8, 18]. In [8], a hierarchical optimization structure combines supply function bidding and tâtonnement process. In [18], a bidding of parameters which locally approximate the supply function was used to apply the Newton method for price updating rule, which accelerates the convergence of tâtonnement process.

7.4 Strategic Behavior of Aggregator

This section formulates strategic behavior of aggregators. We suppose that the social welfare will be maximized by using any one optimization process through pricing described in Sect. 7.3. Under this assumption, the aggregator will strategically choose its parameter r_i. In the first setting, the aggregator will try to pursue its benefit as well as market power. In the second setting, the aggregator is supposed to be equipped with battery storage, and it will try to optimize the storage specifications to increase the benefit.

7.4.1 Market Power Optimization

Market power is defined as the ability to affect the market price profitably from competitive levels [13]. If the market power of the utility company is too large compared to the aggregators, the price will be increased and most of the benefit may be brought to the utility company. On the other hand, if the market powers

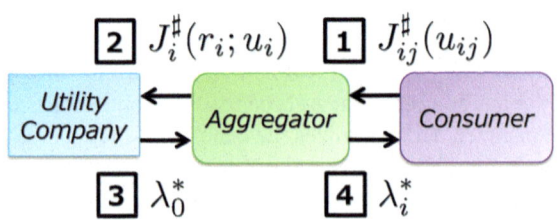

Fig. 7.3 Optimization process via exchanges of information. Modified from [15]

of the aggregators are large, the price will be decreased and the benefit may be properly allocated to the consumers. We suppose that the aggregators make decisions strategically and choose parameter r_i in the cost function in order to pursue their benefit as well as market power. We propose a specific index of market power and define the cost function to evaluate the benefit as well as market power.

In this subsection, we consider the cost function

$$J_i(r_i; u_i) = r_i(u_i^\top Q_i u_i + R_i u_i + C_i), \tag{7.10}$$

of aggregator A_i, which includes the parameter $r_i \in \mathbb{R}$. Essentially, small r_i reduces the operational cost and makes it possible for aggregator A_i to purchase a large amount of u_i. On the other hand, by choosing a large r_i, aggregator A_i may benefit even if only a small amount of u_i is purchased.

We define the market power of aggregator A_i as

$$\mathrm{MP}_i = \sum_{t=1}^{P} \sum_{\tau=1}^{P} \left\| \frac{\partial (p_0)_\tau}{\partial (u_i)_t} \right\| u_i^\top u_i.$$

Note here that $\sum_{t=1}^{P} \sum_{\tau=1}^{P} \left\| \frac{\partial (p_0)_\tau}{\partial (u_i)_t} \right\|$ is the sum of the all elements of the Jacobian matrix $\frac{\partial p_0}{\partial u_i}$, and $\left\| \frac{\partial (p_0)_\tau}{\partial (u_i)_t} \right\|$ represents the sensitivity of the purchased amount of energy by aggregator A_i to the price decided by the utility company A_0. We multiply the term $u_i^\top u_i$ in order to normalize the unit with the cost function. The market power of the utility company can be defined in a similar manner

$$\mathrm{MP}_0 = \sum_{t=1}^{P} \sum_{\tau=1}^{P} \left\| \frac{\partial (p_0)_\tau}{\partial (u_0)_t} \right\| u_0^\top u_0.$$

We also define the market power ratio as

$$\mathrm{RMP}_i = \frac{\mathrm{MP}_i}{\mathrm{MP}_0 + \sum_{j \in N} \mathrm{MP}_j},$$

which is the ratio of market power of aggregator A_i relative to the sum of market powers of the other participants.

We suppose that the aggregators will try to pursue their own benefit as well as market power. We consider "$\mathrm{RMP}_i \times$ benefit" as the cost function for this purpose, and the strategic aggregator will choose r_i and try to increase this cost function. Because the aggregators will try to maximize their cost under the social welfare maximization process, the KKT conditions of the social welfare maximization problem (7.4) becomes constraints. Thus, optimal r_i will solve the following optimization problem, see Appendix C for details,

$$\max_{r_i} \quad \mathrm{RMP}_i \times \left[J_i(r_i; u_i) + (p_i - p_0)^\top u_i \right] \tag{7.11a}$$

$$\text{subject to} \quad \frac{\partial J_0^\sharp(u_0)}{\partial u_0} + p_0 = 0 \tag{7.11b}$$

$$-u_0 + \sum_{i \in N} u_i = 0 \tag{7.11c}$$

$$\frac{\partial J_i(u_i)}{\partial u_i} - p_0 + p_i = 0 \quad i \in N \tag{7.11d}$$

$$-u_i + \sum_{j \in N_i} u_{ij} = 0 \quad i \in N \tag{7.11e}$$

$$\frac{\partial J_{ij}^\sharp(u_{ij})}{\partial u_{ij}} - p_i = 0 \quad j \in N_i, \ i \in N. \tag{7.11f}$$

In order to calculate the market power MP_i of aggregator A_i or to find an optimal solution to (7.11), information on the cost functions of the utility company and all the other aggregators needed to be gathered. The proposed market power index and the optimization procedure in (7.11) may not be useful for optimal design of individual aggregator, since the cost functions of the other aggregators could be a private information and may not be disclosed to the other participants. The proposed strategic optimization of the aggregator is still useful for a qualitative analysis such as how oligopoly impacts on the benefit of the consumers, as we can see in the numerical examples in [15] and Sect. 7.5.

Another possible market power index for the aggregators could be MC_i (Market Centrality) defined by

$$\mathrm{MC}_i = \sum_{t=1}^{P} \sum_{\tau=1}^{P} \left(\left\| \frac{\partial (p_0)_\tau}{\partial (u_i)_t} \right\| + \left\| \frac{\partial (p_i)_\tau}{\partial (u_i)_t} \right\| \right).$$

The market centrality index tries to evaluate an impact of the purchased amount of energy by aggregator A_i not only on the buying price p_0 but also on selling price p_i. A numerical example that considers the market centrality index can be found in [4].

Several indices of market power in electricity market have been considered [9, 21], but the market power of the aggregators has not been discussed yet. The well-known market power indices which can be found in literature may not be applicable to the problem considered here, because the social welfare optimization process considered in Sect. 7.3 uses the optimization-based pricing. The discussions on the market power of the aggregators here are specialized to the market which optimized through optimization-based pricing. The optimization-based pricing for energy demand network including aggregators was also considered in [8, 18], and the market power index proposed here is possibly applicable to the optimization processes considered in [8, 18].

7.4.2 Battery Storage Operation

This subsection considers the strategic operation of battery storage by aggregators. The aggregator is equipped with storage and may choose the total capacity of storage to be installed as well as the ratio of high to low performance storage, where high and low performance means that it has low volatility and is expensive, and vice versa. We suppose that the aggregator will optimize the total capacity and ratio of high to low performance storage to increase its own benefit.

We start with considering the dynamics of high performance storage

$$x_{ibH}[t+1] = \eta_{iH} x_{ibH}[t] + \frac{u_{ibH}[t]}{H_{iH}},$$

where $x_{ibH}[t]$ is state of charge at time t, η_{iH} indicates the volatility of the battery, H_{iH} is the capacity of the battery, and $u_{ibH}[t]$ is the amount of energy charged into the battery at time t. The dynamics of low performance storage are given in a similar manner

$$x_{ibL}[t+1] = \eta_{iL} x_{ibL}[t] + \frac{u_{ibL}[t]}{H_{iL}}.$$

We consider the cost function associated with storage operation as

$$J_{ibH}(u_{ibH}) = \sum_{t=1}^{P} \left\{ -\frac{z_{i1}}{H_{iH}^2} u_{ibH}^2[t] - z_{i2} \left(x_{ibH}[t] - x_{ibH}^{\mathrm{ref}} \right)^2 \right\} - H_{iH} \eta_{iH}^2 z_{i3},$$

$$(7.12a)$$

$$J_{ibL}(u_{ibL}) = \sum_{t=1}^{P} \left\{ -\frac{z_{i1}}{H_{iL}^2} u_{ibL}^2[t] - z_{i2} \left(x_{ibL}[t] - x_{ibL}^{\mathrm{ref}} \right)^2 \right\} - H_{iL} \eta_{iL}^2 z_{i3}, \quad (7.12b)$$

where

$$H_{iH} = \alpha H_{iT}, \quad H_{iL} = (1 - \alpha_i) H_{iT},$$

and given constants x_{ibH}^{ref} and x_{ibL}^{ref} indicate the desired set points of the state of charge for high and low performance storage, respectively, and z_{i1}, z_{i2} and z_{i3} are weighting coefficients.

The first term in the cost function represents the cost for fast charge or discharge. In order to normalize the unit, the cost is divided by the quadratic of capacity. The second term represents the penalty to the deviation from the desired set-point. The last term represents the installation cost of the battery. We assume that the price of the battery is proportional to its capacity and quadratic of its quality, which is represented by the volatility parameter η_{iH} or η_{iL}. Aggregator A_i will choose the total capacity of the batteries H_{iT} and ratio of the high performance batteries α_i. Thus, the strategic parameter is defined by

$$r_i = \{H_{iT}, \ \alpha_i\}.$$

We define the total cost function of aggregator A_i by

$$J_i(r_i; u_i, u_{ibH}, u_{ibL}) = J_i(u_i) + J_{ibH}(r_i; u_{ibH}) + J_{ibL}(r_i; u_{ibL}), \qquad (7.13)$$

where u_i and J_i denote the amount of energy purchased by aggregator A_i and its operational cost, respectively. The storage has dynamics, but the cost function (7.13) can be written in a quadratic form of u_i, u_{ibH} and u_{ibL} as similar to the cost function of the consumers having dynamic appliances, see Appendix A for details.

For given prices p_0 and p_i, since aggregator A_i sells $u_i - u_{ibH} - u_{ibL}$ to the consumers, the benefit-maximization problem (7.2) can be written as

$$\max_{u_i, u_{ibH}, u_{ibL}} \quad J_i(r_i; u_i, u_{ibH}, u_{ibL}) + p_i^\top (u_i - u_{ibH} - u_{ibL}) - p_0^\top u_i.$$

Numerical examples in [16] and Sect. 7.6 illustrate the impact of strategic storage operation by aggregators.

7.5   Case Studies: Strategic Optimization of Market Power-Related Cost Function

We see numerical case studies for strategic behavior of aggregators, where the aggregator tries to maximize market power as well as its own benefit. Other numerical examples can also be found in [15]. The examples in [15] consider, for example, the sequential strategic optimization by the multiple aggregators and the convergence of strategic parameters to the Nash equilibrium.

7.5.1   Optimization of Market Power-Related Cost Function

We consider the network having a single utility company, three aggregators, and ten consumers for each aggregator. The time horizon for optimization is 24 h and is divided into $P = 24$ time slots. The three aggregators have the cost function in the form of (7.10) and the network will be optimized through pricing by one of the optimization processes stated in Sect. 7.3. Under this assumption, this example considers strategic behavior of aggregator A_1, while the other aggregator has fixed parameters as $r_2 = r_3 = 10$, respectively.

Figure 7.4 shows the cost function of aggregator A_1 such as $\text{RMP}_1 \times \left[J_1(r_1; u_1) + (p_1 - p_0)^\top u_1 \right]$ versus parameter r_1. It can be seen that the cost is maximized at around $r_1 = 3.4$. Figure 7.5 shows the optimized demand u_1^* of aggregator A_1 for the cases $r_1 = 3.4$ and 10, respectively. Figure 7.4 implies that there exists a reasonable portion of market power that maximizes the cost of the strategic aggregator. The strategic aggregator can purchase more amount of energy u_1 if it is

Fig. 7.4 Strategy r_1 versus profit of aggregator A_1

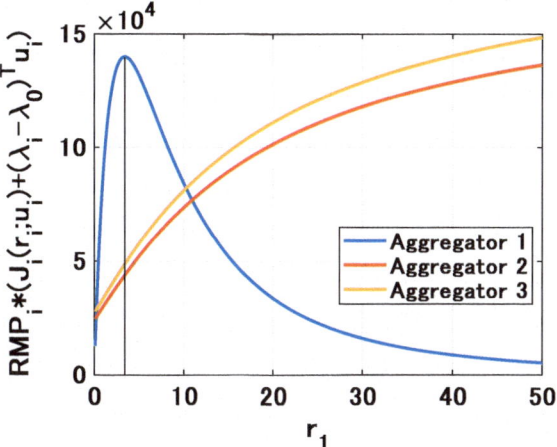

Fig. 7.5 Demand u_1 of aggregator A_1

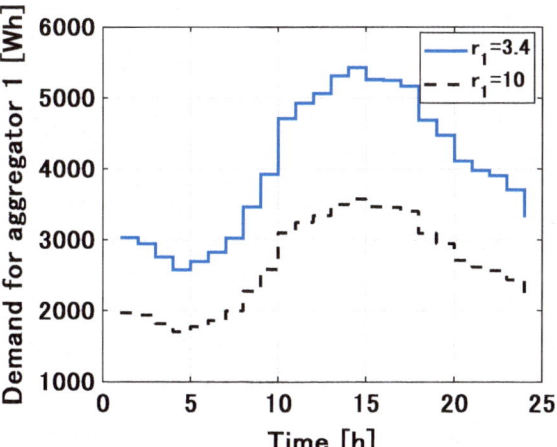

compared to non-strategic case $r_1 = 10$, as it can be seen in Fig. 7.5. This means that the strategic behavior of the aggregator would also be beneficial to the associated consumers, because more amount of energy can be allocated to the consumers.

7.5.2 Optimal Strategy and the Number of Holding Consumers

We consider the network having a single utility company, three aggregators, and 200 consumers in the total. Only aggregator A_1 decides r_1 strategically. At the initial step, we suppose that $n_1 = 50$ and $n_2 = n_3 = 75$. In the next step, aggregator A_1 gets the consumers from the other aggregators such as $n_1 = 52$ and $n_2 = n_3 = 74$

Fig. 7.6 Cost versus the number of holding consumers n_1. Modified from [15]

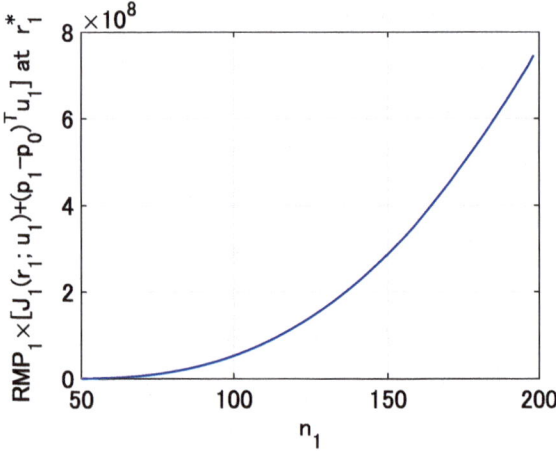

Fig. 7.7 Average of allocated energy to the consumers. Modified from [15]

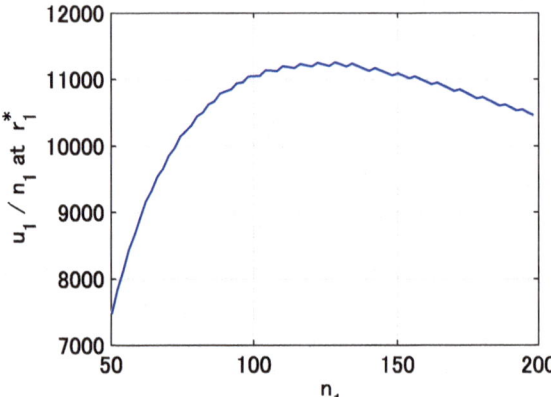

and recalculates the optimal strategy r_1. We continue this process and aggregator A_1 has gradually more and more consumers until $n_1 = 198$. We demonstrate how the number of holding consumers affects the strategic behavior of aggregator.

Figure 7.6 shows the cost function versus n_1. Figure 7.7 shows the average of allocated energy to each consumer that is $(\sum_{\tau=1}^{24}(u_1)_\tau)/n_1$ versus n_1. If the aggregator holds the more and more consumers, it can increase the benefit as well as market power as it can be seen in Fig. 7.6. However, it is not necessarily beneficial to the associated consumers because the average of allocated energy to each consumer shown in Fig. 7.7 becomes decreasing if the aggregator holds an extremely large number of consumers. This example indicates that oligopoly by the strategic aggregator may not be beneficial to the consumers. More details on numerical case studies can be found in [15].

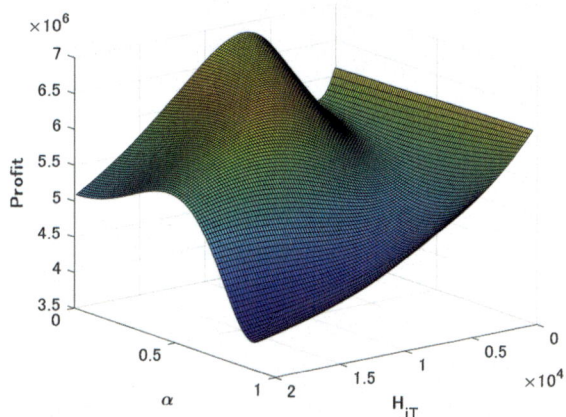

Fig. 7.8 Profit of aggregator A_1. Modified from [16]

7.6   Case Studies: Strategic Operation of Battery Storage

This section considers strategic operation of battery storage by aggregators. We consider the network having a single utility company, three aggregators, and ten consumers for each aggregator. The network will be optimized through pricing by one of the optimization processes stated in Sect. 7.3. Under this assumption, this example supposes that only aggregator A_1 is equipped with storage and strategically choose $r_1 = \{H_{1T}, \alpha_1\}$, where H_{1T} is the total capacity of the storage and α_1 is the ratio of high performance storage. The other parameters in the cost function (7.12) were set to $\eta_{1H} = 0.9$, $\eta_{1L} = 0.5$, $x_{1bH}^{ref} = 0.7$, $x_{1bL}^{ref} = 0.7$, $z_{11} = 50{,}000$, $z_{12} = 50{,}000$ and $z_{13} = 100$, respectively.

Figure 7.8 shows the 3D-plot of the cost function (7.13) of aggregator A_1 versus strategic parameter $r_1 = \{H_{1T}, \alpha_1\}$. Figures 7.9, 7.10, 7.11, and 7.12 show the cut planes of Fig. 7.8 for $H_{1T} = 8000, 10{,}000, 13{,}000, 20{,}000$, respectively.

Figures 7.9, 7.10, 7.11, and 7.12 show that, when the total capacity H_{1T} is given, the aggregator A_1 can maximize its benefit by choosing appropriate α_1, the ratio of high performance storage. From the figures, it can be seen that the best strategy for aggregator A_1 should be $r_1 = \{7200, 0\}$ that maximizes the cost function in this example.

Figure 7.13 shows the 3D-plot of the social welfare (7.4a) of the network. It can be seen that installing the storage with some specific combination of H_{1T} and α_1 can actually increase the social welfare of the network. This means that the network or, in other words, the society may accept the strategic behavior of aggregator A_1. However, the combination of H_{1T} and α_1 that maximize the social welfare may not necessarily maximize aggregator A_1's personal benefit, as we saw in Fig. 7.8. The strategic aggregator may try to increase its own benefit not the social welfare, and it is not a desired behavior for the network. From the market or society design point of view, it may be important to align the maximization of aggregator's personal benefit and the social welfare. If it was realized, the society should accept the strategic behavior of

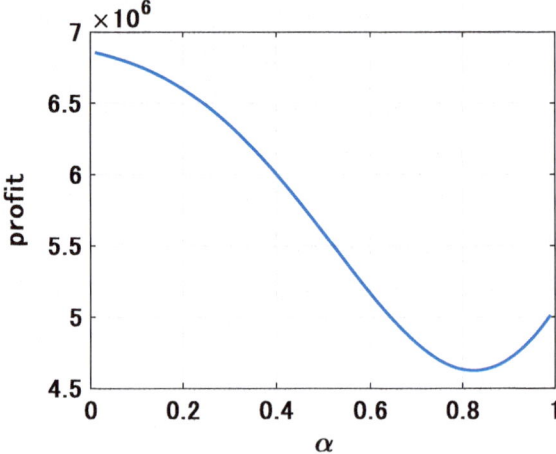

Fig. 7.9 Profit of aggregator A_1 for the total capacity $H_{1T} = 8000$. Modified from [16]

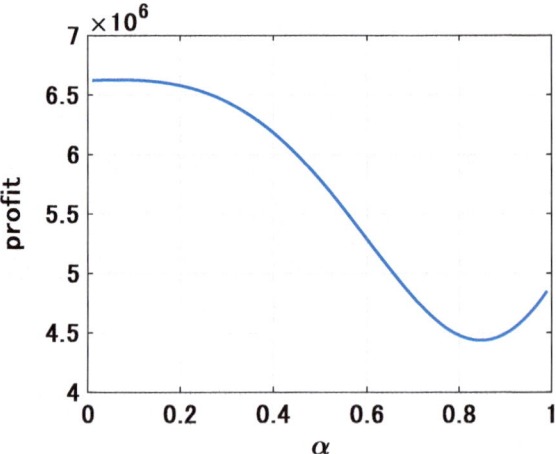

Fig. 7.10 Profit of aggregator A_1 for the total capacity $H_{1T} = 10,000$. Modified from [16]

aggregators. This problem, aligning the maximization of the aggregator's personal benefit and the social welfare, may be formulated as the problem of mechanism design [5, 14, 17], where a suitable transfer cost or, in other words, incentive should be designed to alter aggregator's decision making so as to align the selfish optimization and the social welfare optimization.

7.7 Conclusion

This chapter investigated strategic behavior of aggregators in the three-layered energy demand network optimization problem. We supposed that the network is optimized through price response-based optimization process. Under this assumption, we for-

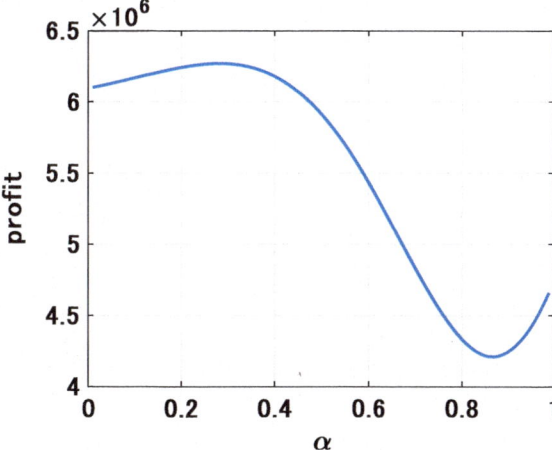

Fig. 7.11 Profit of aggregator A_1 for the total capacity $H_{1T} = 13,000$. Modified from [16]

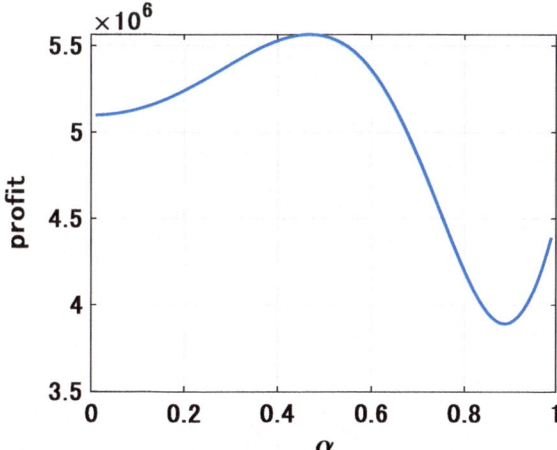

Fig. 7.12 Profit of aggregator A_1 for the total capacity $H_{1T} = 20,000$. Modified from [16]

mulated market power maximization and strategic battery storage operation problems by the aggregators. The strategic decision making by the aggregator could provide useful insights in qualitative analysis of a large energy demand network. Numerical case studies for market power optimization actually indicated that, for example, oligopoly by the aggregator may not be beneficial to the consumers. In case of storage managements, a benefit maximization by the aggregator might not conclude the social welfare maximization, and this encourages researches in other directions, for example, an incentive design problem for the future energy market.

Fig. 7.13 Social welfare. Modified from [16]

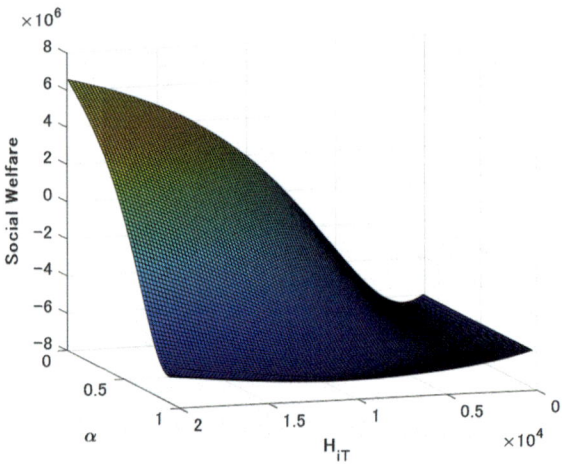

Appendix A: Consumer Model Descriptions

Consumer A_{ij} reports the following utility function in a quadratic form to aggregator A_i:

$$J_{ij}^{\sharp}(u_{ij}) = u_{ij}^{\top} Q_{ij} u_{ij} + R_{ij} u_{ij} + C_{ij}.$$

We denote $u_{ij}[t] \in \mathbb{R}$, $t = 1, \ldots, P$, the consumption of consumer A_{ij} at tth time slot. The Consumers use electricity to utilize appliances, such as PCs and air conditioners. We consider PCs and air conditioners as representative example of the demand which do not have dynamics or have dynamics, respectively.

For the appliances with dynamics, we denote $x_{ij}[t]$ the state, $d_{ij}[t]$ disturbance, and $u_{ijd}[t]$ input. In a representative example, the state variable is room temperature, disturbance is outside air temperature, and input is the electricity consumption of the air conditioner. With these notations, we suppose that the dynamics are described as a linear discrete-time system:

$$x_{ij}[t + 1] = A_{ij} x_{ij}[t] + B_{ij} u_{ijd}[t] + D_{ij} d_{ij}[t].$$

To describe the dynamics including all time slots, we use following expression:

$$x_{ij} = \Phi_{ij} u_{ij} + s_{ij}, \tag{7.14}$$

where

$$s_{ij} = \Psi_{ij} x_{ij}[1] + \Theta_{ij} d_{ij},$$

$$x_{ij} = \begin{bmatrix} x_{ij}[2] \\ \vdots \\ x_{ij}[P+1] \end{bmatrix}, \quad u_{ijd} = \begin{bmatrix} u_{ijd}[1] \\ \vdots \\ u_{ijd}[P] \end{bmatrix}, \quad d_{ij} = \begin{bmatrix} d_{ij}[1] \\ \vdots \\ d_{ij}[P] \end{bmatrix},$$

$$\Psi_{ij} = \begin{bmatrix} A_{ij} \\ A_{ij}^2 \\ \vdots \\ A_{ij}^P \end{bmatrix}, \quad \Phi_{ij} = \begin{bmatrix} B_{ij} & 0 & \cdots & 0 \\ A_{ij}B_{ij} & B_{ij} & & 0 \\ \vdots & \vdots & \ddots & \\ A_{ij}^{P-1}B_{ij} & A_{ij}^{P-2}B_{ij} & \cdots & B_{ij} \end{bmatrix},$$

$$\Theta_{ij} = \begin{bmatrix} D_{ij} & 0 & \cdots & 0 \\ A_{ij}D_{ij} & D_{ij} & & 0 \\ \vdots & \vdots & \ddots & \\ A_{ij}^{P-1}D_{ij} & A_{ij}^{P-2}D_{ij} & \cdots & D_{ij} \end{bmatrix}.$$

The utility function of dynamic demand is formed by a quadratic function of the state x_{ij} and u_{ijd}, but using equation (7.14), the function can be written as a quadratic function of u_{ijd}:

$$J_{ijd}^{\sharp\sharp}(u_{ijd}) = u_{ijd}^T Q_{ijd} u_{ijd} + R_{ijd} u_{ijd} + C_{ijd}.$$

The utility function of the appliances without dynamics is also written as a quadratic function of u_{ijs}:

$$J_{ijs}^{\sharp\sharp}(u_{ijs}) = u_{ijs}^T Q_{ijs} u_{ijs} + R_{ijs} u_{ijs} + C_{ijs}.$$

The utility function of consumer A_{ij} is given as

$$J_{ij}^{\sharp\sharp}(u_{ijd}, u_{ijs}) = J_{ijd}^{\sharp\sharp}(u_{ijd}) + J_{ijs}^{\sharp\sharp}(u_{ijs})$$
$$= u_{ijd}^\top Q_{ijd} u_{ijd} + R_{ijd} u_{ijd} + C_{ijd} + u_{ijs}^\top Q_{ijs} u_{ijs} + R_{ijs} u_{ijs} + C_{ijs}.$$

We now rewrite the function $J_{ij}^{\sharp\sharp}(u_{ijd}, u_{ijs})$ as the function of u_{ij} and obtain $J^{\sharp}(u_{ij})$. We suppose that the consumer optimally allocates u_{ijd} and u_{ijs} when u_{ij} is specified. Then, the utility function is given as

$$J_{ij}^{\sharp}(u_{ij}) = \max_{u_{ijd}, u_{ijs}} J_{ij}^{\sharp\sharp}(u_{ijd}, u_{ijs}) \quad \text{subject to} \quad u_{ij} = u_{ijd} + u_{ijs}$$

or

$$J_{ij}^{\sharp}(u_{ij}) = \min_{\lambda_{ij}} \max_{u_{ijd}, u_{ijs}} J_{ij}^{\sharp\sharp}(u_{ijd}, u_{ijs}) - \lambda_{ij}^\top \left(u_{ijd} + u_{ijs} - u_{ij} \right).$$

The KKT conditions of the above problems are given by

$$\begin{bmatrix} 2Q_{ijd} & 0 & -E \\ 0 & 2Q_{ijs} & -E \\ E & E & 0 \end{bmatrix} \begin{bmatrix} u_{ijd} \\ u_{ijs} \\ \lambda_{ij} \end{bmatrix} = \begin{bmatrix} -R_{ijd}^{\top} \\ -R_{ijs}^{\top} \\ u_{ij} \end{bmatrix},$$

where E denote identify matrix. Then, we get the optimal allocation of u_{ijd} and u_{ijs}:

$$\begin{bmatrix} u_{ijd} \\ u_{ijs} \\ \lambda_{ij} \end{bmatrix} = \begin{bmatrix} 2Q_{ijd} & 0 & -E \\ 0 & 2Q_{ijs} & -E \\ E & E & 0 \end{bmatrix}^{-1} \begin{bmatrix} -R_{ijd}^{\top} \\ -R_{ijs}^{\top} \\ u_{ij} \end{bmatrix}.$$

Here we define

$$G_{ij} = \begin{bmatrix} 2Q_{ijd} & 0 & -E \\ 0 & 2Q_{ijs} & -E \\ E & E & 0 \end{bmatrix}^{-1}$$

and G_{ij} is give as

$$G_{ij} = \begin{bmatrix} Q_{ij}^{inv} & -Q_{ij}^{inv} & 2Q_{ijs}Q_{ij}^{inv} \\ -Q_{ij}^{inv} & Q_{ij}^{inv} & E - 2Q_{ijs}Q_{ij}^{inv} \\ -2Q_{ijs}Q_{ij}^{inv} & -E + 2Q_{ijs}Q_{ij}^{inv} & 2Q_{ijs} - 4Q_{ijs}^2 Q_{ij}^{inv} \end{bmatrix},$$

where

$$Q_{ij}^{inv} = \left(2Q_{ijd} + 2Q_{ijs}\right)^{-1}.$$

We can write $J_{ij}^{\sharp}(u_{ij})$ as

$$J_{ij}^{\sharp}(u_{ij}) = \begin{bmatrix} -R_{ijd}^{\top} \\ -R_{ijs}^{\top} \\ u_{ij} \end{bmatrix}^{\top} G_{ij}^{\top} \begin{bmatrix} Q_{ijd} & 0 & 0 \\ 0 & Q_{ijs} & 0 \\ 0 & 0 & 0 \end{bmatrix} G_{ij} \begin{bmatrix} -R_{ijd}^{\top} \\ -R_{ijs}^{\top} \\ u_{ij} \end{bmatrix}$$

$$+ \begin{bmatrix} R_{ijd} & R_{ijs} & 0 \end{bmatrix} G_{ij} \begin{bmatrix} -R_{ijd}^{\top} \\ -R_{ijs}^{\top} \\ u_{ij} \end{bmatrix} + C_{ijd} + C_{ijs}.$$

Appendix B: Aggregator Model Descriptions

Consumer A_{ij}, who belongs to the subnetwork integrated by aggregator A_i, reports its cost function

$$J_{ij}^{\sharp}(u_{ij}) = u_{ij}^{\top} Q_{ij} u_{ij} + R_{ij} u_{ij} + C_{ij},$$

to aggregator A_i. We suppose that aggregator A_i has a quadratic cost function as similar to Sect. 7.4.1

$$J_i(r_i; u_i) = r_i(u_i^\top Q_i u_i + R_i u_i + C_i).$$

The utility function of the subnetwork is given as $J_i(r_i; u_i) + \sum_{j \in N_i} J_{ij}^\sharp(u_{ij})$, and this needs to be written as a function of $u_i = \sum_{j \in N_i} u_{ij}$. We assume that aggregator A_i optimally allocates u_{ij}, $j \in N_i$ when u_i is specified. Then, the utility function of the subnetwork is written as:

$$J_i^\sharp(r_i; u_i) = J_i(r_i; u_i) + \max_{u_{ij} \; j \in N_i} \sum_{j \in N_i} J_{ij}^\sharp(u_{ij}) \quad \text{subject to} \quad u_i = \sum_{j \in N_i} u_{ij}$$

or

$$J_i^\sharp(r_i; u_i) = J_i(r_i; u_i) + \min_{\lambda_i} \max_{u_{ij} \; j \in N_i} \sum_{j \in N_i} J_{ij}^\sharp(u_{ij}) - \lambda_i^\top \left(-u_i + \sum_{j \in N_i} u_{ij} \right).$$

The KKT conditions of the above problems are given by

$$
\begin{bmatrix}
Q_{i1} + Q_{i1}^\top & 0 & 0 & -E \\
0 & \ddots & 0 & -E \\
\cdots & 0 & Q_{in_i} + Q_{in_i}^\top & -E \\
E & \cdots & E & 0
\end{bmatrix}
\begin{bmatrix}
u_{i1} \\
\vdots \\
u_{in_i} \\
\lambda_i
\end{bmatrix}
=
\begin{bmatrix}
-R_{i1}^\top \\
\vdots \\
-R_{in_i}^\top \\
u_i
\end{bmatrix}.
$$

Here, we define

$$
\text{Gai} =
\begin{bmatrix}
Q_{i1} + Q_{i1}^\top & 0 & 0 & -E \\
0 & \ddots & 0 & -E \\
\cdots & 0 & Q_{in_i} + Q_{in_i}^\top & -E \\
E & \cdots & E & 0
\end{bmatrix},
$$

and

$$RAt = [R_{i1} \; \cdots \; R_{in_i}].$$

Then we get

$$
\text{Gai}
\begin{bmatrix}
u_{i1} \\
\vdots \\
u_{in_i} \\
\lambda_i
\end{bmatrix}
=
\begin{bmatrix}
-RAt^\top \\
u_i
\end{bmatrix}.
$$

We also define

$$\text{Gam} = \text{blockdiag} \begin{bmatrix} Q_{i1} \; \cdots \; Q_{in_i} & 0 \end{bmatrix}.$$

For the shake of simplicity, we assume that $R_i = 0$ and $C_i = 0$, which means

$$J_i(r_i; u_i) = r_i u_i^\top Q_i u_i.$$

Then, the utility function of the subnetwork integrated by aggregator A_i can be rewritten as

$$J_i^\sharp(r_i; u_i) = J_i(r_i; u_i) + \sum_{j \in N_i} J_{ij}^\sharp(u_{ij})$$

$$= r_i u_i^\top Q_i u_i + \begin{bmatrix} u_{i1}^\top & \cdots & u_{in_i}^\top & \lambda_i^\top \end{bmatrix} \text{Gam} \begin{bmatrix} u_{i1} \\ \vdots \\ u_{in_i} \\ \lambda_i \end{bmatrix} + \begin{bmatrix} RAt & 0 \end{bmatrix} \begin{bmatrix} u_{i1} \\ \vdots \\ u_{in_i} \\ \lambda_i \end{bmatrix} + \sum_{j \in N_i} C_{ij}$$

$$= r_i u_i^\top Q_i u_i + \begin{bmatrix} -RAt & u_i^\top \end{bmatrix} (\text{Gai}^{-1})^\top \text{GamGai}^{-1} \begin{bmatrix} -RAt^\top \\ u_i \end{bmatrix}$$

$$+ \begin{bmatrix} RAt & 0 \end{bmatrix} \text{Gai}^{-1} \begin{bmatrix} -RAt^\top \\ u_i \end{bmatrix} + \sum_{j \in N_i} C_{ij}.$$

By using notations $GAs = (\text{Gai}^{-1})^\top \text{GamGai}^{-1}$ and $GAt = \begin{bmatrix} RAt & 0 \end{bmatrix} \text{Gai}^{-1}$, we get

$$J_i^\sharp(r_i; u_i) = \begin{bmatrix} -RAt & u_i^\top \end{bmatrix} GAs \begin{bmatrix} -RAt^\top \\ u_i \end{bmatrix} + GAt \begin{bmatrix} -RAt^\top \\ u_i \end{bmatrix} + \sum_{j \in N_i} C_{ij} + r_i u_i^\top Q_i u_i.$$

By using notations

$$GAs = \begin{bmatrix} GAs1 & GAs2 \\ GAs3 & GAs4 \end{bmatrix}, \quad GAt = [GAt1 \ GAt2],$$

we can write $J_i^\sharp(r_i; u_i)$ as

$$J_i^\sharp(r_i; u_i) = u_i^\top GAs4 u_i - RAtGAs2 u_i + RAtGAs1RAt^\top$$
$$- u_i^\top GAs3RAt^\top - GAt1RAt^\top + GAt2 u_i + \sum_{j \in N_i} C_{ij} + r_i u_i^\top Q_i u_i,$$

which means that

$$Q_i^\sharp = GAs4 + r_i Q_i,$$
$$R_i^\sharp = -RAtGAs2 - RAtGAs3^\top + GAt2,$$
$$C_i^\sharp = RAtGAs1RAt^\top - GAt1RAt^\top + \sum_{j \in N_i} C_{ij}.$$

Appendix C: Details on Market Power Calculation

Market power of aggregator A_i is defined as

$$\mathrm{MP}_i = \sum_{t=1}^{P} \sum_{\tau=1}^{P} \left\| \frac{\partial (\lambda_0)_\tau}{\partial (u_i)_t} \right\| u_i^\top u_i.$$

By using Jacobian matrix

$$\frac{\partial \lambda_0}{\partial u_i} = \begin{bmatrix} \frac{\partial \lambda_0[1]}{\partial u_i[1]} & \cdots & \frac{\partial \lambda_0[1]}{\partial u_i[P]} \\ \vdots & & \vdots \\ \frac{\partial \lambda_0[P]}{\partial u_i[1]} & \cdots & \frac{\partial \lambda_0[P]}{\partial u_i[P]} \end{bmatrix},$$

MP_i is expressed as

$$\mathrm{MP}_i = \mathbf{1}^\top \frac{\partial \lambda_0}{\partial u_i} \mathbf{1} u_i^\top u_i,$$

where $\mathbf{1}$ is a vector whose all elements are 1.

Here, we show how to calculate $\dfrac{\partial (\lambda_0)_\tau}{\partial (u_i)_t}$. The primal problem of the social welfare maximization can be written as

$$\max_{\substack{u_0 \in \mathbb{R}^P \\ u_i \in \mathbb{R}^P \ i \in N}} \quad J_0^\sharp(u_0) + \sum_{i \in N} J_i^\sharp(r_i; u_i)$$

$$\text{subject to} \quad u_0 = \sum_{i \in N} u_i.$$

The dual problem can be written as

$$\min_{\lambda_0 \in \mathbb{R}^P} \max_{\substack{u_0 \in \mathbb{R}^P \\ u_i \in \mathbb{R}^P \ i \in N}} \quad J_0^\sharp(u_0) + \sum_{i \in N} J_i^\sharp(r_i; u_i) - \lambda_0^\top \left(-u_0 + \sum_{i \in N} u_i \right),$$

the KKT conditions are

$$\frac{\partial J_0^\sharp(u_0)}{\partial u_0} + \lambda_0 = 0$$

$$\frac{\partial J_i^\sharp(r_i; u_i)}{\partial u_i} - \lambda_0 = 0, \quad \forall i \in N$$

$$- u_0 + \sum_{i \in N} u_i = 0.$$

We want to have the derivative of $\lambda_0^*[\tau]$ at the optimal solution $u_i^*[t]$. In order to get the derivative of $\lambda_0^*[\tau]$, we fix $u_i[t]$ as a given constant value and then get the optimal $\lambda_0[\tau]$ as a function of $u_i[t]$. The KKT conditions when $u_i[t]$ is fixed are written as

$$\frac{\partial J_0^{\sharp}(u_0)}{\partial u_0} + \lambda_0 = 0$$

$$\frac{\partial J_k^{\sharp}(r_k; u_k)}{\partial u_k} - \lambda_0 = 0, \quad k \neq i$$

$$\frac{\partial J_i^{\sharp}(r_i; u_i)}{\partial u_i^{(-t)}} - \lambda_0^{(-t)} = 0$$

$$u_0 - \sum_{i \in N} u_i = 0,$$

where $a^{(-t)}$ denotes the vector that t-th element is removed from a. In the problem considered in this chapter, we can get the LQ form:

$$\left(Q_0^{\sharp} + Q_0^{\sharp T} \right) u_0 + R_0^{\sharp T} + \lambda_0 = 0$$

$$\left(Q_k^{\sharp} + Q_k^{\sharp T} \right) u_k + R_k^{\sharp T} - \lambda_0 = 0, \quad k \neq i$$

$$p_t^{\top} \left\{ \left(Q_i^{\sharp} + Q_i^{\sharp T} \right) u_i + R_i^{\sharp T} - \lambda_0 \right\} = 0$$

$$u_0 - \sum_{i \in N} u_i = 0,$$

where p_t denotes the matrix that t-th column from $P \times P$ identity matrix. Note that $Q_i^{\sharp}$ and $R_i^{\sharp}$ include parameter r_i as described in Appendix 7.7.

To get $\dfrac{\partial (\lambda_0)_{\tau}}{\partial (u_i)_t}$, we differentiate each term of KKT conditions:

$$\frac{\partial^2 J_0^{\sharp}(u_0)}{\partial u_0^2} \frac{du_0}{du_i[t]} + \frac{d\lambda_0}{du_i[t]} = 0$$

$$\frac{\partial^2 J_k^{\sharp}(r_k; u_k)}{\partial u_k^2} \frac{du_k}{du_i[t]} - \frac{d\lambda_0}{du_i[t]} = 0, \quad k \neq i$$

$$\frac{\partial}{\partial u_i[t]} \left(\frac{\partial J_i^{\sharp}(r_i; u_i)}{\partial u_i^{(-t)}} \right) + \frac{\partial^2 J_i^{\sharp}(r_i; u_i)}{\partial u_i^{(-t)2}} \frac{du_i^{(-t)}}{du_i[t]} - \frac{d\lambda_0^{(-t)}}{du_i[t]} = 0$$

$$\frac{du_0}{du_i[t]} - \sum_{k \in N} \frac{du_k}{du_i[t]} = 0.$$

The LQ form of these equations are

$$\left(Q_0^\sharp + Q_0^{\sharp\top}\right)\frac{du_0}{du_i[t]} + \frac{d\lambda_0}{du_i[t]} = 0$$

$$\left(Q_k^\sharp + Q_k^{\sharp\top}\right)\frac{du_k}{du_i[t]} - \frac{d\lambda_0}{du_i[t]} = 0, \quad k \neq i$$

$$p_t^\top\left(Q_i + Q_i^{\sharp\top}\right)p_{t2} + p_t^\top\left(Q_i^\sharp + Q_i^{\sharp\top}\right)p_t\frac{du_i^{(-t)}}{du_i[t]} - \frac{d\lambda_0^{(-t)}}{du_i[t]} = 0$$

$$\frac{du_0}{du_i[t]} - p_{t2} - p_t\frac{du_i^{(-t)}}{du_i[t]} - \sum_{k\neq i}\frac{du_k}{du_i[t]} = 0,$$

where p_{t2} is the vector that the t-th element of P-dimension column zero vector is changed to 1. This means that, for each r_i, market power can be determined by solving a linear equation, which can be solved easily.

References

1. Boyd S, Vandenberghe L (2004) Convex optimization. Cambridge University Press
2. Chai B, Chen J, Yang Z, Zhang Y (2014) Demand response management with multiple utility companies: A two-level game approach. IEEE Trans Smart Grid 5(2):722–731
3. Gkatzikis L, Koutsopoulos I, Salonidis T (2013) The role of aggregators in smart grid demand response markets. IEEE Trans Sel Areas Commun 31(7):1247–1257
4. Hirata K, Okajima Y, Nishizawa K, Uchida K (2017) Decentralized optimization and strategic bidding of aggregators in energy demand networks. In: Proceedings of the 60th Japan joint automatic control conference, SaA1–4 (in Japanese)
5. Jackson MO (2003) Mechanism theory. EOLSS Publishers
6. Jokić A, Lazar M, van den Bosch PPJ (2009) Real-time control of power systems using nodal prices. Int J Electr Power Energy Syst 31(9):522–530
7. Kiani A, Annaswamy A (2011) Wholesale energy market in a smart grid: dynamic modeling and stability. In: Proceedings of the 50th IEEE conference on decision and control, pp 2202–2207
8. Kim H, Thottan M (2011) A two-stage market model for microgrid power transactions via aggregators. Bell Labs Tech J 16(3):101–108
9. Lee YY, Hur J, Baldick R, Pineda S (2011) New indices of market power in transmission-constrained electricity markets. IEEE Trans Power Syst 26(2):681–689
10. Li N, Chen L, Low SH (2011) Optimal demand response based on utility maximization in power networks. In: Proceedings of the 2011 IEEE power and energy society general meeting, pp 1–8
11. Li T, Shahidehpour M (2005) Strategic bidding of transmission-constrained GENCOs with incomplete information. IEEE Trans Power Syst 20(1):437–447
12. Lin CE, Viviani GL (1984) Hierarchical economic dispatch for piecewise quadratic cost functions. IEEE Trans Power Apparus Syst PAS-103(6):1170–1175
13. Mas-Colell A, Whinston MD, Green JR (1995) Microeconomic theory. Oxford University Press
14. Murao T, Hirata K, Okajima Y, Uchida K (2018) Real-time pricing for LQG power networks with independent types: a dynamic mechanism design approach. Eur J Control 39:95–105
15. Okajima Y, Hirata K, Murao T, Hatanaka T, Gupta V, Uchida K (2017) Strategic behavior and market power of aggregators in energy demand networks. In: Proceedings of the 56th IEEE conference on decision and control, pp 694–701

16. Okajima Y, Hirata K, Gupta V, Uchida K (2018) Strategic battery storage management of aggregators in energy demand networks. In: Proceedings of the 2nd IEEE conference on control technology and applications, pp 444–449

17. Okajima Y, Murao T, Hirata K, Uchida K (2015) Integration of day-ahead energy market using VCG type mechanism under equality and inequality constraints. In: Proceedings of the 2015 IEEE multi-conference on systems and control, pp 187–194

18. Papavasiliou A, Hindi H, Greene D (2010) Market-based control mechanisms for electric power demand response. In: Proceedings of the 49th IEEE conference on decision and control, pp 1891–1898

19. Samadi P, Mohsenian-Rad H, Schober R, Wong VMS, Jatskevich J (2010) Optimal real-time pricing algorithm based on utility maximization for smart grid. In: Proceedings of the 2010 first IEEE international conference on smart grid communications, pp 415–420

20. Stoft S (2002) Power system economics. IEEE Press

21. Wang P, Xiao Y, Ding Y (2004) Nodal market power assessment in electricity markets. IEEE Trans Power Syst 19(3):1373–1379

22. Wen F, David AK (2001) Optimal bidding strategies and modeling of imperfect information among competitive generators. IEEE Trans Power Syst 16(1):15–21

Chapter 8
Incentive-Based Economic and Physical Integration for Dynamic Power Networks

Yasuaki Wasa, Toshiyuki Murao and Ken-Ichi Akao

Abstract This chapter describes an incentive-based market that integrates the economic layer and the physical layer to guarantee high-quality physical ancillary services in dynamic power networks. Toward the full liberalization of the electricity markets, it is indispensable to develop a socially optimal regulation market with a high-speed market-clearing mechanism while preventing maliciously strategic biddings of market participants called agents. To realize the markets, the utility functioning as a system operator prepares an appropriate incentive mechanism to elicit the agents' private information. We refer to this real-time incentive-enabled market as an incentivizing market. After illustrating the issues of the agents' strategic biddings without incentive mechanisms, we mainly propose incentivizing market mechanisms for moral hazard problems and adverse selection problems in dynamic power networks, respectively. We also discuss features, possibilities, and limitations of our proposed mechanisms through simulations for linear quadratic Gaussian (LQG) power networks.

8.1 Introduction

Toward large-scale integration of distributed energy resources (DERs), including renewables, demand response, and storage devices, in today's era of electricity deregulation, one of the control problems for dynamic power networks is to achieve quality

Y. Wasa (✉)
Department of Electrical Engineering and Bioscience, Waseda University,
Tokyo 169-8555, Japan
e-mail: wasa@aoni.waseda.jp

T. Murao
Department of Robotics, Kanazawa Institute of Technology,
Ishikawa 921-8501, Japan
e-mail: murao@neptune.kanazawa-it.ac.jp

K.-I. Akao
Graduate School of Social Sciences, Waseda University,
Tokyo 169-8050, Japan
e-mail: akao@waseda.jp

© Springer Nature Singapore Pte Ltd. 2020   181
T. Hatanaka et al. (eds.), *Economically Enabled Energy Management*,
https://doi.org/10.1007/978-981-15-3576-5_8

assurance of frequency, voltage, and power—known as ancillary services—in real time [2]. Because the DERs will be owned and operated by different agents, the control drivers of ancillary services have been migrating from traditional electricity suppliers (e.g., [18, 22, 31]) to participants in competitive power-balancing markets. These participants comprise agents and a utility: The agents control their physical systems including generators and demand response units strategically according to their own criterion, whereas the utility is a market operator that economically integrates all the agent controls with the whole physical system stable. Such market and control designs are known as market-based control and transactive control [36].

To establish the market-based controls, various market mechanisms have been proposed in the past decade [11, 12, 14, 15, 20, 46]. The fundamental approach that enables distributed decision making of market participants is a real-time pricing mechanism operated by the utility in dynamic power networks as pioneered by Berger and Schweppe [6]. Kiani et al. [20] and Wasa et al. [46] have presented a transactive control integrating this pricing mechanism into a three-layer market architecture based on timescale control decomposition (see, e.g., [17] and Chaps. 1 and 2 of this book). Hirata et al. [14, 15] have proposed a gradient-based pricing strategy tracking to a socially optimal solution under distributed decision making. Ela et al. [11, 12] have proposed a market and pricing mechanism implementing the primary frequency response. However, these papers [11, 12, 14, 15, 20, 46] assume non-strategic agents. To address agents' *strategic* behavior under electricity liberalization, a market mechanism is required to achieve the following distributed implementation; each agent bids *truthfully*, i.e., *non-strategically*, based on its private information in response to a one-shot market-clearing price, while the utility clears the market based on the bidding and sets the prices in real time. From the information asymmetry between the utility and the agents, the transient states led by the agents' strategic behavior may conflict with the utility's objective of achieving high-quality ancillary services. Consequently, a market mechanism needs to contain some incentive mechanisms for eliciting agents' private information, achieving the utility's objective, and securing agents' strategic behavior. In economics, the incentive mechanisms are divided into two types: *moral hazard problem* for preventing the strategic behavior of the agents' control policy called hidden action and *adverse selection problem* to make the agents report their truthful state and type information called hidden information [7]. In this chapter, we refer to this incentive-enabled balancing market as an *incentivizing market* and provide formulas and optimal conditions for the two incentive design problems, respectively.

The technical papers addressing the moral hazard problem in power networks are relatively few [41, 42, 44]. Moral hazard incentive induces the agents to select their controls minimizing a public cost; this is the utility's objective, but it cannot adjust the agents' control directly in real time out of respect for individual decision making. This chapter introduces a problem formulation proposed in our papers [41, 42], which is motivated by principal–agent problems in contract theory [9, 16, 21, 33, 34]. However, the optimal incentive condition presented in the economic papers [9, 16, 21, 33, 34] cannot be applied to the dynamic power network models with the limited control dimension, which is smaller than those of the whole states. Thus, we present a novel parameterization of the reward functional as an optimal inventive model and an incentive synthesis problem to derive the public cost minimization, which we sometimes call public optimality.

On the other hand, to overcome the adverse selection incentivizing problem, several price-based mechanisms for rational agents have been proposed based on the mechanism design theory in economics [5, 19]. Silva et al. [35] have provided a solution to economic dispatch while considering *incentive compatibility* (IC) constraints and *individual rationality* (IR) constraints.[1] The IC and IR constraints are essential properties to design a mechanism that agents will voluntarily and truthfully have access to the power network. Samadi et al. [32] have designed an incentive mechanism for demand-side management, which holds efficiency, user truthfulness, and nonnegative transfer. However, the methodologies reported in [32, 35] are limited to the framework of static mechanism design. Langbort et al. [23, 37] have proposed some real-time pricing methods based on dynamic mechanism design. Taylor et al. [39] have investigated dynamic real-time pricing with a tax for a regulation service in a time-invariant linear quadratic (LQ) system. This chapter introduces an incentive mechanism based on a dynamic mechanism design presented in our paper [26]. This incentivizing scheme corresponds to the Groves mechanism [19] in the mechanism design literature.

This chapter has been organized as follows: In Sect. 8.2, we introduce a general problem formulation constituting an incentivizing market mechanism integrating an economic layer and a physical layer. In Sect. 8.3, we discuss the influence of the agents' strategic biddings over the whole dynamical system and the necessity of incentive-based market mechanisms through simulations with an intrinsic linear quadratic Gaussian (LQG) model. In Sect. 8.4, we present an optimal incentive condition for a moral hazard incentivizing market and an adverse selection incentivizing market. The proposed market mechanism with incentives can achieve public optimality based on private optimal controls subjected to IC and IR constraints. In Sect. 8.5, we illustrate the physical and economic performance of the presented market mechanisms via simulations with the Institute of Electrical Engineers of Japan (IEEJ) East 30-machine system [40]. In Sect. 8.6, we conclude our discussion.

8.2 Problem Formulation

This chapter handles a real-time electricity regulation market. As shown in Chap. 2 of this book, the conventional power system control is divided into three control time cycles termed primary control (voltage and frequency stabilization), secondary control (quasi-stationary load power imbalance control), and tertiary control (economic dispatch) [17, 20, 46]. In particular, we focus on a market-based transactive control

---

[1]The definition of IR and IC constraints is slightly different between moral hazard problems and adverse selection problems. The IC and IR are formally defined in Sect. 8.4.

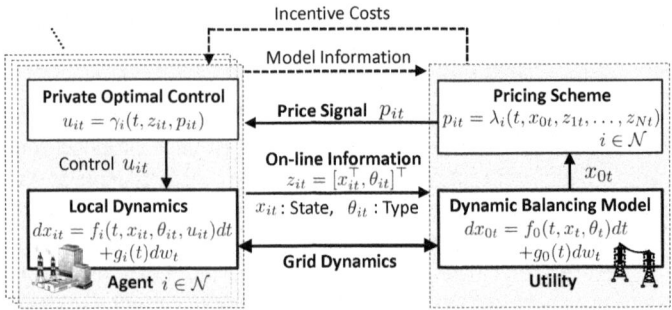

Fig. 8.1 Framework of an incentivizing market

problem, which integrates the control objectives of both the primary control and the secondary control.

This chapter considers two kinds of market players, *utility* and *agents* (Fig. 8.1). We also adopt the average system frequency model [6] or its general model [26], which is known as generic power dynamics of high-speed response for ancillary service control problems. The power dynamics indicate the state deviations around the nominal trajectory planned over a future time interval $[t_0, t_f]$ in a spot energy market. Assume a power network consists of N-agents (areas) including generators and consumers and the utility is a market operator that economically integrates all the agent controls with the physical system stable.[2] Unless noted otherwise, we suppose that the parameters and functions in the system models and the cost models are smooth on time, state, and control regions and have boundedness and polynomial growth properties.[3] For readers who are interested in a mathematically rigorous treatment, please see our companion paper [26, 43].

As mentioned in Chap. 1 of this book, the systems and control engineers' standpoint for problem formulation is very different from that of economists. This chapter advances a discussion based on the viewpoint of multi-agent systems (MASs) and their control theory. In particular, the conventional MAS in systems and control field is considered in the absence of the utility functioning as a leader or a principal, which is different from the framework of the principal–agent problems and the market design problems in economics. Hence, we interpret some economic policies of the proposed market mechanisms while extending the framework of the MAS to the electricity market design problem integrating the economic layer into the physically dynamic power network.

---

[2]Following the custom in the economics literature, we define that the utility is a feminine noun and the agent is a masculine noun.

[3]The function $f(x)$ is *polynomial growth* if there exist some positive value $K > 0$ and a positive integer k such that $|f(x)| \leq K(1 + |x|^k)$ for all the elements x.

8.2.1 Grid Model and Information Exchange

Agent Model Each agent $i \in \mathcal{N} := \{1, \ldots, N\}$ has an independent subsystem with the local state $x_i \in \mathbb{R}^{n_i}$, the operation-type parameter $\theta_i \in \Theta_i := \{1, 2, \ldots, s_i\}$, and the local control input $u_i \in \mathbb{R}^{m_i}$, where the set $\mathbb{R}^n$ indicates n-dimensional real space. The parameters x_i, θ_i, and u_i depend on time t. Afterward, for $x.$, $\theta.$, and $u.$, we use an abbreviation like $x_{it} := x_i(t)$, $x_t := x(t)$. The state evolution of the subsystem i over a future time interval $[t_0, t_f]$ is described by

$$dx_{it} = f_i(t, x_{it}, \theta_{it}, u_{it})dt + g_i(t)dw_t, \quad t_0 \le t \le t_f, \ i \in \mathcal{N}, \tag{8.1}$$

where $w_t \in \mathbb{R}^q$ is the disturbance including the stochastic nature of renewables and is modeled by a white Gaussian random process with zero mean and unity variance defined on $[t_0, t_f]$.[4] The system representation (8.1) can be regarded as a generic component model [17], and a portion of the local states x_i is physically and informationally connected to the whole power network. For the specific models of the system functions f_i and g_i, refer to, e.g., [17, 22]. The states x_i indicate typically the deviation of power generation or consumption from the scheduled trajectory; the typical examples of the type parameter θ_i are the operation method of the turbine generators and the plug-and-play level in demand-side management; the inputs u_i denote control variables to compensate for the deviation.

Given initial conditions $x_i(t) = x_i$ and $\theta_i(t) = \theta_i$, the agent i's private economic cost functional along the local dynamics (8.1) during $[t, t_f]$ is given by

$$C_i(t, x_i, \theta_i) = \varphi_i(t_f, x_{it_f}, \theta_{it_f}) + \int_t^{t_f} l_i(\tau, x_{i\tau}, \theta_{i\tau}, u_{i\tau})d\tau, \ i \in \mathcal{N}, \tag{8.2}$$

where l_i is an integral cost functional and φ_i is a terminal cost functional. The private cost functional (8.2) means the loss for energy generation or consumption from the scheduled nominal values over the time period $[t, t_f]$. As the future type is undecided at any current time t, the type parameter θ_i is modeled as a Markov step process

$$\Pr\{\theta_i(\tau + \Delta) = l \mid \theta_i(\tau) = k\} = \begin{cases} \pi_{kl}^i \Delta + o(\Delta), & \text{if } k \ne l \\ 1 + \pi_{kk}^i \Delta + o(\Delta), & \text{if } k = l \end{cases},$$

$$\pi_{kk}^i := -\sum_{l \ne k} \pi_{kl}^i, \ t \le \tau \le \tau + \Delta, \ i \in \mathcal{N}, \tag{8.3}$$

where the parameter $\pi_{kl}^i \ge 0$, $k \ne l$ means a transition probability from type k to type l and $o(\cdot)$ indicates a high-order term [8, 26].

---

[4]Our problem formulation admits that each control u_i affects only a part of all the states x. For instance, the control u_i of the agent i cannot operate the public state x_0 and the others' state $x_j, j \in \mathcal{N} \setminus \{i\}$ instantly. Meanwhile, most of the economic literature (e.g., [9, 16, 21, 33, 34]) implicitly assumes the direct control for all the states.

Utility Model The utility manages the power and/or frequency imbalance caused by the state deviations of the physically connected subsystems and other deviations from physical constraints as well. The balancing equation with the interaction state variables $x_0 \in \mathbb{R}^{n_0}$ obeys the following equation:

$$dx_{0t} = f_0(t, x_t, \theta_t)dt + g_0(t)dw_t, \quad t_0 \le t \le t_f, \tag{8.4}$$

and is evaluated by the power balance cost functional:

$$C_0(t, x_0) = \varphi_0(t_f, x_{0t_f}) + \int_t^{t_f} l_0(\tau, x_{0\tau})d\tau, \tag{8.5}$$

where $x = [x_0^\top, x_1^\top, \ldots, x_N^\top]^\top \in \mathbb{R}^n$, $n := n_0 + n_1 + \ldots + n_N$, is the collection of the states of the utility dynamics (8.4) and the agents' dynamics (8.1); $\theta = [\theta_1, \ldots, \theta_N]^\top \in \Theta := \Theta_1 \times \ldots \times \Theta_N$ is the collection of their type parameters; and $u = [u_1^\top, \ldots, u_N^\top]^\top \in \mathbb{R}^m$, $m := m_1 + \ldots + m_N$, is the aggregated control input. The function φ_0 is a terminal cost functional of the utility, and l_0 is her integral cost functional.

We formulated the grid model with the evaluation functionals (8.2) and (8.5) from the current time t to the terminal time t_f[5]; namely, our evaluation functionals are of a model-predictive type. Afterward, the aggregated grid model is described as

$$dx_t = f(t, x_t, \theta_t, u_t)dt + g(t)dw_t, \quad t_0 \le t \le t_f, \tag{8.6}$$

$$f(t, x_t, \theta_t, u_t) := \begin{bmatrix} f_0(t, x_t, \theta_t) \\ f_1(t, x_{1t}, \theta_{1t}, u_{1t}) \\ \vdots \\ f_N(t, x_{Nt}, \theta_{Nt}, u_{Nt}) \end{bmatrix}, \quad g(t) := \begin{bmatrix} g_0(t) \\ g_1(t) \\ \vdots \\ g_N(t) \end{bmatrix},$$

and we sometimes use $\theta := [\theta_i, \theta_{-i}]$ and $u := [u_i, u_{-i}]$ for $i \in \mathcal{N}$ with $\theta_{-i} := [\theta_1, \ldots, \theta_{i-1}, \theta_{i+1}, \ldots, \theta_N]^\top$ and $u_{-i} := [u_1^\top, \ldots, u_{i-1}^\top, u_{i+1}^\top, \ldots, u_N^\top]^\top$.

Information Exchange and Decision-making Process Let us introduce the framework for information exchange and decision-making process among the utility and the agents in the real-time power network. First, each agent $i \in \mathcal{N}$ obtains the state information $x_{i\tau}$ at time $\tau \in [t_0, t_f]$ and reports the information $z_{i\tau} = [x_{i\tau}^\top, \theta_{i\tau}]^\top$ to the utility.[6] Next, using the information $z_\tau = [x_\tau^\top, \theta_\tau^\top]^\top$ reported from all the agents, the utility calculates command signals $p_{i\tau} = \lambda_i(\tau, z_\tau)$, $i \in \mathcal{N}$, to achieve not only public cost minimization but also two requisite mechanism properties: IC and IR. We call

---

[5]From the time-consistency property of dynamic programming (DP), we can handle the cost functionals for any time $t \in [t_0, t_f]$, which is not limited to the initial time t_0, without loss of generality.

[6]Even if the utility knows the above functions and dynamics, the utility cannot execute a public optimal control rigorously without obtaining the actual state information $x(\tau)$ and the type information $\theta(\tau)$ along the stochastic dynamics (8.6) and the type transition (8.3). Hence, the state and type information $z_{i\tau} = [x_{i\tau}^\top, \theta_{i\tau}]^\top$ at time τ is also one of the agents' online private information.

the scheme $\lambda := [\lambda_1^\top, \ldots, \lambda_N^\top]^\top$ *pricing scheme* and call the command signal $p_{i\tau}$ *price* for agent i. Here, it is assumed that the utility knows all the system functions, the cost functionals, and the type transition probabilities π_{kl}^i of every agent.[7] Then, the utility transmits the command signal $p_{i\tau}$ to each agent $i \in \mathcal{N}$. Finally, agent i chooses his control input $u_{i\tau} = \gamma_i(\tau, z_{i\tau}, p_{i\tau})$ to achieve an individual objective, using the command signal $p_{i\tau}$ received from the utility.

Here, the pricing scheme $\lambda_i(\tau, z_\tau)$ for each agent $i \in \mathcal{N}$ is regarded as a utility's policy scheme so that the utility, as well as a market system operator, decides economical taxes and subsidies depending on physical state variables x. The utility can select a suitable pricing scheme λ_i for agent i from an appropriate class Λ_i. On the other hand, each agent i also has a bidding parameter $z_{i\tau}$ reported to the utility and the right to decide his control scheme $\gamma_i(\tau, z_{i\tau}, p_{i\tau})$ strategically. Figure 8.1 shows a diagram of the framework for information exchange, the decision-making processes of pricing scheme and controls, and incentives mentioned later.

Admissible Controls and Implementations Let us next define *admissible controls* and *implementations* in our framework. An admissible control law of agent $i \in \mathcal{N}$, denoted as $\tilde{\gamma}_i \in \Gamma_i$, provides a feedback control $u_i(\tau) = \tilde{\gamma}_i(\tau, z_\tau) \in U_i \subseteq \mathbb{R}^{m_i}, i \in \mathcal{N}$ that assures the existence of the unique solution of the state trajectory $x_\tau, t \leq \tau \leq t_f$ along the dynamics (8.6), $\tilde{\gamma}_i$ is continuous on $[t_0, t_f]$ and Lipschitz-continuous in $\mathbb{R}^n$, and U_i is compact. As stated in the above framework, the control input $u_{i\tau}$ of agent i must be designed by using only a state and type $z_{i\tau} = [x_{i\tau}^\top, \theta_{i\tau}]^\top$ and the price $p_{i\tau}$ given by the utility. For an admissible control law $\tilde{\gamma}_i$, when the function γ_i and the pricing scheme λ_i are such that $\tilde{\gamma}_i(\tau, z_\tau) = \gamma_i(\tau, z_{i\tau}, \lambda_i(\tau, z_\tau))$, the function γ_i is called the *implementation* of the feedback control law $\tilde{\gamma}_i$ for the pricing scheme λ_i.[8]

8.2.2 Control Objectives and Market Model

Control Objectives The publicly economic cost for evaluating the whole power network efficiency is defined by, for a given $\tilde{\gamma} = [\tilde{\gamma}_1^\top, \ldots, \tilde{\gamma}_N^\top]^\top$ in $\Gamma := \Gamma_1 \times \ldots \times \Gamma_N$,

$$C(t, x, \theta; \tilde{\gamma}) = C_0(t, x_0; \tilde{\gamma}) + \sum_{i \in \mathcal{N}} C_i(t, x_i, \theta_i; \tilde{\gamma}_i), \tag{8.7}$$

which is the sum of all the agent costs (8.2) and the power balance cost (8.5). *The main problem of the utility is to make all the agents take a public optimal control which minimizes the expectation of the public cost* (8.7) under a given initial condition

---

[7]On a real-time power electricity market, it is preferable to determine the price without iterative calculation, which is known as tâtonnement-type calculation. In this case, a profit-neutral agent needs some information. This assumption is also set in other papers (see, e.g., [3, 39]). Meanwhile, some papers have tackled to relax this assumption [13, 27].

[8]For the sufficient condition for the existence of γ_i, see, e.g., [43, Lemma 1] and (8.27) in this chapter.

(x, θ) at time t, i.e.,

$$\min_{\tilde{\gamma}_i \in \Gamma_i, i \in \mathcal{N}} \mathbb{E}\left[C(t, x, \theta; \tilde{\gamma}) \,\middle|\, x(t) = x, \theta(t) = \theta, \tilde{\gamma} \right], \tag{8.8}$$

where $\mathbb{E}[\tilde{a} \mid \tilde{b}]$ denotes the conditional expectation of a cost $\tilde{a}$ given some conditions $\tilde{b}$. Note that the public cost minimization problem (8.8) is equivalent to the maximization problem of *social welfare*.

Meanwhile, *each agent $i \in \mathcal{N}$ aims at minimizing his expected private cost* (8.2) strategically, i.e.,

$$\min_{\tilde{\gamma}_i \in \Gamma_i} \mathbb{E}\left[C_i(t, x_i, \theta_i; \tilde{\gamma}_i) \,\middle|\, x(t) = x, \theta(t) = \theta, \tilde{\gamma}_i, \tilde{\gamma}_{-i} \right], \tag{8.9}$$

where $\tilde{\gamma}_{-i} := [\tilde{\gamma}_1^\top, \dots, \tilde{\gamma}_{i-1}^\top, \tilde{\gamma}_{i+1}^\top, \dots, \tilde{\gamma}_N^\top]^\top$.[9]

Here, a rational agent, which is defined as Homo economicus in economics, has his own preference and tries to minimize his individual cost strategically. Meanwhile, the utility, whose dynamics is derived from the interplay among the agents' dynamics, intends to minimize public cost. We emphasize that each agent has not only the control input u_i but also the information parameter $z_i = [x_i^\top, \theta_i]^\top$ as a part of bidding parameters, but the above situation ensures that agents behave strategically and *the utility cannot operate their bidding parameters (u_i, z_i) compellingly at any time* in electricity liberalization. Through bidding information reported from all the agents, the utility decides a price scheme λ_i based on a market mechanism and reports the price to all the agents. Due to the interplay of the local states x_i on the utility dynamics (8.4) with the power balance cost (8.5) and the asymmetric information between the utility and the agents in the above situation, the public optimal control is generally different from the private optimal control. Then, the system behavior led by the agents' selfish behavior destabilizes the system at worse. In the following, we discuss optimal controls for each agent (we call *private optimal controls*) and for the whole power network system (which is called *public optimal control*) separately.

Private Optimal Control Policy We first consider an optimal control problem for each agent $i \in \mathcal{N}$, which is described as (8.9). The solution to the private control problem (8.9) is denoted by $u_i^\sharp(\tau) = \tilde{\gamma}_i^\sharp(\tau, z_\tau)$, $t \leq \tau \leq t_f$, and is called the *private optimal control* of agent $i \in \mathcal{N}$. The notations $\gamma_i^\sharp$, $\lambda_i^\sharp$, and $p_i^\sharp$ denote the corresponding implementation, pricing scheme, and price for the private optimal control law $\tilde{\gamma}_i^\sharp$, respectively.

The private control problem for the admissible control is equivalent to a standard optimal control problem with a jump parameter [48]. Applying the dynamic programming (DP) approach to the private control problem of the agent i, we find the following optimal control,

---

[9]In economics, the cost minimization problem (8.9) is often solved as the maximization problem of the agent's reward $\mathbb{E}\left[-C_i\right]$. From the consistency of the problem formulation, we discuss (8.9) throughout this chapter.

$$u_i^{\sharp}(\tau) = \tilde{\gamma}_i^{\sharp}(\tau, z_{\tau}) = \arg \min_{u_i \in U_i} H_i \left(\tau, x_i, \theta_i, \left(\frac{\partial V_i(\tau, x_i, \theta_i)}{\partial x_i} \right)^{\top}, u_i \right) \Bigg|_{\substack{x_i = x_{i\tau} \\ \theta_i = \theta_{i\tau}}}, \quad (8.10)$$

where

$$H_i(\tau, x_i, \theta_i, p_i, u_i) = l_i(\tau, x_i, \theta_i, u_i) + p_i^{\top} f_i(\tau, x_i, \theta_i, u_i) \quad (8.11)$$

and $V_i(\tau, x_i, \theta_i)$ obeys the following Hamilton–Jacobi–Bellman (HJB) equation:

$$-\frac{\partial V_i(\tau, x_i, \theta_i)}{\partial \tau} = \min_{u_i \in U_i} \left\{ H_i \left(\tau, x_i, \theta_i, \left(\frac{\partial V_i(\tau, x_i, \theta_i)}{\partial x_i} \right)^{\top}, u_i \right) \right.$$
$$\left. + \sum_{\tilde{\theta}_i = 1}^{s_i} \pi_{\theta_i \tilde{\theta}_i}^i V_i(\tau, x_i, \tilde{\theta}_i) + \frac{1}{2} \mathrm{tr} \left(g_i(\tau) g_i^{\top}(\tau) \frac{\partial^2 V_i(\tau, x_i, \theta_i)}{\partial x_i^2} \right) \right\},$$
$$V_i(t_f, x_i, \theta_i) = \varphi_i(t_f, x_i, \theta_i), \quad (8.12)$$

where $\mathrm{tr}(\cdot)$ indicates the trace of the square matrix. In order to introduce the price $p_i(\tau) \in \mathbb{R}^{n_i}$ designed by the utility, if the agent i's private optimal control is implementable, then the control is characterized by

$$u_i^{\sharp}(\tau) = \gamma_i^{\sharp}(\tau, z_{i\tau}, p_i^{\sharp}(\tau)) = \arg \min_{u_i \in U_i} H_i(\tau, x_{i\tau}, \theta_{i\tau}, p_i^{\sharp}(\tau), u_i) \quad (8.13)$$

at any time τ in the interval $[t, t_f]$. Note that the partial derivative of the value function V_i regarding the state gives the pricing signal for agent i: $p_i^{\sharp} = (\partial V_i / \partial x_i)^{\top}$, which achieves the private cost minimization. This adjoint variable p_i is sometimes called the *shadow price* of the state vector [10].

Public Optimal Control Policy In the same way as the private optimal control policy, we derive the solution to the public control problem (8.8), which is called *public optimal control*. The public optimal control is denoted by $u_i^*(\tau) = \tilde{\gamma}_i^*(\tau, z_{\tau})$, $t \leq \tau \leq t_f, i \in \mathcal{N}$, and the corresponding implementation, pricing scheme, and price for the public optimal control law $\tilde{\gamma}_i^*$ are denoted by γ_i^*, λ_i^*, and p_i^*, respectively. Then, the public optimal control is given by

$$u^*(\tau) = \tilde{\gamma}^*(\tau, z_{\tau}) = \arg \min_{u_i \in U_i, i \in \mathcal{N}} H \left(\tau, x, \theta, \left(\frac{\partial V(\tau, x, \theta)}{\partial x} \right)^{\top}, u \right) \Bigg|_{[x^{\top}, \theta^{\top}]^{\top} = z_{\tau}},$$
$$(8.14)$$

where

$$H(\tau, x, \theta, p, u) = H_0(\tau, x, \theta, p_0) + \sum_{i \in \mathcal{N}} H_i(\tau, x_i, \theta_i, p_i, u_i) \qquad (8.15)$$

$$H_0(\tau, x, \theta, p_0) = l_0(\tau, x_0) + p_0^\top f_0(\tau, x, \theta) \qquad (8.16)$$

and $V(\tau, x_\tau, \theta_\tau)$ obeys the following HJB equation:

$$-\frac{\partial V(\tau, x, \theta)}{\partial \tau} = \min_{u_j \in U_j, j \in \mathcal{N}} \left\{ H\left(\tau, x, \theta, \left(\frac{\partial V(\tau, x, \theta)}{\partial x}\right)^\top, u\right) \right.$$

$$\left. + \sum_{i \in \mathcal{N}} \sum_{\tilde{\theta}_i=1}^{s_i} \pi^i_{\theta_i \tilde{\theta}_i} V(\tau, x, \tilde{\theta}_i, \theta_{-i}) + \frac{1}{2} \mathrm{tr}\left(g(\tau)g^\top(\tau)\frac{\partial^2 V(\tau, x, \theta)}{\partial x^2}\right) \right\},$$

$$V(t_f, x, \theta) = \varphi_0(t_f, x) + \sum_{i \in \mathcal{N}} \varphi_i(t_f, x_i, \theta_i). \qquad (8.17)$$

If the public optimal control is implementable, then the optimal implementation $\gamma^* = [\gamma_1^*, \ldots, \gamma_N^*]^\top$ is characterized by

$$u^*(\tau) = \gamma^*(\tau, z_\tau, p^*(\tau)) = \arg \min_{u_i \in U_i, i \in \mathcal{N}} H(\tau, x_\tau, \theta_\tau, p^*(\tau), u) \qquad (8.18)$$

at each time τ in the interval $[t, t_f]$ for the price $p^* = (\partial V/\partial x)^\top$, which is just a pricing scheme minimizing the public cost (8.7).

Incentivizing Market Mechanism As shown in Sect. 8.3, under the asymmetric information between the utility and the agents, the agents behaving strategically have no incentive to take public optimal controls in some cases. Consequently, a market mechanism is needed that contains some incentive mechanisms with rewards for eliciting agents' private information while achieving the public objective and securing agents' strategic behavior. In this chapter, we refer to this real-time incentive-enabled market as an *incentivizing market* and provide formulas and fundamental conditions for their design.

When the utility is regarded as a market planner, the utility designs an incentivizing market mechanism and makes an auction rule on the basis of the evaluation functionals and the grid model information. The auction rule takes place in the following steps [41]:

Step 1   Utility tells agents the auction system, and the agents decide participation.
Step 2   Agent offers his bidding parameter made from his private information.
Step 3   Based on agents' bids, price is fixed so as to minimize public cost.
Step 4   Agent decides his control to minimize his own cost based on price.
Step 5   Utility pays rewards to agents.

Note that steps 2, 3, and 4 will be executed continuously over a time interval. In Sect. 8.4, we propose two market mechanisms so that all the agents participate in the incentivizing market, which is the main challenge of this chapter.

8.3   Lessons from Non-incentive Strategic Bidding for LQG Models

Afterward, we confine the aforementioned power network model to an LQG model and consider the influence of the agents' strategic biddings over the whole dynamical system through the LQG model.

8.3.1   LQG Model and Optimal Control Policy

LQG Model We introduce a dynamical model linearized around an equilibrium state, which is scheduled in a spot energy market in advance. To be concrete, we consider $U_i = \mathbb{R}^{m_i}$, $i \in \mathcal{N}$, and replace general nonlinear system models in (8.1) and (8.4) with linearized system models

$$f_i(t, x_{it}, \theta_{it}, u_{it}) = A_{ii}(t, \theta_{it})x_{it} + B_{ii}(t, \theta_{it})u_{it}, \quad i \in \mathcal{N}, \tag{8.19}$$

$$f_0(t, x_t, \theta_t) = A_{00}(t)x_{0t} + \sum_{i \in \mathcal{N}} A_{0i}(t, \theta_{it})x_{it}, \tag{8.20}$$

and cost functionals in (8.2) and (8.5) with quadratic-form cost functionals

$$l_i(t, x_{it}, \theta_{it}, u_{it}) = x_{it}^\top Q_i(t, \theta_{it})x_{it} + u_{it}^\top R_i(t, \theta_{it})u_{it}, \quad i \in \mathcal{N}, \tag{8.21}$$

$$\varphi_i(t_f, x_{it_f}, \theta_{it_f}) = x_{it_f}^\top Q_{if}(t_f, \theta_{it_f})x_{it_f}, \qquad i \in \mathcal{N}, \tag{8.22}$$

$$l_0(t, x_{0t}) = x_{0t}^\top Q_0(t)x_{0t} \tag{8.23}$$

$$\varphi_0(t_f, x_{0t_f}) = x_{0t_f}^\top Q_{0f}(t_f)x_{0t_f}, \tag{8.24}$$

where $Q_j \geq 0$, $Q_{jf} \geq 0$, $j \in \{0\} \cup \mathcal{N}$ and $R_i > 0$, $i \in \mathcal{N}$. Therefore, the aggregated grid model and the public cost are substituted for

$$f(t, x_t, u_t, \theta_t) = A(t, \theta_t)x_t + B(t, \theta_t)u_t,$$

$$A = \begin{bmatrix} A_{00} & A_{01} & \cdots & A_{0N} \\ 0 & A_{11} & & 0 \\ \vdots & & \ddots & \\ 0 & 0 & & A_{NN} \end{bmatrix}, \quad B = \begin{bmatrix} 0 & \cdots & 0 \\ B_{11} & & 0 \\ & \ddots & \\ 0 & & B_{NN} \end{bmatrix},$$

$$C(t, x; u, \theta) = x_{t_f}^\top Q_f(t_f, \theta_{t_f}) x_{t_f} + \int_t^{t_f} \left(x_\tau^\top Q(\tau, \theta_\tau) x_\tau + u_\tau^\top R(\tau) u_\tau \right) d\tau,$$

$$Q = \begin{bmatrix} Q_0 & 0 & \cdots & 0 \\ 0 & Q_1 & & 0 \\ \vdots & & \ddots & \\ 0 & 0 & & Q_N \end{bmatrix}, \quad Q_f = \begin{bmatrix} Q_{0f} & 0 & \cdots & 0 \\ 0 & Q_{1f} & & 0 \\ \vdots & & \ddots & \\ 0 & 0 & & Q_{Nf} \end{bmatrix}, \quad R = \begin{bmatrix} R_i & & 0 \\ & \ddots & \\ 0 & & R_N \end{bmatrix}.$$

Private Optimal Control Policy Under the above LQG model, the value function of the agent $i \in \mathcal{N}$ is given by $V_i(\tau, x_i, \theta_i) = x_i^\top P_i(\tau, \theta_i) x_i + b_i(\tau, \theta_i)$ if there are a matrix $P_i(\tau, \theta_i)$ and a scalar value $b_i(\tau, \theta_i)$, $t \le t \le t_f$, led by the solution to the following differential equations:

$$-\frac{\partial P_i(\tau, \theta_i)}{\partial \tau} = P_i(\tau, \theta_i) A_{ii}(\tau, \theta_i) + A_{ii}^\top(\tau, \theta_i) P_i(\tau, \theta_i)$$

$$-P_i(\tau, \theta_i) B_{ii}(\tau, \theta_i) R_i^{-1}(\tau, \theta_i) B_{ii}^\top(\tau, \theta_i) P_i(\tau, \theta_i)$$

$$+Q_i(\tau, \theta_i) + \sum_{\tilde{\theta}_i=1}^{s_i} \pi_{\theta_i \tilde{\theta}_i}^i P_i(\tau, \tilde{\theta}_i), \quad P_i(t_f, \theta_i) = Q_{if}(t_f, \theta_{it_f}), \quad (8.25)$$

$$-\frac{\partial b_i(\tau, \theta_i)}{\partial \tau} = \sum_{\tilde{\theta}_i=1}^{s_i} \pi_{\theta_i \tilde{\theta}_i}^i b_i(\tau, \tilde{\theta}_i) + \mathrm{tr}\left(P_i(\tau, \theta_i) g_i(\tau) g_i^\top(\tau) \right), \quad b_i(t_f, \theta_i) = 0. \quad (8.26)$$

In addition, the private optimal control of agent i is implementable and given by

$$u_i^\sharp(\tau) = \gamma_i^\sharp(\tau, z_{i\tau}, p_i^\sharp(\tau)) = -\frac{1}{2} R_i^{-1}(\tau, \theta_{i\tau}) B_{ii}^\top(\tau, \theta_{i\tau}) p_i^\sharp(\tau), \quad t \le \tau \le t_f, \quad (8.27)$$

for the price $p_i^\sharp(\tau) = 2 P_i(\tau, \theta_{i\tau}) x_i(\tau)$ (see [26, 48]).

Public Optimal Control Policy In a manner that parallels the private optimal control policy, the value function corresponding to the public optimal control policy is obtained by $V(\tau, x, \theta) = x^\top P(\tau, \theta) x + b(\tau, \theta)$ if there are a symmetric matrix $P(\tau, \theta) \ge 0$ and a scalar value $b(\tau, \theta)$, $t \le t \le t_f$, led by the solution to the following differential equations:

$$-\frac{\partial P(\tau, \theta)}{\partial \tau} = P(\tau, \theta) A(\tau, \theta) + A^\top(\tau, \theta) P(\tau, \theta)$$

$$-P(\tau, \theta) B(\tau, \theta) R^{-1}(\tau, \theta) B^\top(\tau, \theta) P(\tau, \theta)$$

$$+Q(\tau, \theta) + \sum_{i \in \mathcal{N}} \sum_{\tilde{\theta}_i=1}^{s_i} \pi_{\theta_i \tilde{\theta}_i}^i P(\tau, \tilde{\theta}_i, \theta_{-i}), \quad P(t_f, \theta) = Q_f(t_f, \theta_{t_f}),$$

$$(8.28)$$

$$-\frac{\partial b(\tau, \theta)}{\partial \tau} = \sum_{i \in \mathcal{N}} \sum_{\tilde{\theta}_i = 1}^{s_i} \pi^i_{\theta_i \tilde{\theta}_i} b(\tau, \tilde{\theta}_i, \theta_{-i}) + \text{tr} \left(P(\tau, \theta) g(\tau) g^\top(\tau) \right), \ b(t_f, \theta) = 0.$$

$$(8.29)$$

By considering the derivative of the value function V regarding the state x, which becomes the price signal for agents, we obtain the public optimal control. The public optimal control u^* is given by

$$u_i^*(\tau) = \gamma_i^*(\tau, z_{i\tau}, p_i^*(\tau)) = -\frac{1}{2} R_i^{-1}(\tau, \theta_{i\tau}) B_{ii}^\top(\tau, \theta_{i\tau}) p_i^*(\tau), \ t \le \tau \le t_f, \ i \in \mathcal{N},$$

$$(8.30)$$

for the price p_i^* defined by

$$p^*(\tau) := \begin{bmatrix} p_0^*(\tau) \\ p_1^*(\tau) \\ \vdots \\ p_N^*(\tau) \end{bmatrix} = 2 \underbrace{\begin{bmatrix} P_{00}(\tau, \theta_\tau) & P_{01}(\tau, \theta_\tau) & \cdots & P_{0N}(\tau, \theta_\tau) \\ P_{10}(\tau, \theta_\tau) & P_{11}(\tau, \theta_\tau) & \cdots & P_{1N}(\tau, \theta_\tau) \\ \vdots & \vdots & \ddots & \vdots \\ P_{N0}(\tau, \theta_\tau) & P_{N1}(\tau, \theta_\tau) & \cdots & P_{NN}(\tau, \theta_\tau) \end{bmatrix}}_{=P(\tau, \theta_\tau)} \begin{bmatrix} x_0(\tau) \\ x_1(\tau) \\ \vdots \\ x_N(\tau) \end{bmatrix}.$$

$$(8.31)$$

Note that the optimal price p^* is given as the gradient of the value function regarding the current state. The price is called shadow price in the economics literature. Then, the optimal implementation γ^* is characterized by (8.18) and

$$u_i^*(\tau) = \gamma_i^*(\tau, z_{i\tau}, p_{i\tau}^*) = \arg \min_{u_i \in U_i} H_i(\tau, x_{i\tau}, \theta_{i\tau}, p_{i\tau}^*, u_i) \qquad (8.32)$$

at each time τ in the interval $[t, t_f]$.

Here, we emphasize from the above results on the LQG models that the feedback control N-tuple $\left[u_{1\tau}^{*\top}, \ldots, u_{N\tau}^{*\top} \right]^\top = \left[\gamma_1^{*\top}(\tau, z_{1\tau}, p_{1\tau}^*), \ldots, \gamma_N^{*\top}(\tau, z_{N\tau}, p_{N\tau}^*) \right]^\top$ given by (8.30) with the price p_i^*, which the utility computes according to (8.31), is *person-by-person optimal*, i.e., a *Nash equilibrium*, for the public cost functional (8.8) defined over the interval $t \le \tau \le t_f$ as well as for the Hamiltonian $H(\tau, x_\tau, \theta_\tau, p_\tau^*, \gamma^*)$ at each time τ. See [26] for more details.

8.3.2 Drawbacks in Non-incentive Strategic Bidding Through Examples

For simplicity, let us consider LQ problem with two agents ($N = 2$), whose local dynamics (8.1) and cost functional (8.2) for type $\theta_i = 1, i = 1, 2$, are given by

$$dx_{it} = (-x_{it} + u_{it})dt, \ x_i(0) = 1, \quad C_i = \int_0^{10} (x_{it}^2 + u_{it}^2)dt, \quad i = 1, 2, \qquad (8.33)$$

that is $n_i = m_i = 1$, $A_{ii}(t, 1) = -1$, $B_{ii}(t, 1) = 1$, $g_i(t) = 0$, $Q_i(t, 1) = 1$, Q_{if}
$(t_f, 1) = 0$ and $R_{t,1} = 1$, $i = 1, 2, 0 = t_0 \leq t \leq t_f = 10$. Here, we assume that each
agent always reports the type $\theta_i = 1$ during the period $[t_0, t_f]$, i.e., $\pi_{kl}^i = 0$ for all k,
l in (8.3). In other words, we ignore a Markov step process of the type parameters
defined by (8.3) and the type parameter obeys a deterministic (fixed) process during
the time period. The utility's dynamics (8.4) and cost functional (8.5) are also given
by

$$dx_{0t} = (x_{1t} + x_{2t})dt, \quad x_0(0) = 0, \quad C_0 = \int_0^{10} x_{0t}^2 dt, \tag{8.34}$$

that is $n_0 = 1$, $A_{00}(t) = 0$, $A_{01} = A_{02} = 1$, $g_0(t) = 0$, $Q_0(t) = 1$ and $Q_{0f}(t_f) = 0$,
$0 \leq t \leq 10$.

The primary objective of the utility, as well as a system operator, is to execute
the public optimal control u_i^*, $i = 1, 2$ (8.30) minimizing the public cost (8.7) and
solved in a centralized fashion. The resulting behavior for the public optimal control
is shown in Fig. 8.2, and all the states converge to zero smoothly. Then, the resulting
costs are $C = 2.4148$ and $C_1 = C_2 = 0.8835$.

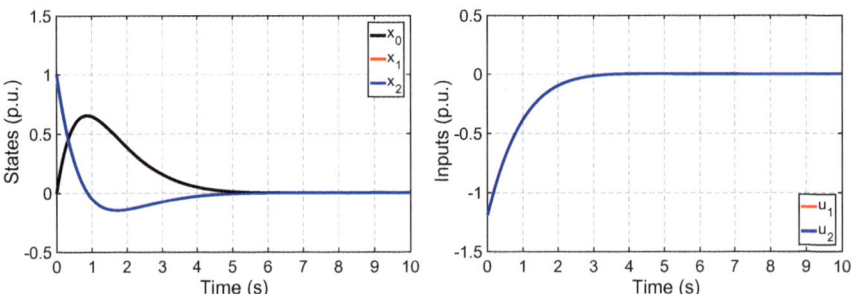

Fig. 8.2 Time evolution for the public optimal control with true bidding $\theta_1 = 1$ (*left* states x_i,
$i = 0, 1, 2$, *right* inputs $u_i = u_i^*$, $i = 1, 2$). Resulting costs are $C = 2.4148$, $C_1 = C_2 = 0.8835$

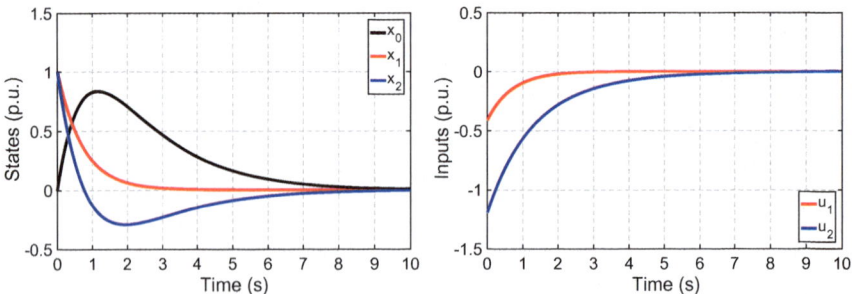

Fig. 8.3 Time evolution for agent 1's private optimal bidding with true bidding $\theta_1 = 1$ (*left* states
x_i, $i = 0, 1, 2$, *right* inputs $u_1 = u_1^\sharp$ and $u_2 = u_2^*$). Resulting costs are $C = 3.3656$, $C_1 = 0.4172$,
$C_2 = 1.3759$, and the behavior except x_1 is worsened, compared with Fig. 8.2

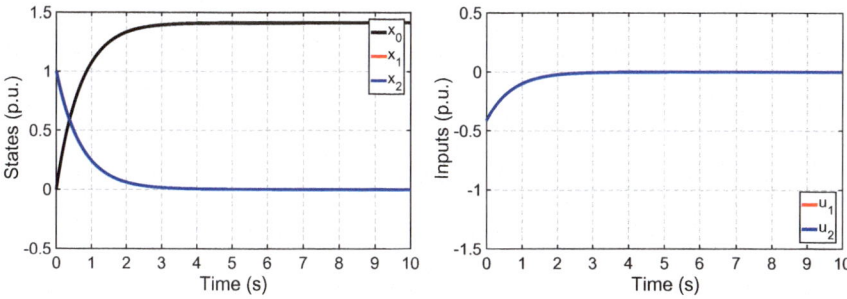

Fig. 8.4 Time evolution for private optimal bidding (*left* states x_i, $i = 0, 1, 2$, *right* inputs $u_i = u_i^*$, $i = 1, 2$). Resulting costs are $C = 18.7180$, $C_1 = C_2 = 0.4172$. The state x_0 cannot converge to 0

Meanwhile, the utility cannot control the agents' inputs u_i directly without agreement of all the participants including the agents and each agent aims at minimizing his private cost (8.2) strategically. Even if all the agents report the true type information θ_i in advance, they have the right to select freely their own control inputs u_i without supervised by the utility. Hence, let us now consider the case that the utility operates the price p_i^* given by (8.31) but agent 1 operates the private optimal control $u_1^\sharp$ (8.27) without the usage of the utility's guidance on public optimal price information p_i^*. Note that agent 2 follows the public optimal control $u_2 = u_2^*$ (8.30). Then, the resulting time evolution is shown in Fig. 8.3 and the corresponding costs are $C = 3.3656$, $C_1 = 0.4172$, and $C_2 = 1.3759$. As the cost $C_1 = 0.4172$ with the private optimal control input $u_1^\sharp$ is lower than the cost 0.8835 led by the controls u_i^* minimizing the public cost, each agent i has the potential to change the bidding parameter of controls u_i strategically. We also see from Figs. 8.2 and 8.3 that the strategic input of agent 1 makes the convergence performance of x_0 and x_2 worsen. Moreover, Fig. 8.4 indicates the results for the case that both the agents execute their private optimal controls $u_i^\sharp$ and the corresponding private costs $C_1 = C_2 = 0.4172$ are less than those of the public optimal controls. However, the balancing state x_0 cannot converge to zero, which does not meet the desired objective of the utility. It is called a *moral hazard* problem, and a promising mechanism to the moral hazard problem will be introduced in Sect. 8.4.1.

The other strategic bidding of the agent is to misreport his online information $z_{it} = [x_{it}^\top, \theta_{it}]^\top$ intentionally. Here, agent $i = 1$ has a different dynamics, denoted by a type $\theta_1 = 2$, which is

$$dx_{1t} = (-2x_{1t} + u_{1t})dt, \tag{8.35}$$

i.e., $A_{11}(t, 2) = -2$, and let us consider the case that he $i = 1$ reports an untruthful bidding parameter $\theta_1 = 2$ to the utility in advance instead of the actual type $\theta_1 = 1$. Then, the utility computes a public price p_i obeying (8.31) for the types $(\theta_1, \theta_2) = (2, 1)$ and each agent behaves by following the public optimal control $u_i(t) = \gamma_i^*(t, z_t, p_{it})$. The time evolution of the corresponding states and inputs is shown in Fig. 8.5, and the costs are $C = 2.4794$, $C_1 = 0.6888$, and $C_2 = 1.0466$.

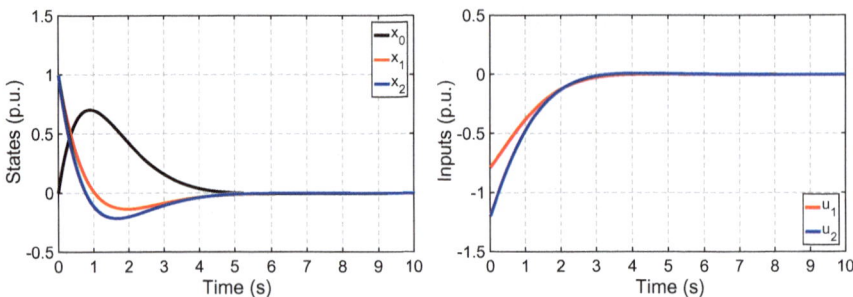

Fig. 8.5 Time evolution for the public optimal control with an agent 1's strategic bidding $\theta_1 = 2$ (*left* states x_i, $i = 0, 1, 2$, *right* inputs $u_i = u_i^*$, $i = 1, 2$). Agent 1's resulting cost is $C_1 = 0.6888$, which is reduced from the cost 0.8835 in case of the public cost minimization (Fig. 8.2), while the public cost and agent 2's cost are worsened to $C = 2.4794$ and $C_2 = 1.0466$

We see from Figs. 8.2 and 8.5 that agent 1 has a potential for reducing his own cost C_1 strategically by misreporting the type parameter θ_1. On the other hand, the public cost and the private cost of the other agent are worsened as a side effect. To prevent such a strategic and/or untruthful type information bidding, an incentive-based market mechanism is required and it is known as an *adverse selection* problem. One of the promising solutions will be introduced in Sect. 8.4.2.

8.4 Incentivizing Market Design

To prevent rational agents from selecting strategic bidding not realizing the public objective, the primary objective of the utility functioning as system operator is to design an incentive mechanism achieving *public optimality* based on private optimal controls subjected to IC and IR constraints. This section aims at proposing dynamic integration mechanisms in incentivizing markets in order to solve moral hazard problems and adverse selection problems with grid dynamics.

Before starting technical discussions on incentive designs, we introduce some common formulations and definitions. The incentive functional C_i^{mt} with which the utility provides an agent $i \in \mathcal{N}$ is composed of the online price signal $p_i(\tau) = \lambda_i(\tau, z(\tau)) \in \mathbb{R}^{n_i}$ and a reward functional W_i specified later. In a manner that parallels the cost functional (8.2) without incentives and that the admissible controls are limited to implementable ones, the net cost functional I_i for implemented control laws $\gamma = [\gamma_1^\top, \ldots, \gamma_N^\top]^\top$ and pricing schemes $\lambda = [\lambda_1^\top, \ldots, \lambda_N^\top]^\top$ is defined as

$$I_i(t, z; \gamma_i, \gamma_{-i}, \lambda) = \mathbb{E}\left[C_i(t, z_i; \gamma_i) \,\middle|\, z(t) = z, \gamma \right] + C_i^{mt}(t, z; \gamma_i, \gamma_{-i}, \lambda), \quad i \in \mathcal{N}, \tag{8.36}$$

and each rational agent basically aims at minimizing the net cost (8.36) instead of (8.9). We denote by $u_i^b(\tau) = \gamma_i^b(\tau, z_{i\tau}, \lambda_i(\tau, z_\tau)), t \leq \tau \leq t_f$ a private, implemented, optimal control for the net private cost functional (8.36).[10] For pricing schemes $\lambda = [\lambda_1^\top, \ldots, \lambda_N^\top]^\top$, let $\gamma^b = [\gamma_1^{b\top}, \ldots, \gamma_N^{b\top}]^\top$ be an N-tuple of implementable optimal control laws, i.e., a *Nash equilibrium* in Γ, defined by

$$u_i^b(t) = \gamma_i^b(t, z_{it}, \lambda_i(t, z_t)) = \arg\min_{\gamma_i \in \Gamma_i} I_i(t, z_t; \gamma_i, \gamma_{-i}^b, \lambda), \quad i \in \mathcal{N}. \tag{8.37}$$

We here define three properties of the integration mechanisms in incentivizing markets as below.

Definition 1 A mechanism is *publicly optimal* if private optimal controls u_i^b given by (8.37) are equivalent to the public optimal controls u_i^* minimizing the public cost functional (8.14).

Definition 2 Given information z, prices p, and the others' optimal controls u_{-i}^b, a mechanism is *incentive compatible* (IC) if the controls u_i^b given by (8.37) are optimal for private cost functional I_i (8.36).

Definition 3 We denote by $k_i(t, z)$ a prescribed cost level of agent $i \in \mathcal{N}$ under a condition $z(t) = [x(t)^\top, \theta(t)^\top]^\top = [x^\top, \theta^\top]^\top$, which is an important threshold to determine whether he participates in an incentivizing market as shown in step 1 of incentivizing market mechanism described in Sect. 8.2.2. Then, a mechanism is *individually rational* (IR) if the expected cost of each agent with the private optimal controls is lower than his prescribed cost $k_i(t, z)$.

8.4.1 Moral Hazard Incentivizing Market

Given the true state and type information $z_{it} = [x_{it}^\top, \theta_{it}]^\top$ reported by each agent $i \in \mathcal{N}$, we first consider an optimal incentive for the strategic bidding of hidden actions, i.e., the agents' control policies u_i, in order to achieve the utility's net cost minimization. For simplicity of discussion, we suppose that each agent $i \in \mathcal{N}$ deterministically and truly reports the type parameter θ_i in LQG models. Hence, without loss of generality, we consider the deterministic case of $s_i \equiv 1$ and $\pi_{kl}^i \equiv 0$ for all $i \in \mathcal{N}$.

---

[10]The private control $u_i^\sharp$ used in the previous sections is optimal for each agent in the case of the private cost functional C_i defined in (8.2). Meanwhile, the control u_i^b used in this section is defined as the private optimal control for private costs *adding an incentive cost*.

Moral Hazard Incentive Design Problem To incentivize agents' behavior in the incentivizing market mechanism, the utility as a market planner uses a reward (salary) functional W_i, $i \in \mathcal{N}$.[11] The incentive functional C_i^{mt} in (8.36) is redesigned by the expected reward functional W_i for the agent $i \in \mathcal{N}$, which is parameterized by a parameter ξ_i in a class Ξ_i and an initial reward h_i in a class $\mathcal{H}$, i.e.,

$$C_i^{mt}(t, z; \gamma_i, \gamma_{-i}, \lambda) = -\mathbb{E}\left[W_i(t, z; \gamma_i, \gamma_{-i}, \lambda) \,\Big|\, z(t) = z, \gamma_i, \gamma_{-i}, \lambda \right]. \quad (8.38)$$

As the utility pays rewards W_i to each agent $i \in \mathcal{N}$, her net cost functional is described as:

$$I(t, z; \gamma, \lambda) = \mathbb{E}\left[C_0(t, x_0; \gamma, \lambda) \,\Big|\, z(t) = z, \gamma, \lambda \right] + \mathbb{E}\left[\sum_{i \in \mathcal{N}} W_i(t, z; \gamma, \lambda) \,\Big|\, z(t) = z, \gamma, \lambda \right]. \quad (8.39)$$

The design problem of our market is reduced to a reward design problem, which minimizes the utility's net cost I subjected to IC and IR. The reward design problem is formulated as follows: For a given initial condition $z(t) = [x(t)^{\top}, \theta(t)^{\top}]^{\top} = [x^{\top}, \theta^{\top}]^{\top}$, $(t, x, \theta) \in [t_0, t_f] \times \mathbb{R}^n \times \Theta$, agents' prescribed cost level $k_i(t, z)$, $i \in \mathcal{N}$, and model information, the utility finds an optimal reward functional about W_i with parameters ξ_i and h_i that implements optimal controls $u_i(t) = \gamma_i(t, z_{it}, \lambda_i(t, z_t))$, $t \in [t_0, t_f]$, $i \in \mathcal{N}$ such that

$$\min_{\substack{\gamma_i \in \Gamma_i, \lambda_i \in \Lambda_i, \\ \xi_i \in \Xi_i, h_i \in \mathcal{H}, i \in \mathcal{N}}} I(t, z; \gamma, \lambda) \quad (8.40)$$

$$\text{subject to } I_i(t, z; \gamma_i, \gamma_{-i}, \lambda) = \min_{\mu_i \in \Gamma_i} I_i(t, z; \mu_i, \gamma_{-i}, \lambda), \quad i \in \mathcal{N}, \quad (8.41)$$

$$I_i(t, z; \gamma_i, \gamma_{-i}, \lambda) \le k_i(t, z), \quad i \in \mathcal{N}, \quad (8.42)$$

where Ξ_i is an appropriate class of reward parameters ξ_i derived from the necessary condition of the form of the optimal reward functionals and $\mathcal{H}$ is that of h_i as shown in (8.43) (see [43] for more details). Constraints (8.41) and (8.42) are just IC constraint and IR constraint, respectively. The IC constraint (8.41) guarantees that the designed reward incentivizes each agent's behavior to adopt the optimal control that minimizes his own cost, so that the agents' controls compose a Nash equilibrium (8.37). The constraint (8.41) also intends that, even if the control profiles are not bidden, the utility can reconstruct them by using the bidden model information. Meanwhile, the IR constraint (8.42) assures a prescribed level of each agent's cost that incentivizes the agent to participate in the market and send the utility his model information in advance. Now, if the agents' bidding is done, the utility can decide a market-clearing price immediately by performing the above optimization and send the price

---

[11] We can prove rigorously that the form of the optimal reward functional W_i for the dynamics (8.6) and the private cost functionals (8.36) is limited to (8.43). See [43] for the technical details.

to each agent together with the reward payment in real time. Thus, we obtain a non-iterative/one-shot incentivizing market mechanism.

The reward design problem (8.40)–(8.42) reformulates the conventional contract problems in economics [7, 9, 16, 21, 33, 34] in order to adapt the contract problem to an incentivizing market mechanism in dynamical physical systems (8.6) with agents' controls. To synthesize the reward design problem, it is required to parameterize the reward functional different from previously reported types [9, 16, 21, 33, 34].

Optimal Contract Condition To solve the reward design problem, we begin by using constraints (8.41) and (8.42) to specify a form of the reward functional. Hereinafter, we focus the fundamental results of our mechanism and the detailed technical description is shown in [43]. Suppose that there is a Nash equilibrium u^b defined by (8.37) for a tuple of reward functionals $[W_1, \ldots, W_N]^\top$. Then, consequently, given an initial condition $z(t) = z$ and optimal controls u^b_{-i} of all the agents except agent $i \in \mathcal{N}$, the reward functional W_i of the agent i with arbitrary implementable control law $\gamma_i \in \Gamma_i$ has the form:

$$
W_i(t, z; \gamma_i, \bar{\gamma}^b_{-i}, \lambda) = \varphi_i(t_f, x_{it_f}) - h_i(t, z)
$$
$$
+ \int_t^{t_f} \left[\xi_i^\top(\tau, z_\tau) f(\tau, z_\tau, \gamma^b_i(\tau, z_{i\tau}, \lambda_i(\tau, z_{i\tau})), u^b_{-i}(\tau)) \right.
$$
$$
\left. + l_i(\tau, x_{it}, \gamma^b_i(\tau, z_{i\tau}, \lambda_i(\tau, z_{i\tau}))) \right] d\tau - \int_t^{t_f} \xi_i^\top(\tau, z_\tau) dx_\tau, \qquad (8.43)
$$

along $dx_\tau = f(\tau, z_\tau, \gamma_i(\tau, z_{i\tau}, \lambda_i(\tau, z_\tau)), u^b_{-i}(\tau)) d\tau + g(\tau) dw_\tau$, where the controls γ^b_i constitute the Nash equilibrium. Equation (8.43) is defined by $\xi_i := [\xi_{i0}^\top, \xi_{i1}^\top, \ldots, \xi_{iN}^\top]^\top$ in $\Xi_i = \Xi_{i0} \times \Xi_{i1} \times \ldots \times \Xi_{iN}$ for $\xi_{ij} = 0, j \in \{0\} \cup \mathcal{N} \setminus \{i\}$, and for any ξ_{ii} in the class Ξ_{ii} and any h_i in the class $\mathcal{H}$. For the proof of (8.43), see, e.g., [43]. Note that ξ_i takes its value in $\mathbb{R}^n$ and ξ_{ij} takes its value in $\mathbb{R}^{n_j}$. Then, by taking the reward parameter $\xi_{ii} \in \Xi_{ii}$ as a pricing scheme λ_i, which is provided by the utility, we have

$$
\gamma^b_i(\tau, z_i, p_i) = \arg \min_{u_i \in U_i} H_i(\tau, z_i, p_i, u_i), \quad i \in \mathcal{N}, \qquad (8.44)
$$
$$
p_i(\tau) = \lambda_i(\tau, z_\tau), \quad \lambda_i \in \Lambda_i, \quad i \in \mathcal{N}, \qquad (8.45)
$$
$$
u^b_i(\tau) = \gamma^b_i(\tau, z_i, p_i(\tau)), \quad i \in \mathcal{N}. \qquad (8.46)
$$

From the above observation on (8.43)–(8.46), we limit the class of the reward parameters Ξ_i to Λ_i so that we take $\xi_{ij} = 0, i \neq j$, and $\xi_{ii} = \lambda_i$ in the following discussion. Note that the basic idea of (8.43) is motivated from [16] and the above dynamic optimization for market-clearing has an overlapped structure of a dynamic game and an optimal control problem. The paper [43] developed a contract theory approach to this optimization based on DP.

Using (8.43), we next obtain Proposition 1 regarding the relationship between the reward parameters $(\lambda_i, h_i) \in \Lambda_i \times \mathcal{H}$ and the constraints (8.41) and (8.42).

Proposition 1 *(a) For the reward functionals (8.43) with reward parameters* $\lambda = [\lambda_1^\top, \ldots, \lambda_N^\top]^\top \in \Lambda_1 \times \ldots \times \Lambda_N =: \Lambda$ *and* $h = [h_1, \ldots, h_N]^\top \in \mathcal{H} \times \ldots \times \mathcal{H} =: \mathcal{H}^N$, *a tuple of control laws* $[\gamma_1^\top, \ldots, \gamma_N^\top]^\top \in \Gamma$ *is a Nash equilibrium if and only if it has the form defined by (8.44).*
(b) For the reward functionals (8.43) with reward parameters $\lambda \in \Lambda$ *and* $h \in \mathcal{H}^N$ *and the Nash equilibrium (8.44), the IR constraint (8.42) is fulfilled if and only if the initial reward* h_i *is specified such as* $h_i(t, z) \leq k_i(t, z)$, $i \in \mathcal{N}$.

Proposition 1 indicates that the IC constraint (8.41) and the IR constraint (8.42) are fulfilled for any reward parameters $[\lambda_i^\top, h_i]^\top \in \Lambda_i \times \mathcal{H}$, which are also regarded as controls of the reward design problem for the utility. Proposition 1 (a) also shows that our complex reward design problem (8.40)–(8.42) is converted to the single optimal control problem in which the utility's net cost functional is minimized by the price p and the initial reward h_i. Therefore, the utility can replace the reward design problem (8.40)–(8.42) with an optimal control problem of the parameters $[\lambda_i^\top, h_i]^\top$, which is summarized in Theorem 1.

Theorem 1 *The reward design problem (8.40)–(8.42) with the reward parameters* $\lambda = [\lambda_1^\top, \ldots, \lambda_N^\top]^\top \in \Lambda$ *and* $h = [h_1, \ldots, h_N]^\top \in \mathcal{H}^N$ *is equivalent to the optimal control problem by* λ *and* h *for the utility's net cost functional:*

$$\inf_{\lambda \in \Lambda, h \in \mathcal{H}^N} I(t, x; \gamma^\flat, \lambda) \tag{8.47}$$

subject to initial rewards satisfying $h_i(t, z) \leq k_i(t, z)$, $i \in \mathcal{N}$, *private optimal controls* $\gamma^\flat(\tau, z, \lambda(\tau, z_\tau)) := [\gamma_1^{\flat\top}(\tau, z_1, \lambda_1(\tau, z_\tau)), \ldots, \gamma_N^{\flat\top}(\tau, z_N, \lambda_N(\tau, z_\tau))]^\top$ *defined by (8.44), the reward functional* W_i *given by (8.43) and the stochastic state equation:* $dx_\tau = f(\tau, z_\tau, \gamma^\flat(\tau, z_\tau, \lambda(\tau, z_\tau)))d\tau + g(\tau)dw_\tau$, $\tau \in [t, t_f]$.

From Proposition 1 and Theorem 1, (8.39) results in

$$I(t, z; \gamma^\flat, \lambda) = \mathbb{E}\left[C_0(t, x_0; \gamma^\flat, \lambda) \,\middle|\, z(t) = z, \gamma^\flat, \lambda\right]$$
$$+ \sum_{i \in \mathcal{N}} \mathbb{E}\left[C_i(t, z_i; \gamma_i^\flat, \lambda_i) \,\middle|\, z_i(t) = z_i, \gamma_i^\flat, \lambda_i\right] - \sum_{i \in \mathcal{N}} h_i(t, z). \tag{8.48}$$

We see from Theorem 1 and (8.48) that $h_i(t, z) \equiv k_i(t, z)$ is optimal for any $(t, x, \theta) \in [t_0, t_f] \times \mathbb{R}^n \times \Theta$. Then, if we assume zero costs for all the agents, i.e., $k_i(t, z) \equiv 0$, $i \in \mathcal{N}$, we see that the utility's net cost derived from (8.48) is essentially equivalent to the public cost (8.7). Hence, the optimal control policy (8.44) and the optimal price solved by the optimal control problem (8.47) with $k_i(t, z) \equiv 0$ are just the public optimal control u^* (8.18) and the corresponding price $p^\flat = (\partial V / \partial x)^\top$ at (τ, z_τ), which is just (8.31). We emphasize that the incentive mechanism with the optimal reward functional W_i prevents the strategic behavior of non-public optimal controls as shown in Fig. 8.3.[12] In addition, the optimal solution $\lambda^* = [\lambda_1^{*\top}, \ldots, \lambda_N^{*\top}]^\top \in \Lambda$ of the optimal control problem in (8.47) leads to the Nash equilibrium $[\gamma_1^{\flat}(\tau, z_1, \lambda_1(\tau, z))^\top, \ldots, \gamma_N^{\flat}(\tau, z_N, \lambda_N(\tau, z))^\top]^\top$ combining the private optimal controls, which satisfies the IC constraint, and $h^* = [h_1^{*\top}, \ldots, h_N^{*\top}]^\top \in \mathcal{H}^N$ guarantees the IR constraint. Finally, the incentive mechanism with (8.38) and (8.43) based on the contract theory approach achieves the balanced budget whereas in general most incentive mechanisms do not satisfy the budget balance condition [29, 38]. In other words, if the utility operates the public optimal reward parameters λ^* and h^* and each rational agent executes his private optimal control based on (8.36) with (8.38) and (8.43), the sum of costs $I + I_1 + \ldots + I_N$ is always the lowest cost $\mathbb{E}\left[C \mid z(t) = z, \tilde{\gamma}, \lambda\right] = \mathbb{E}\left[C_0 + C_1 + \ldots + C_N \mid z(t) = z, \tilde{\gamma}, \lambda\right]$.

We see from Theorem 1 that the utility has a selection option of the optimal reward parameters λ and h in the sense of (8.47) while modifying her net cost functional I. The fact means that the parameters λ and h can function as policy variables of the utility in economics.

Remark 1 Because the utility and the agents act as leader and followers, respectively, the structure of our contract problem (8.40)–(8.42) is close to the games that derive closed-loop Stackelberg (hierarchical) solutions. However, the classical Stackelberg solution does not admit the time-consistency property of the leader's strategy without a rather restrictive precommitment and/or conditions including LQ differential games [25]. We see from Theorem 1 that our problem has the time-consistency property of the utility's strategy (λ, h) and the agents' strategy $\gamma^{\flat}$ by interposing the reward functionals W_i between the utility and the agents.

Remark 2 The necessary condition of the optimal reward functional (8.43) holds for any dynamical systems satisfying some mathematical assumptions. Proposition 1 (a), especially (8.44), holds for any control-affine systems including LQG models. The stability analysis for the long-term average cost is discussed in [47]. In economics, Benchekroun and Long [4] investigate the optimal policy between pollution and tax (incentive) for single agent and infinite-time horizon case, and Akao [1] considers an optimal tax and subsidy scheme for a simple dynamical model from the viewpoint of environmental economics.

8.4.2 Adverse Selection Incentivizing Market

Adverse Selection Incentive Design Problem We see from (8.13) and (8.27) that the private optimal control of each agent is limited to the form $\gamma_i^{\sharp}$ (8.13) or $\gamma_i^{\flat}$ (8.37) and the honest utility executes the price (policy variable) p^* led by the pricing scheme

---

[12]Strictly speaking, the solution to the reward design problem presented in this chapter does not give a defense policy against cooperative strategic controls of all the agents as shown in Fig. 8.4. The paper [45] presents a defense policy with another incentive design inspired by the contract theory approach.

λ^* such as (8.31) to minimize the public cost. We call p^* the public optimal price. Thus, each rational agent $i \in \mathcal{N}$ designs his control law taking account of his net cost I_i with γ_i, that is

$$I_i(t, z; \tilde{\gamma}_i, \tilde{\gamma}^*_{-i}) = \mathbb{E}\left[C_i(t, z_i; \tilde{\gamma}_i) \mid z_i(t) = z_i, \gamma_i^b, \lambda_i\right] + C_i^{mt}(t, z; \tilde{\gamma}_i, \tilde{\gamma}^*_{-i}), \quad (8.49)$$

where $\tilde{\gamma}^*(\tau, z) = \gamma_i^{\sharp}(\tau, z_i, p^*(\tau))$ and the public optimal price $p^*(\tau) = \lambda^*(\tau, z_\tau)$ are calculated by the utility and they are based on online information $z(\tau)$ reported from the agents. We next consider an optimal incentive for the strategic bidding of hidden information, i.e., the agents' online information $z_i = [x_i^\top, \theta_i]^\top$, when each agent executes his private optimal control $\gamma_i^b(\tau, z_i, p_i^*(\tau))$.

In our framework for information exchange and control determination shown in Sect. 8.2, agent $i \in \mathcal{N}$ can strategically report fictitious information about his own state and type, such as $\sigma_{ii}(z_{i\tau}) = [\sigma_{iix}(x_{i\tau}^\top)^\top, \sigma_{ii\theta}(\theta_{i\tau})]^\top$, to the utility at each time τ in the period $[t, t_f]$. In this case, the utility will calculate the price $p_i(\tau)$ on the basis of the data $\sigma_i(z_\tau) = \{x_{0\tau}, \sigma_{iix}(x_{i\tau}), x_{-i\tau}, \sigma_{ii\theta}(\theta_{i\tau}), \theta_{-i\tau}\}$ with the result that $p_i(\tau) = \lambda_i(\tau, \sigma_i(z_\tau))$. As an implementation based on the pricing scheme $\lambda_i(\tau, \sigma_i(z))$ cannot certify public optimality for power networks as shown in Fig. 8.5, an integration mechanism with C_i^{mt} needs to prevent each agent from misreporting strategically. The IC constraint means that a truthful reporting is optimal from the viewpoint of agent i's cost functional, and the mechanism is IR if each agent who participates in the power network and minimizes his private cost functional is less than his prescribed cost level $k_i(t, z)$.

Optimal Contract Condition To make a mechanism possessing the above properties, we introduce an incentive cost functional C_i^{mt} that is paid to each agent at the ex post stage as an economic (monetary) transfer. Thus, each rational agent designs his control law taking account of the incentive cost functional C_i^{mt} imposed in the form of an additional cost. Since the private cost functional is redesigned as (8.49), an incentive cost functional C_i^{mt} for agent i is given by

$$C_i^{mt}(t, z; \tilde{\gamma}_i, \tilde{\gamma}^*_{-i}) = \mathbb{E}\left[C_0(t, x_0) + \sum_{j \in \mathcal{N}\setminus\{i\}} C_j(t, z_j) \,\Big|\, z(t) = z, \tilde{\gamma}_i, \tilde{\gamma}^*_{-i}\right]$$
$$- \min_{\tilde{\gamma}_{-i}} \mathbb{E}\left[C(t, z; 0, \tilde{\gamma}_{-i}) \,\Big|\, z(t) = z, \tilde{\gamma}_{-i}\right] + h_i(t, z), \quad (8.50)$$

where $h_i(t, z) \in \mathcal{H}$ is an initial reward. The form of incentive cost functional (8.50) is inspired by the pivot mechanism (Vickrey–Clarke–Groves (VCG) mechanism) [5, 19].[13]

---

[13]The term $C(t, z; 0, \tilde{\gamma}_{-i})$ means the public cost in the absence of the agent i's control (strategic action), i.e., $u_i \equiv 0$. However, to avoid an instability condition led by $u_i = 0$, there is a case to replace a traditional control policy similar to $u_i = 0$ as $u_i = \bar{u}_i \neq 0$.

Here, the notation $u_i^b(\tau) = \tilde{\gamma}_i^b(\tau, z_\tau)$, $t \leq \tau \leq t_f$ denotes an optimal private control law for the private cost functional (8.49). By using the private cost functional (8.49) with incentive cost (8.50), Theorem 2 holds.

Theorem 2 *[26] (a) For the private cost functional (8.49) with incentive cost functional (8.50), the private optimal controls $u_i^b(\tau) = \tilde{\gamma}_i^b(\tau, z_\tau)$, $t \leq \tau \leq t_f$, $i \in \mathcal{N}$ coincide with the public optimal control (8.30).*
(b) The mechanism given by the incentive cost functional (8.50) and the private optimal control $u_i^b(\tau)$, $t \leq \tau \leq t_f$, $i \in \mathcal{N}$ is IC and IR, i.e.,

$$I_i(t, z; \tilde{\gamma}_i^b(t, z), \tilde{\gamma}_{-i}^*(t, z)) \leq I_i(t, z; \tilde{\gamma}_i^b(t, \sigma_i(z)), \tilde{\gamma}_{-i}^*(t, \sigma_i(z))), \qquad (8.51)$$

$$I_i(t, z; \tilde{\gamma}_i^b, \tilde{\gamma}_{-i}^*) = h_i(t, z) \leq k_i(t, z). \qquad (8.52)$$

Generally, it is appropriate that a dynamic integration mechanism can allow each agent to determine whether he is connected with a power network at any time. The next mechanism based on the Hamiltonian permits each agent to select whether to join the mechanism for each sampling period. If every agent agrees to an evaluation of his loss through the Hamiltonian, then each agent can select whether to join the mechanism every time and he reports information. Considering the characterization of the temporal and spatial decompositions (8.32), we introduce the private cost functional

$$J_i(t, z, u_i, \tilde{\gamma}_{-i}^*(t, z)) = H_i(t, z_i, p_i^*(t), u_i) + H_i^{mt}(t, z, p^*(t), \tilde{\gamma}_{-i}^*(t, z)) \quad (8.53)$$

with

$$H_i^{mt}(t, z, p^*, \tilde{\gamma}_{-i}^*) = H_0(t, z, p_0^*(t)) + \sum_{j \in \mathcal{N}\setminus\{i\}} H_j(t, z_j, p_j^*(t), \tilde{\gamma}_j^*(t, z))$$
$$-H(t, z, p^*(t), 0, \tilde{\gamma}_{-i}^\dagger(t, z)), \qquad (8.54)$$

where $\tilde{\gamma}_{-i}^\dagger$ is the optimal control law that minimizes the cost functional $\mathbb{E}\big[C(t, z; 0, \tilde{\gamma}_{-i}) \mid z(t) = z, \tilde{\gamma}_{-i}\big]$. Here, as we see from the discussion in Sect. 8.4.1 that the relation between k_i and h_i is independent of the agents' private economic cost under the optimality condition of u_i, we set $h_i = k_i = 0$ without loss of generality.

Similar to the mechanism based on the cost functional, we present Theorem 3 for the case of the dynamic integration mechanism based on the Hamiltonian.

Theorem 3 *[26] (a) For the private cost functional (8.53) with incentive cost functional (8.54), the private optimal controls $u_i^b(t) = \tilde{\gamma}_i^b(t, z)$, $i \in \mathcal{N}$ coincide with the public optimal control (8.30).*
(b) The mechanism given by the incentive cost functional (8.54) and the private optimal control $u_i^b(t)$, $i \in \mathcal{N}$ is IC and IR, that is

$$J_i(t, z, \tilde{\gamma}_i^b(t, z), \tilde{\gamma}_{-i}^*(t, z)) = \min_{u_i \in U_i} J_i(t, z, u_i, \tilde{\gamma}_{-i}^*(t, z))$$

$$\leq J_i(t, z, \tilde{\gamma}_i^b(t, \sigma_i(z)), \tilde{\gamma}_{-i}^*(t, \sigma_i(z))), \qquad (8.55)$$

$$J_i(t, z, \tilde{\gamma}_i^b(t, z), \tilde{\gamma}_{-i}^*(t, z)) \leq 0. \qquad (8.56)$$

Both Theorems 2 and 3 show that the proposed dynamic integration mechanism ensures public optimality from private optimal controls, IC and IR constraints. These properties are very important for rational agents and the utility that aims to minimize the public cost. We emphasize that *the utility can achieve the global control objective by taking advantage of the price and the incentive cost despite having no direct controls.*

Remark 3 Assuming that the online parameter z_i is always reported truthfully, the VCG mechanism used in (8.50) is just one of the reward designs in moral hazard problems, but it does not consider budget balance constraints. Hence, we imply that the incentive mechanism with (8.38) and (8.43) based on the contract theory approach can satisfy the budget balance constraints whereas the incentive model (8.50) based on (VCG) mechanism design approach does not guarantee its constraints. The additional rewards will be paid from the social welfare budget. More general incentivizing schemes of this type, especially the schemes realizing the budget balance, are sought in [28].

Remark 4 One of the practical issues of the model-based approach used in this chapter is the compression of the model information and the online information. We have tried to compress the model information using randomized models [30], a model reduction method [27]. One of the options reducing the online information is to use the output feedback strategy with Kalman filter [24], which helps privacy protection and reduction of communication loads.

Remark 5 This chapter and the related control literature [24, 26, 27, 30] assume that, under the adverse selection problems, the policy variables (λ_i, h_i), $i \in \mathcal{N}$, of the utility are an optimal scheme for the public cost $\mathbb{E}[C]$, i.e., (8.31), and the resulting incentive mechanism (8.50) satisfies the IC and IR constraints. Meanwhile, the economists mainly focus on the design and economic interpretation of the policy variables (λ_i, h_i) including taxes and subsidies under IC and IR constraints. To design the policy variables in the adverse selection incentivizing markets, one of the promising mechanisms is to combine the moral hazard incentive mechanism and the adverse selection incentive mechanism handled in this section. Bolton and Dewatripont [7, Sect. 6.3] investigate the feasibility of the combined mechanism in the static model. It is one of the future challenge topics to design the mechanism integrating both of the incentives for the dynamical models as in Sect. 8.2.

8.5 Simulation

We finally demonstrate the effectiveness of the proposed incentive mechanisms through two simulations.

Performance Evaluation for Motivating Example Let us first consider the performance of the proposed incentive schemes through the motivating example shown in Sect. 8.3.2. Given a prescribed cost level $k_i \equiv 0$ for all $i \in \mathcal{N}$ and moral hazard incentive models (8.38) with (8.43), Table 8.1 shows the net costs I in (8.39), I_1 and I_2 in (8.36), and the incentives C_i^{mt} for the cases used in Figs. 8.2–8.5. First, we see from Table 8.1 that all the cases satisfy $\mathbb{E}[C] = I + I_1 + I_2$, which is the evidence that the moral hazard incentivizing market guarantees budget balance constraint. Next, comparing the case for Fig. 8.2 and that of Fig. 8.3, agent 1 pays only the prescribed cost $I_1 = 0$ unless he executes an irrational control $u_1 \neq \tilde{\gamma}_1^{\flat}$, and the irrational action of agent 1 reduces his incentive C_1^{mt} and the net public cost I. Third, although the moral hazard incentivizing market based on Theorem 3 does not strictly guarantee the case for Fig. 8.4, we see from the result for Fig. 8.4 that preventing the agents' malicious action in the moral hazard problem with the proposed incentive mechanism is strengthened. Finally, from the case of Fig. 8.5, the moral hazard incentivizing market cannot prevent the agents' strategic action for the adverse selection problems.

Table 8.2 shows the performance of the adverse selection incentivizing market. Comparing agent 1's net cost I_1 for each case, in particular, Figs. 8.2 and 8.5, the cost $I_1 = -1.9185$ for the public optimal control case (Fig. 8.2) is lower than the other cases that agent 1 takes an irrational action. The results indicate that the adverse selection incentivizing market is going as planned.

Table 8.1 Net costs I, I_1, I_2 and incentives C_1^{mt}, C_2^{mt} for moral hazard incentive (8.38) with (8.43)

Cases	Fig. 8.2	Fig. 8.3	Fig. 8.4	Fig. 8.5
C	2.4148	3.3656	18.7180	2.4794
I	2.4148	0.1022	−31.3234	2.4794
I_1	0	0.9551	25.0207	0
I_2	0	2.3083	25.0207	0
C_1^{mt}	−0.8835	0.5379	24.6036	−0.6888
C_2^{mt}	−0.8835	0.9324	24.6036	−1.0466

Table 8.2 Net costs I, I_1, I_2 and incentives C_1^{mt}, C_2^{mt} for adverse selection incentive (8.50)

Cases	Fig. 8.2	Fig. 8.3	Fig. 8.4	Fig. 8.5
I_1	−1.9185	−0.9677	−0.9677	−1.7649
I_2	−1.9185	−1.4003	−0.9677	−2.2337
C_1^{mt}	−2.8020	−1.3848	−1.3848	−2.4537
C_2^{mt}	−2.8020	−2.7762	−1.3848	−3.2803

Performance Evaluation for IEEJ East 30-machine System Finally, we numerically verify the performance of our incentives through a dynamic power network model based on the IEEJ 30-machine system [40]. The IEEJ system imitates actual power network systems in East Japan area, and it is divided into five areas as shown in Fig. 8.6 in order to be applied to our framework. Each area is regarded as an agent with an individual subsystem; all generators and consumers in each area are represented by some equivalent generators. The details of the models used in this simulation are shown in Appendix. Note that the agent dynamics (8.1) are stable systems. For simplicity of discussion, we set $k_i(t, z) \equiv 0$. Similar to Sect. 8.3.2, we assume that each agent always operates the type $\theta_i = 1$ and the type parameter obeys a deterministic (fixed) process during the time period $[t_0, t_f] = [0, 30]$ s.

We first consider the agents' strategic control and its performance in the proposed moral hazard incentivizing market. The left of Fig. 8.7 shows the time evolution of the state x when all the rational agents execute their own private optimal control, i.e., consequently the public optimal controls $u_i = \tilde{\gamma}_i^b = \tilde{\gamma}_i^*$. We see from the figure that each state converges sufficiently to zero at the terminal time $t_f = 30$ s. The right of

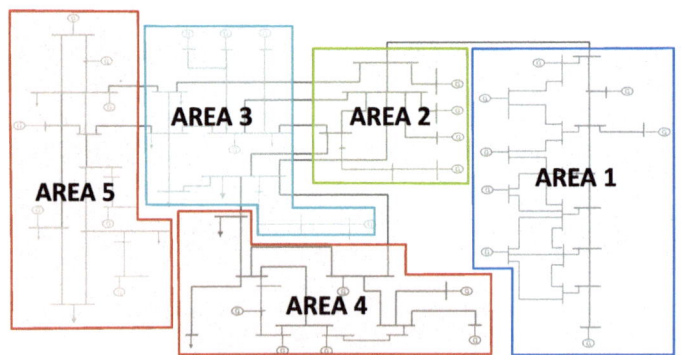

Fig. 8.6 IEEJ East 30-machine system divided into five areas

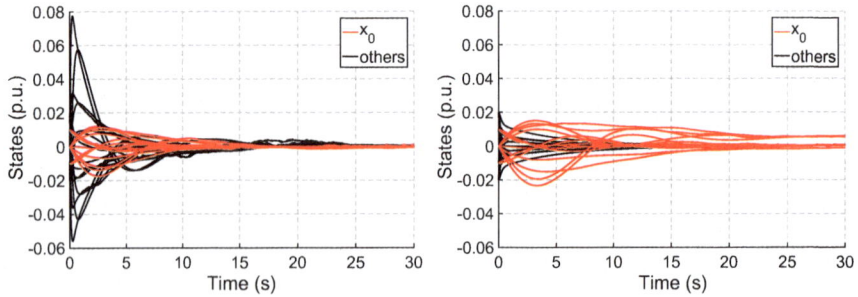

Fig. 8.7 Time evolution of state x from $t_0 = 0$ s to $t_f = 30$ s for a sample noise in case of moral hazard incentive (8.38) with (8.43) (*left* public optimal controls $u_i = \tilde{\gamma}_i^b = \tilde{\gamma}_i^*$, *right* private optimal controls $u_i = \tilde{\gamma}_i^{\sharp}$)

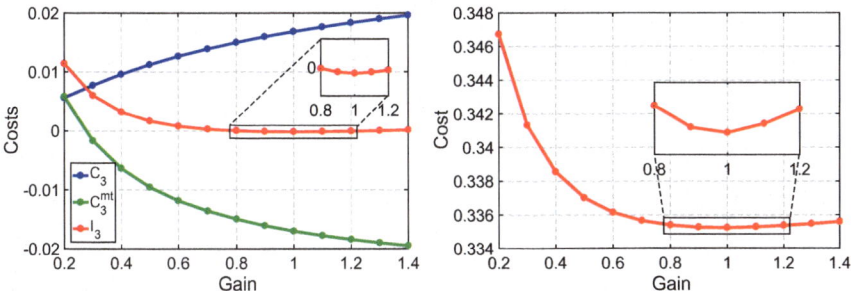

Fig. 8.8 Averaged costs for agent 3's strategic control $u_3 = k_p \tilde{\gamma}_3^{\flat}$ with gain $k_p \in [0.2, 1.4]$ (*left* C_3, C_3^{mt}, and I_3, *right* I)

Table 8.3 Average costs for control law, reported state, and type of agent 3. The fictitious report information is $\sigma_{33x}(x_3) = 0$ and $\sigma_{33\theta}(\theta_3) = 2$

	Control law	Reported state	Reported type	Cost I_3
(a)	$\tilde{\gamma}_3^{\flat}$	x_3	θ_3	-0.0561
(b)	$\tilde{\gamma}_3^{\flat}$	$\sigma_{33x}(x_3)$	θ_3	0.0190
(c)	$\tilde{\gamma}_3^{\flat}$	x_3	$\sigma_{33\theta}(\theta_3)$	-0.0380
(d)	$\tilde{\gamma}_3^{\flat}$	$\sigma_{33x}(x_3)$	$\sigma_{33\theta}(\theta_3)$	0.0191
(e)	$\tilde{\gamma}_3^{\sharp}$	x_3	θ_3	-0.0166

Fig. 8.7 is the result when each agent takes his private optimal control without the incentive, i.e., $u_i = \tilde{\gamma}_i^{\sharp}$. We see from Fig. 8.7 that the agents' selfish optimal behavior cannot lead the state x_0 to zero. In other words, thanks to the reward W_i, the utility can achieve the public optimality.

We next verify the requirement of IC and IR. Suppose that agent 3's control u_3 is not always optimal, i.e., $u_3 = \beta \tilde{\gamma}_3^{\flat}$ with a gain parameter $\beta > 0$, whereas the others execute the private optimal control with incentive $u_i = \tilde{\gamma}_i^{\flat}$ rationally. Then, the averaged cost indexes C_3, C_3^{mt}, I_3 and C of 1000 noise samples for each β are shown in Fig. 8.8. We see from the left of Fig. 8.8 that the averaged cost of agent 3 for the rational control $\beta = 1$ is approximately zero, which is the same as the theoretical result, i.e., $I_3 = h_3 = k_3 = 0$. We also see from Fig. 8.8 that the rational control case ($\beta = 1$) minimizes the agent 3's own net cost I_3 and the public cost C. These results are evidenced that the proposed moral hazard incentivizing market guarantees IC and IR.

We finally verify the performance of the strategic information bidding in the proposed adverse selection incentivizing market. Table 8.3 shows the average cost for several control laws, reported states, and types of agent 3. We see from Table 8.3 that the cost $I_3 = -0.0561$ for the true information case (a) is less than the others. Moreover, the optimal cost $I_3 = -0.0561$ is lower than zero and the honest agent 3 gains some benefit from the utility. The results indicate that the proposed adverse selection incentivizing market is a successful mechanism. The other simulations considering a Markov step process in (8.3) are shown in [26].

8.6 Conclusion

This chapter has investigated an optimal contract mechanism between a utility and
multi-agents in dynamic power systems while taking into account the physical con-
flicts among subsystems operated by the agents and economic optimalities of all the
market players. In particular, to prevent the rational agents from selecting a strategic
bidding, we have presented an optimal incentive condition for a moral hazard incentivizing market and an adverse selection incentivizing market. We have also discussed
and demonstrated the physical and economic performance of the proposed market
mechanisms with the two incentive mechanisms via simulations of IEEJ models.

Appendix: Detailed Network Model

We describe the detailed models of the system parameters and the cost parameters
used in Sect. 8.5.

As introduced in the papers related to electricity systems (e.g., [17], this book and
reference therein), there is a lot of options to model the agents' physical dynamics
including generators and demand response units. Considering the time constant of
each equipment, in this chapter, we introduce an agent model based on a generating
unit with a reheat steam turbine, one of the turbine-governing systems, described
in [22, Sect. 11.1.4]. To be concrete, the matrices A_{ii} and B_{ii} described in (8.19) are
given by time-invariant models:

$$A_{ii}(t, \theta_i) = \begin{bmatrix} \frac{-1}{T_G(\theta_i)} & 0 & 0 \\ \frac{1}{T_{CH}(\theta_i)} & -\frac{1}{T_{CH}(\theta_i)} & 0 \\ \frac{F_{HP}(\theta_i)}{T_{CH}(\theta_i)} & \frac{1}{T_{RH}(\theta_i)} - \frac{F_{HP}(\theta_i)}{T_{CH}(\theta_i)} & -\frac{1}{T_{RH}(\theta_i)} \end{bmatrix}, \quad B_{ii}(t, \theta_i) = \begin{bmatrix} \frac{1}{T_G(\theta_i)} \\ 0 \\ 0 \end{bmatrix},$$

(8.57)

where the parameters of the standard type, denoted by $\theta_i = 1$, are $T_G(1) = 0.2$ s,
$T_{CH}(1) = 0.3$ s, $T_{RH}(1) = 7.0$ s, and $F_{HP}(1) = 0.3$, and the variant case, denoted by
θ_2, is $T_G(2) = 0.2$ s, $T_{CH}(2) = 0.5$ s, $T_{RH}(2) = 5.5$ s, and $F_{HP}(2) = 0.4$. Hence, the
dimension of the local state x_i is $n_i = 3$ and that of control u_i is $m_i = 1$. Note that the
local dynamics led by (8.57) is stable. We see from A_{ii} in (8.57) that each element
of the steady-state x_i is the same value. The matrices in the cost functional in (8.21)
and (8.22) are given by

$$Q_i(t, \theta_i) = Q_{if}(t_f, \theta_i) = \begin{bmatrix} 3 & 0 & 0 \\ 0 & 3 & 0 \\ 0 & 0 & 3 \end{bmatrix}, \quad R_i(t, \theta_i) = 1.$$

for any type $\theta_i \in \Theta_i$.

We next introduce the balancing equation (8.20). Table 8.4 shows the typical
parameters, where M_i is the net inertia constant of the aggregated equipment in area

Table 8.4 Typical parameters of the dynamical model used in the simulation

Area i	M_i	D_i	Line e	In (e)	Out (e)	Y_e
1	9.0645	3.8575	1	1	2	1.4642
2	8.0333	1.9183	2	2	3	3.3842
3	9.0253	6.8935	3	3	4	3.3708
4	9.0444	4.5166	4	3	5	1.1668
5	8.3354	5.4479				

$i \in \mathcal{N}$, D_i is load damping constant in area i, and Y_e is the inverse of the inductance from area in(e) $\in \mathcal{N}$ to area out(e) $\in \mathcal{N}$ on the transmission line e. Let us denote by ω_i, δ_i, and ΔP_i the frequency deviation, the phase deviation, and the power imbalance in area i, respectively. Then, following [22, 46], the dynamics are given by

$$d\delta_{it} = \omega_{it}dt, \ d\omega_{it} = (M_i^{-1}\Delta P_{it} - M_i^{-1}D_i\omega_{it})dt, \ i \in \mathcal{N},$$
$$\Delta P_{1t} = \begin{bmatrix} 0 & 0 & 1 \end{bmatrix} x_{1t} + s_{1t} + w_{1t},$$
$$\Delta P_{2t} = \begin{bmatrix} 0 & 0 & 1 \end{bmatrix} x_{2t} - s_{1t} + s_{2t} + w_{2t},$$
$$\Delta P_{3t} = \begin{bmatrix} 0 & 0 & 1 \end{bmatrix} x_{3t} - s_{2t} + s_{3t} + s_{4t} + w_{3t},$$
$$\Delta P_{4t} = \begin{bmatrix} 0 & 0 & 1 \end{bmatrix} x_{4t} - s_{3t} + w_{4t},$$
$$\Delta P_{5t} = \begin{bmatrix} 0 & 0 & 1 \end{bmatrix} x_{5t} - s_{4t} + w_{5t},$$
$$ds_{1t} = Y_1(\omega_{2t} - \omega_{1t})dt, \ ds_{2t} = Y_2(\omega_{3t} - \omega_{2t})dt,$$
$$ds_{3t} = Y_3(\omega_{4t} - \omega_{3t})dt, \ ds_{4t} = Y_4(\omega_{5t} - \omega_{3t})dt.$$

From the above equations, we define $x_0 = [\delta_1, \ldots, \delta_5, \omega_1, \ldots, \omega_5, s_1, \ldots, s_4]^\top$ and obtain A_{00}, A_{0i} in (8.20) and g in (8.6). We also set $Q_0 = Q_{0f} = \text{diag}(30, \ldots, 30, 1, \ldots, 1, 100, \ldots, 100)$, and the variance of w_t is $\text{diag}(0.8977, 0.1436, 5.7684, 2.1604, 3.3137) \times 10^{-3}$, where $\text{diag}(\cdot)$ indicates the elements of the diagonal matrix. The initial condition of the state $x(t_0)$ is as follows: $\delta_i = 0$, $i \in \mathcal{N}$, $\omega_1 = 0.01$, $\omega_2 = -0.01$, $\omega_3 = 0.01$, $\omega_4 = -0.01$, $\omega_5 = 0.01$, $s_e = 0$, $e = 1, 2, 3, 4$, and $x_1 = 0.01$, $x_2 = -0.02$, $x_3 = 0.02$, $x_4 = 0$, $x_5 = -0.01$, which means that all the elements of x_i are the same.

References

1. Akao K (2008) Tax schemes in a class of deferential games. Econ Theory 35(1):155–174
2. Amin M, Annaswamy AM, DeMarco CL, Samad T (2013) IEEE vision for smart grid controls: 2030 and beyond. IEEE Press
3. Baldick R (2002) Electricity market equilibrium models: the effect of parametrization. IEEE Trans Power Syst 17(4):1170–1176
4. Benchekroun H, Long NV (1998) Efficiency inducing taxation for polluting oligopolists. J Public Econ 70(2):325–342

5. Bergemann D, Välimäki J (2010) The dynamic pivot mechanism. Econometrica 78(2):771–789
6. Berger AW, Schweppe FC (1989) Real time pricing to assist in load frequency control. IEEE Trans Power Syst 4(3):920–926
7. Bolton P, Dewatripont M (2005) Contract theory. The MIT Press
8. Costa OLV, Fragoso MD, Todorov MG (2013) Continuous-time Markov jump linear systems. Springer-Verlag
9. Cvitanic J, Zhang J (2013) Contract theory in continuous-time models. Springer
10. Dockner E, Jørgensen S, Long NV, Sorger G (2000) Differential games in economics and management science. Cambridge University Press
11. Ela E, Gevorgian V, Tuohy A, Kirby B, Milligan M, O'Malley M (2014) Market designs for the primary frequency response ancillary service - Part I: motivation and design. IEEE Trans Power Syst 29(1):421–431
12. Ela E, Gevorgian V, Tuohy A, Kirby B, Milligan M, O'Malley M (2014) Market designs for the primary frequency response ancillary service - Part II: case studies. IEEE Trans Power Syst 29(1):432–440
13. Farokhi F, Johansson KH (2015) Optimal control design under limited model information for discrete-time linear systems with stochastically-varying parameters. IEEE Trans Automtic Control 60(3):684–699
14. Hirata K, Hespanha JP, Uchida K (2014) Real-time pricing leading to optimal operation under distributed decision makings. In: Proceedings of 2014 American control conference, pp 1925–1932
15. Hirata K, Hespanha JP, Uchida K (2015) Real-time pricing and distributed decision makings leading to optimal power flow of power grids. In: Proceedings of 2015 American control conference, pp 2284–2291
16. Holmstrom B, Milgrom P (1987) Aggregation and linearity in the provision of intertemporal incentives. Econometrica 55(2):303–328
17. Ilic MD (2016) Toward a unified modeling and control for sustainable and resilient electric energy systems. Found Trends Electr Energy Syst 1(1–2):1–141
18. Ilic MD, Liu SX (1996) Hierarchical power systems control—Its value in a changing industry. Springer
19. Jackson MO (2003) Mechanism theory. In: Derigs U (ed) Encyclopedia of life support systems. EOLSS Publishers
20. Kiani A, Annaswamy A, Samad T (2014) A hierachical transactive control architecture for renewables integration in smart grids: Analytical modeling and stability. IEEE Trans Smart Grid 5(4):2054–2065
21. Koo HK, Shim G, Sung J (2008) Optimal multi-agent performance measures for team contracts. Math Financ 18(4):649–667
22. Kundur P (1994) Power system stability and control, McGraw-Hill Professional
23. Langbort C (2012) A mechanism design approach to dynamic price-based control of multi-agent systems. In: Johansson R, Rantzer A (eds) Distributed decision making and control. 113–129, Springer-Verlag
24. Matsui S, Murao T, Hirata K, Uchida K (2015) A dynamic output integration mechanism for LQG power networks with random type parameters. Proc Asian Control Conf 2015:1–6
25. Mehlmann A (1988) Applied differential games. Springer
26. Murao T, Hirata K, Okajima Y, Uchida K (2018) Real-time pricing for LQG power networks with independent types: A dynamic mechanism design approach. Eur J Control 39:95–105
27. Murao T, Hirata K, Uchida K (2016) An approximate dynamic Integration mechanism for LQ power networks with multi-time scale structures. In: Proceedings of 2016 European control conference, pp 202–209
28. Murao T, Okajima Y, Hirata K, Uchida K (2014) Dynamic balanced integration mechanism for LQG power networks with independent types. In: Proceedings of 53rd IEEE Conference on decision and control, pp 1395–1402
29. Nisan N, Roughgarden T, Tardos É, Vazirani VV (eds) (2007) Algorithmic game theory, Cambridge University Press

30. Okajima Y, Murao T, Hirata K, Uchida K (2013) A dynamic mechanism for LQG power networks with random type parameters and pricing delay. In: Proceedings of 52nd IEEE conference on decision and control, pp 2384–2390

31. Rebours YG, Kirschen DS, Trotignon M, Rossignol S (2007) A survey of frequency and voltage control ancillary services - Part I: technical features. IEEE Trans Power Syst 22(1):350–357

32. Samadi P, Mohsenian-Rad H, Schober R, Wong VWS (2012) Advanced demand side management for the future smart grid using mechanism design. IEEE Trans Smart Grid 3(3):1170–1180

33. Sannikov Y (2013) Contracts: the theory of dynamic principal-agent relationships and the continuous-time approach. In: Acemoglu D, Arellano M, Dekel E (eds) Advances in economics and econometrics, 10th world congress of the econometric society. Cambridge University Press

34. Schattler H, Sung J (1993) The first-order approach to the continuous time principal-agent problem with exponential utility. J Econ Theory 61:331–371

35. Silva C, Wollenberg BF, Zheng CZ (2001) Application of mechanism design to electric power markets (Republished). IEEE Trans Power Syst 16(4):862–869

36. Stoustrup J, Annaswamy AM, Chakrabortty A, Qu Z (2018) Smart gird control: Overview and research opportunities. Springer-Verlag

37. Tanaka T, Cheng AZW, Langbort C (2012) A dynamic pivot mechanism with application to real time pricing in power systems. In: Proceedings of 2012 American control conference, pp 3705–3711

38. Tanaka T, Li N, Uchida K (2018) On the relationship between the VCG mechanism and market clearing. In: Proceedings of 2018 American control conference, pp 4597–4603

39. Taylor JA, Nayyar A, Callaway DS, Poolla K (2013) Consolidated dynamic pricing of power system regulation. IEEE Trans Power Syst 28(4):4692–4700

40. The Institute of Electrical Engineers of Japan (1999) Standard models for Japanese power system. IEEJ Tech. Report 754 (in Japanese)

41. Uchida K, Hirata K, Wasa Y (2018) Incentivizing market and control for ancillary services in dynamic power grids. In: Gird Smart (ed) Stoustrup J, Annaswamy AM, Chakrabortty A, Qu Z. Overview and research opportunities, Springer-Verlag, Control, pp 47–58

42. Wasa Y, Hirata K, Uchida K (2017) A dynamic contract mechanism for risk-sharing management on interdependent electric power and gas supply networks. In: Proceedings of 2017 Asian control conference, pp 1222–1227

43. Wasa Y, Hirata K, Uchida K (2018) Contract theory approach to incentivizing market and control design. arXiv:1709.09318

44. Wasa Y, Hirata K, Uchida K (2019) Optimal agency contract for incentive and control under moral hazard in dynamic electric power networks. IET Smart Grid 2(4):594–601. https://digital-library.theiet.org/content/journals/10.1049/iet-stg.2018.0256

45. Wasa Y, Murao T, Tanaka T, Uchida K (2019) Strategic bidding of private information for principal-agent type dynamic LQ networks. In: Proceedings of 2019 European control conference, pp 3383–3389

46. Wasa Y, Sakata K, Hirata K, Uchida K (2017) Differential game-based load frequency control for power networks and its integration with electricity market mechanisms. In: Proceedings of 1st IEEE conference on control technology and applications, pp 1044–1049

47. Wasa Y, Uchida K (2019) Optimal dynamic incentive and control contract among principal and agents with moral hazard and long-term average reward. Proc Asian Control Conf 2019:31–36

48. Wonham WM (1970) Random differential equations in control theory. In: Bharucha-Reid AT (ed) Probabilistic methods in applied mathematics, (2). Academic Press, pp 131–212

Chapter 9
Distributed Dynamic Pricing in Electricity Market with Information Privacy

Toru Namerikawa and Yoshihiro Okawa

Abstract In the future power network, power consumers as well as power generators participate in electricity market trading as selfish market players. Dynamic pricing is one of the useful tools to manage such networks in a distributed manner by changing electricity prices appropriately. In this chapter, we show distributed price decision procedures regarding the dynamic pricing to maximize social welfare in a power grid with information privacy of market players. Specifically, we first deal with an electricity market that covers multiple regional areas in a power grid and propose a market trading algorithm to derive the optimal regional electricity prices based on alternating decision making of market players. Subsequently, we deal with an electricity market in one regional area where an aggregator participates in the market trading as a mediator between the market operator and consumers. For this market, we propose a trading algorithm to adjust power demand of consumers depending on their lifestyles in a day-ahead electricity market. This chapter also shows the convergence properties of the proposed market trading algorithms and illustrates that these methods enable us not only to derive the optimal electricity prices but also to improve their convergence speed through numerical simulation.

9.1 Introduction

Dynamic pricing of electricity [10] is a price decision method that changes electricity prices appropriately according to various conditions in a power grid, e.g., time and seasons. Because of this property, the dynamic pricing enables us to use limited energy resources efficiently in the future power network where many kinds of power

T. Namerikawa (✉)
Department of System Design Engineering, Keio University, 3-14-1 Hiyoshi,
Kohokuku, Yokohama, Kanagawa 223-8522, Japan
e-mail: namerikawa@sd.keio.ac.jp

Y. Okawa
Artificial Intelligence Laboratory, Fujitsu Laboratories Ltd., 4-1-1 Kamikodanaka,
Nakahara-ku, Kawasaki, Kanagawa 211-8588, Japan
e-mail: okawa.y@fujitsu.com

© Springer Nature Singapore Pte Ltd. 2020
T. Hatanaka et al. (eds.), *Economically Enabled Energy Management*,
https://doi.org/10.1007/978-981-15-3576-5_9

consumers and generators participate in electricity market trading as market players. On the other hand, to maintain power supply–demand balance in a power grid is significantly important for the stable operation of power systems, even though the above-mentioned power consumers and generators determine their power demand or supply selfishly by only considering their own profits. Therefore, we have to design a market mechanism for the dynamic pricing appropriately so that it achieves not only to maximize social welfare including the profits of the selfish market players participating in its trading but also to satisfy physical constraints of the power grid including power balance at all times.

Similar to the dynamic pricing, regional or nodal price decision methods to minimize the cost of generators in a power grid have been studied as the optimal power flow (OPF) problem or the locational marginal price (LMP) problem since a long time ago, e.g., [4, 9, 25]. More recently, some literature has dealt with the dynamic pricing problems where the market mechanisms are designed by considering the profits of consumers with their utility functions [13, 22] based on concepts from microeconomics. For example, the paper [15] proposed a game-theoretic approach to design a distributed real-time electricity pricing mechanism that guarantees the individual rationality of market players. On the other hand, the papers [3, 21] designed their market mechanisms in which the constraint regarding power balance in a power grid including power flow is set with Kirchhoff's current and voltage laws and so did the paper [18] with the DC approximation. Furthermore, the paper [16] showed the stability of such market mechanisms based on passivity analysis, where the optimal power demand, supply, and electricity prices that maximize social welfare consisting of the utility and cost of each market player are obtained in a distributed manner through market trading. Besides this literature, the paper [20] proposed an incentive design method for power reduction of consumers based on their utility functions to adjust power imbalance in a power grid.

How to reduce the number of iterations until convergence in market trading algorithms with keeping information privacy of market players is another significant issue to design a reasonable market mechanism, since it corresponds to reducing the minimum necessary number of trades among market players in practice. Most of the existing work regarding a distributed market trading algorithm is based on the Uzawa algorithm [1] that adopts dual decomposition and the gradient method. This method enables us to solve primal market problems such as social welfare maximization in a distributed manner without using private information of consumers and generators participating in market trading. That is, information privacy of market players is protected if we use such distributed market trading algorithms. However, in general, the number of iterations until convergence increases as the number of decision variables increases in the original gradient method. The paper [17] focused on this issue and showed that the convergence speed of the market algorithm is improved by using the iterative gradient method with a line search, while this method requires the information of cost functions of generators to update prices. On the other hand, the alternating direction method of multipliers (ADMM) [2, 26] is a fully distributed optimization method that improves its convergence speed by solving augmented Lagrange dual problems associated with the primal optimization ones. Some literature, e.g., [5, 6], solved the OPF problems by using this ADMM algorithm for the minimization of the generating cost in a power grid.

Increasing the number of market players also causes another issue to design a reasonable market mechanism that maximizes the profit of each market player participating in its trading. In particular, we have to design a market mechanism by taking into account that consumers obtain different utility depending on their lifestyles even if they live in the same regional area and consume the same amount of power demand at the same time. However, in order to design such a market mechanism precisely, its mechanism including a trading system becomes complicated. As a countermeasure of this issue, aggregators are expected to participate in market trading as a mediator between some consumers and the market operator to make its trading efficient. Some existing work has focused on the roles of the aggregator and designed their market mechanisms including its participation in trading. For example, the paper [19] proposed a demand adjustment method with which the aggregator allocates power reduction among consumers required from the power grid operator by giving those consumers financial incentives. Also, the paper [7] dealt with the hierarchical market model with the participation of multiple aggregators and proposed a day-ahead market trading algorithm to adjust the demand of consumers in a distributed manner, while these methods are based on one representative behavior model of consumers.

In this chapter, we show distributed price decision procedures for the dynamic pricing of electricity through market trading with alternating decision making of market players. Specifically, we first deal with an electricity market covering multiple regional areas in a power grid and formulate the social welfare maximization problem to be solved in this market based on the representative behavior models of power consumers and generators with the physical constraints of a power grid. For this problem, we propose a distributed market trading algorithm to derive the optimal electricity prices for each market player based on the alternating decision making of market players. This proposed price decision method enables us not only to maximize social welfare including the profits of both consumers and generators in a distributed manner with the satisfaction of the physical constraints of a power grid but also to improve convergence speed of its algorithm, compared with a conventional gradient-based one. Subsequently, we deal with an electricity market regarding one specific regional area in which an aggregator manages some consumers having different lifestyles. For this market model, we design the behavior models of consumers with the utility depending on the types of electric appliances (EAs) to represent their power-consuming behavior more realistically according to their lifestyles. Then, we propose a market trading algorithm to adjust demand of those consumers optimally in a distributed manner by using two kinds of electricity prices in a day-ahead electricity market with the participation of an aggregator. This chapter also shows the convergence properties of the proposed market trading algorithms described above, respectively, and illustrates their validity and effectiveness through numerical simulation.

The rest of this chapter is organized as follows: We first show a price decision procedure for the electricity market covering multiple areas in a power grid in Sect. 9.2. Specifically, we introduce the behavior models of consumers and generators and a power grid model, respectively, as problem formulations in Sect. 9.2.1, and show a distributed market trading algorithm to derive the optimal regional electricity prices based on alternating decision making of market players with its convergence analysis in Sect. 9.2.2. Then, we illustrate the validity and effectiveness of this proposed method through numerical simulation in Sect. 9.2.3. Subsequently, in Sect. 9.3, we deal with the electricity market regarding one specific regional area in which an aggregator participates in market trading as a mediator between consumers and the market operator. In this section, we first introduce the utility obtained by consumers depending on the types of EAs and design the behavior models of market players including the aggregator in Sect. 9.3.1. Then, in Sect. 9.3.2, we show a market trading algorithm that solves demand adjustment problems of consumers in a distributed manner in a day-ahead electricity market. We also illustrate the validity and effectiveness of this proposed algorithm through numerical simulation in Sect. 9.3.3. Finally, we conclude this chapter in Sect. 9.4.

9.2 Distributed Dynamic Pricing with Alternating Decision Making of Market Players Considering Power Flow

In this section, we show a distributed price decision procedure through market trading with alternating decision making of market players. Specifically, we deal with an electricity market covering multiple regional areas in a power grid and formulate the social welfare maximization problem with physical constraints of the power grid including power flow among areas. Then, we propose a distributed market trading algorithm to derive the optimal electricity prices for market players in each regional area based on the alternating decision making of market players. This proposed price decision method enables us not only to maximize social welfare including the profits of both power consumers and generators with their information privacy but also to improve the convergence speed of its algorithm. This section also shows the convergence property of the proposed market trading algorithm and illustrates its validity and effectiveness through numerical simulation.

9.2.1 Problem Formulations Regarding Electricity Market and Power Grid

The model of a power network that contains a power grid and an electricity market is shown in Fig. 9.1. The power grid in this network consists of multiple regional areas where each of them is connected with transmission lines and power flows among areas through those lines. On the other hand, three kinds of market players participate in market trading: power consumers, power generators, and an independent system operator (ISO). Among these players, power consumers and generators selfishly

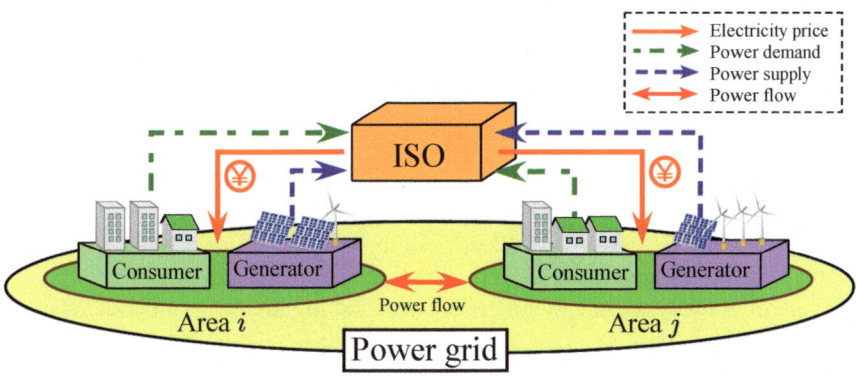

Fig. 9.1 Power grid with electricity market

determine their power demand or supply to maximize their own profits based on an electricity price and their utility or cost functions, while the ISO is a nonprofit organization responsible for operating the electricity market and power grid.

9.2.1.1 Behavior Models of Consumer and Generator in Market Trading

This subsection describes the behavior models of power consumers and generators. Suppose that the power-consuming and generating behaviors of these market players are summarized into their representative models if they live in the same regional area. Specifically, we define the welfare function regarding consumers in area i as

$$\mathcal{W}_{d_i}(x, \lambda_i) := v_i(x) - \lambda_i x, \tag{9.1}$$

where $v_i : \mathbb{R} \to \mathbb{R}$ is a utility function [13, 22] that represents monetary satisfaction obtained by consumers in area i when they consume power demand $x \in \mathbb{R}$, and $\lambda_i \in \mathbb{R}$ is an electricity price in that area. With this welfare function, consumers in area i determine their own power demand $d_i^{\text{opt}} \in \mathbb{R}$ in market trading. That is,

$$d_i^{\text{opt}} = \arg \max_{d_i^{\min} \leq x \leq d_i^{\max}} \mathcal{W}_{d_i}(x, \lambda_i), \tag{9.2}$$

where $d_i^{\min} \in [0, \infty)$ and $d_i^{\max} \in [0, \infty)$ are the lower and upper bounds of power demand of consumers in area i, respectively, and they satisfy $d_i^{\min} \leq d_i^{\max}$.

Similarly, we define the welfare function regarding generators in area i with the price λ_i as

$$\mathcal{W}_{s_i}(x, \lambda_i) := \lambda_i x - c_i(x), \tag{9.3}$$

where $c_i : \mathbb{R} \to \mathbb{R}$ is a cost function to generate power supply $x \in \mathbb{R}$ from their power-generating facilities. With this welfare function, generators in area i determine their own power supply $s_i^{\text{opt}} \in \mathbb{R}$ in market trading. That is,

$$s_i^{\text{opt}} = \arg \max_{s_i^{\text{min}} \leq x \leq s_i^{\text{max}}} \mathcal{W}_{s_i}(x, \lambda_i), \tag{9.4}$$

where $s_i^{\text{min}} \in [0, \infty)$ and $s_i^{\text{max}} \in [0, \infty)$ are the lower and upper bounds of power supply generated from generators in area i, respectively, and they satisfy $s_i^{\text{min}} \leq s_i^{\text{max}}$.

Regarding the above-mentioned utility and cost functions, we introduce the following assumptions.

Assumption 1 The utility functions v_i are in $C^2[0, \infty)$, strictly increasing and strictly concave.

Assumption 2 The cost functions c_i are in $C^2[0, \infty)$, strictly increasing and strictly convex.

9.2.1.2 Power Grid Model

Figure 9.2 shows one of the detailed examples of a power grid model considered in this section. Let us divide such a power grid into L areas and denote the number of nodes in each area by n_i. We also denote the set of areas by $\mathcal{A} := \{1, 2, \ldots, L\}$. The following assumptions are introduced regarding this power grid model, which are commonly used in the OPF and LMP problems to simplify their power flow equations among areas regarding AC power grids.

Assumption 3 The power grid is assumed to satisfy the following properties:

1. Resistance loss in the transmission grid is negligible.
2. The voltage of each node approximately equals to 1 p.u.
3. The voltage phase difference between each node is sufficiently small.

Let $s_i \in [s_i^{\text{min}}, s_i^{\text{max}}]$ and $d_i \in [d_i^{\text{min}}, d_i^{\text{max}}]$ be the power supply from generators and power demand of consumers in area i, respectively. Also, let $\sigma_i \in [\sigma_i^{\text{min}}, \sigma_i^{\text{max}}]$ be the power supply generated from conventional large-scale power-generating facilities in area i such as thermal, hydroelectric, and nuclear power plants, where $\sigma_i^{\text{min}} \in [0, \infty)$ and $\sigma_i^{\text{max}} \in [0, \infty)$ are its lower and upper bounds satisfying $\sigma_i^{\text{min}} \leq \sigma_i^{\text{max}}$. Note that these conventional generating facilities are managed by the ISO as part of the power grid. With Assumption 3, the active power flow equation linearized by using the DC approximation [11] regarding area $i \in \mathcal{A}$ is given by

$$s_i + \sigma_i - d_i = \sum_{j \in \mathcal{A}_i} P_{ij}, \tag{9.5}$$

where $\mathcal{A}_i \subset \mathcal{A}$ is the set of areas that directly connect to area i. In addition, P_{ij} is a linearized active power flow between areas i and j, which is given by

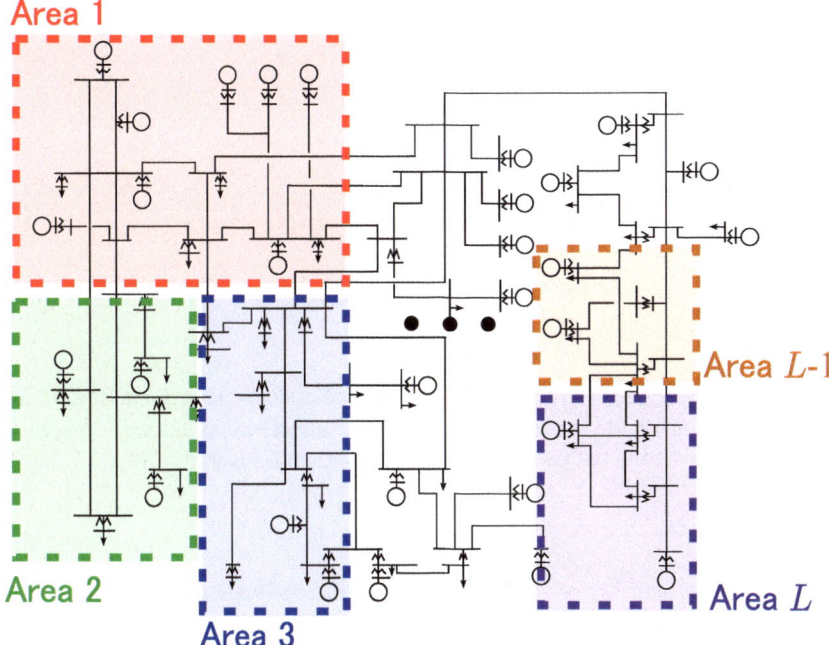

Fig. 9.2 Power grid model with L areas

$$P_{ij} = \sum_{(i_k, j_l) \in \mathcal{N}_{ij}} B_{i_k j_l} (\theta_{i_k} - \theta_{j_l}), \tag{9.6}$$

where $\theta_{i_k} \in (-\pi, \pi)$ and $\theta_{j_l} \in (-\pi, \pi)$ are the voltage phase angles of the node i_k in area i and the node j_l in area j, respectively, while $B_{i_k j_l}$ is an imaginary part of the admittance regarding the transmission line between these two nodes and $\mathcal{N}_{ij}$ is the set of pairs of nodes which connect areas i and j.

By summarizing the power flow equation given in (9.5) for all $i \in \mathcal{A}$, the active power flow equation of the entire power grid becomes

$$s + \sigma + \bar{B}\theta = d, \tag{9.7}$$

where $s, \sigma, d,$ and θ are, respectively, given by

$$s = [s_1, \ldots, s_L]^{\mathrm{T}}, \ \sigma = [\sigma_1, \ldots, \sigma_L]^{\mathrm{T}}, \ d = [d_1, \ldots, d_L]^{\mathrm{T}},$$
$$\theta = \left[\theta_1^{\mathrm{T}}, \ldots, \theta_L^{\mathrm{T}}\right]^{\mathrm{T}}, \ \theta_i = [\theta_{i_1}, \ldots, \theta_{i_{n_i}}]^{\mathrm{T}}, i \in \mathcal{A}.$$

In addition, the ijth sub-matrices of $\bar{B} \in \mathbb{R}^{L \times N}$ ($N := \sum_{i=1}^{L} n_i$) are given by

$$\bar{B}_{ij} := \begin{cases} \begin{bmatrix} \bar{B}_{ii_1} & \dots & \bar{B}_{ii_{n_i}} \end{bmatrix} \in \mathbb{R}^{1 \times n_i} & \text{if } j = i \\ \begin{bmatrix} \bar{B}_{ij_1} & \dots & \bar{B}_{ij_{n_j}} \end{bmatrix} \in \mathbb{R}^{1 \times n_j} & \text{if } j \neq i, \ j \in \mathcal{A}_i \ , \\ \mathbf{0}_{1 \times n_j} & \text{if } j \neq i, \ j \notin \mathcal{A}_i \end{cases} \tag{9.8}$$

where

$$\bar{B}_{ii_k} := - \sum_{j \in \mathcal{A}_i} \sum_{(i_k, j_l) \in \mathcal{N}_{i_k j}} B_{i_k j_l}, \qquad k = 1, 2, \dots, n_i, \tag{9.9}$$

$$\bar{B}_{ij_l} := \sum_{(i_k, j_l) \in \mathcal{N}_{i_k j}} B_{i_k j_l}, \qquad l = 1, 2, \dots, n_j. \tag{9.10}$$

In (9.9) and (9.10), $\mathcal{N}_{i_k j} \subset \mathcal{N}_{ij}$ is the set of pairs of nodes: one is the node i_k in area i and the other is a node in area j that directly connects to the node i_k. We set the upper and lower bounds of power flow between the nodes i_k and j_l as

$$P_{i_k j_l}^{\min} \leq B_{i_k j_l} (\theta_{i_k} - \theta_{j_l}) \leq P_{i_k j_l}^{\max}, \quad (i_k, j_l) \in \mathcal{N}_{ij}, \tag{9.11}$$

where $P_{i_k j_l}^{\max} \in \mathbb{R}$ and $P_{i_k j_l}^{\min} \in \mathbb{R}$. Since we deal with the lossless power grid as mentioned in Assumption 3, the above condition becomes

$$\left| B_{i_k j_l} (\theta_{i_k} - \theta_{j_l}) \right| \leq P_{i_k j_l}^{\max}, \quad (i_k, j_l) \in \mathcal{N}_{ij}. \tag{9.12}$$

On the other hand, let $\theta_{i_k}^{\max}$ and $\theta_{i_k}^{\min}$ be the upper and lower bounds of the voltage phase angles of the node i_k, which satisfy $\theta_{i_k}^{\max} \in (-\pi, \pi)$ and $\theta_{i_k}^{\min} = -\theta_{i_k}^{\max}$. With these upper and lower values, the maximum power flow $P_i^{\max}$ from area i to all of the other connected areas $j \in \mathcal{A}_i$ is given by

$$P_i^{\max} := \sum_{j \in \mathcal{A}_i} \sum_{(i_k, j_l) \in \mathcal{N}_{ij}} B_{i_k j_l} \left(\theta_{i_k}^{\max} + \theta_{j_l}^{\max} \right). \tag{9.13}$$

The following assumption is introduced regarding the upper and lower bounds of the power supply, demand, and the voltage phase angle in each area.

Assumption 4 The upper bound $\theta_{i_k}^{\max}$ of the voltage phase angle strictly satisfies the inequality (9.12) and also satisfies

$$\left(d_i^{\min} - s_i^{\max} - \sigma_i^{\max} < P_i^{\max} \right) \text{ or } \left(s_i^{\min} + \sigma_i^{\min} - d_i^{\max} < P_i^{\max} \right), \ i \in \mathcal{A}. \tag{9.14}$$

The condition (9.14) means that the maximum gap between the power demand and supply in each area is lower than the maximum power flow from or going out to the other directly connected areas, and this assumption is required to guarantee the existence of a solution regarding the optimization problem described in the rest of this section.

9.2.2   Distributed Maximization of Social Welfare Based on Alternating Decision Making in Market Trading

9.2.2.1   Design of Social Welfare Function

We consider maximizing the social welfare that includes the profits of both power consumers and generators participating in market trading as well as operation cost to manage a power grid owned by the ISO. For this purpose, we define the social welfare function as

$$\mathcal{W}(s, \sigma, d, \theta) := \sum_{i=1}^{L} \left\{ v_i(d_i) - c_i(s_i) - c_{L_i}(\sigma_i) - f_i(\theta_i) \right\}, \tag{9.15}$$

where $c_{L_i} : \mathbb{R} \to \mathbb{R}$ and $f_i : \mathbb{R}^{n_i} \to \mathbb{R}$ are the cost functions to operate large-scale power-generating facilities and those to change or adjust the voltage phase angles of nodes in area i, respectively. We introduce the following assumptions regarding these cost functions.

Assumption 5   The cost functions c_{L_i} are in $C^2[0, \infty)$, strictly increasing and strictly convex.

Assumption 6   The cost functions f_i are in $C^2(-\pi, \pi)$ and strictly convex.

Note that the ISO knows the information of these two cost functions since the ISO manages them as part of the power grid.

Based on the social welfare function $\mathcal{W}$ given in (9.15) and the physical constraints regarding the power grid described in the previous subsection, the optimization problem to be solved through market trading is given by

$$\max_{s, \sigma, d, \theta} \mathcal{W}(s, \sigma, d, \theta), \tag{9.16}$$

subject to
$$s + \sigma + \bar{B}\theta = d, \tag{9.17}$$

$$s_i^{\min} \leq s_i \leq s_i^{\max}, \quad \forall i \in \mathcal{A}, \tag{9.18}$$

$$d_i^{\min} \leq d_i \leq d_i^{\max}, \quad \forall i \in \mathcal{A}, \tag{9.19}$$

$$\sigma_i^{\min} \leq \sigma_i \leq \sigma_i^{\max}, \quad \forall i \in \mathcal{A}, \tag{9.20}$$

$$-\theta_{i_k}^{\max} \leq \theta_{i_k} \leq \theta_{i_k}^{\max}, \quad \forall k \in \{1, 2, \ldots, n_i\}, \quad \forall i \in \mathcal{A}, \tag{9.21}$$

$$|B_{i_k j_l}(\theta_{i_k} - \theta_{j_l})| \leq P_{i_k j_l}^{\max}, \quad \forall (i_k, j_l) \in \mathcal{N}_{ij}. \tag{9.22}$$

In this optimization problem, the constraint (9.17) represents the linearized active power flow equation and the other conditions in (9.18)–(9.22) represent the upper and lower bounds regarding each variable and the power flow, respectively.

9.2.2.2 Optimal Electricity Price

It is well known that the optimal price is represented by the optimal Lagrange multiplier of the dual problem associated with the primal optimization one. In this subsection, we show that the optimal regional electricity prices that maximize social welfare
given in (9.15) are also consistent with the optimal solutions of the Lagrange dual
problem associated with the optimization problem (9.16)–(9.22), while power consumers and generators selfishly determine their power demand and supply according
to their behavior models given in (9.2) and (9.4) in market trading.

Let us denote the Lagrange multiplier associated with the constraint (9.17) by
$\lambda_0 := [\lambda_{0_1}, \ldots, \lambda_{0_L}]^T \in \mathbb{R}^L$. With this multiplier, we define the partially Lagrange
dual function $\mathcal{L}(s, \sigma, d, \theta, \lambda_0)$ as

$$
\begin{aligned}
&\mathcal{L}(s, \sigma, d, \theta, \lambda_0) \\
&:= \mathcal{W}(s, \sigma, d, \theta) + \lambda_0^T (s + \sigma + \bar{B}\theta - d) \\
&= \sum_{i=1}^{L} \left\{ v_i(d_i) - c_i(s_i) - c_{L_i}(\sigma_i) - f_i(\theta_i) \right\} + \lambda_0^T (s + \sigma + \bar{B}\theta - d). \quad (9.23)
\end{aligned}
$$

The dual problem associated with the optimization problem (9.16)–(9.22) is given
by

$$
\min_{\lambda_0} \max_{(s, \sigma, d, \theta) \in \mathcal{V}} \mathcal{L}(s, \sigma, d, \theta, \lambda_0), \quad (9.24)
$$

where $\mathcal{V}$ is the set of variables (s, σ, d, θ) satisfying the constraints (9.17)–(9.22).

Let $(s^*, \sigma^*, d^*, \theta^*) \in \mathcal{V}$ be the optimal solution of (9.16)–(9.22) where

$$
s^* = [s_1^*, \ldots, s_L^*]^T, \ \sigma^* = [\sigma_1^*, \ldots, \sigma_L^*]^T, \ d^* = [d_1^*, \ldots, d_L^*]^T,
$$
$$
\theta^* = \left[\theta_1^{*T}, \ldots, \theta_L^{*T}\right]^T, \ \theta_i^* = [\theta_{i_1}^*, \ldots, \theta_{i_{n_i}}^*]^T, i \in \mathcal{A},
$$

and $\lambda_0^* = [\lambda_{0_1}^*, \ldots, \lambda_{0_L}^*]^T \in \mathbb{R}^L$ be the optimal Lagrange multiplier of the dual problem (9.24). Then, we obtain the following theorem.

Theorem 1 *Let Assumption 1–6 be held. If the electricity prices of each area are*
$\lambda_{0_i}^*, \forall i \in \mathcal{A}$, *the power demand d_i^{opt} and power supply s_i^{opt} determined by consumers*
and generators in each area according to (9.2) and (9.4) are consistent with d_i^* and
s_i^*, *respectively.*

Proof Because of the saddle point theorem, the following inequality holds for any
$(s, \sigma, d, \theta) \in \mathcal{V}$ regarding the Lagrange dual function $\mathcal{L}$ given in (9.23):

$$
\mathcal{L}(s^*, \sigma^*, d^*, \theta^*, \lambda_0^*) \geq \mathcal{L}(s, \sigma, d, \theta, \lambda_0^*). \quad (9.25)
$$

In other words, $(s^*, \sigma^*, d^*, \theta^*)$ is the optimal solution of the optimization problem:

$$\max_{(s,\sigma,d,\theta)\in\mathcal{V}} \sum_{i=1}^{L} \left\{ v_i(d_i) - c_i(s_i) - c_{L_i}(\sigma_i) - f_i(\theta_i) \right\} + \lambda_0^{*\mathrm{T}}(s + \sigma + \bar{B}\theta - d).$$

$$(9.26)$$

Since (9.26) and the constraints (9.17)–(9.22) are independent regarding s, σ, d and θ, we obtain

$$d_i^* = \arg \max_{d_i^{\min} \le d_i \le d_i^{\max}} v_i(d_i) - \lambda_{0_i}^* d_i, \quad i \in \mathcal{A}, \tag{9.27}$$

$$s_i^* = \arg \max_{s_i^{\min} \le s_i \le s_i^{\max}} \lambda_{0_i}^* s_i - c_i(s_i), \quad i \in \mathcal{A}. \tag{9.28}$$

By comparing the above two equations with the behavior models of market players given in (9.2) and (9.4), the power demand d_i^{opt} and the power supply s_i^{opt} determined by consumers and generators are consistent with the optimal solutions d_i^* and s_i^* of the optimization problem (9.16)–(9.22), respectively, if the electricity prices λ_i of each area are equal to the optimal Lagrange multipliers $\lambda_{0_i}^*$, $\forall i \in \mathcal{A}$.   □

This result concludes that the Lagrange multipliers can be regarded as the prices in electricity market trading. Therefore, we denote the Lagrange multipliers by λ instead of λ_0 in the rest of this section.

9.2.2.3 Distributed Market Trading Algorithm with Alternating Decision Making of Market Players and Convergence Analysis

Theorem 1 shows that the optimal power demand and supply in each area which maximize the social welfare given in (9.15) are obtained if the ISO sets their electricity prices appropriately, even though power consumers and generators determine their own power demand and supply selfishly in market trading. However, the ISO cannot derive those optimal prices directly since the utility functions and cost functions of the other market players are unknown to the ISO because of their information privacy. As a countermeasure of this issue, we show a distributed market trading algorithm that not only derives the optimal regional electricity prices to maximize social welfare in a distributed manner but also improves the convergence speed of the algorithm. Specifically, we propose a distributed price decision method based on the alternating direction method of multipliers (ADMM) to maximize social welfare which includes the profits of both power consumers and generators with the satisfaction of the physical constraints of the power grid.

We define the augmented Lagrange function $\hat{\mathcal{L}}(s, \sigma, d, \theta)$ associated with the optimization problem (9.16)–(9.22) as

$$\hat{\mathcal{L}}(s, \sigma, d, \theta) := \mathcal{W} + \lambda^{\mathrm{T}}(s + \sigma + \bar{B}\theta - d) - \frac{1}{2}\gamma \left\| s + \sigma + \bar{B}\theta - d \right\|^2. \tag{9.29}$$

With this function, we obtain the partially dual problem:

$$\min_{\lambda} \quad \max_{(s,\sigma,d,\theta)\in\mathcal{V}} \quad \hat{\mathcal{L}}(s,\sigma,d,\theta,\lambda). \tag{9.30}$$

The ADMM solves such dual problems in a distributed manner by updating each decision variable alternately, even though the problem includes a quadratic term which cannot be decomposed with dual decomposition. We propose the price decision method based on the slightly modified ADMM algorithm. Specifically, in our proposed method, power consumers and generators determine their own power demand and supply according to (9.2) and (9.4), while the ISO determines the power supply from large-scale power-generating facilities and voltage phase angles in each area by solving the optimization problem including a quadratic term of the constraint regarding the power balance in a power grid. The detailed price decision procedure of the proposed method is given in Algorithm 1.

As shown in Algorithm 1, the ISO updates the electricity price according to (9.36) with the power demand and supply reported from consumers and generators. That is, the ISO does not require the private information of the other market players such as utility functions or cost functions.

Next, we show the electricity prices updated with Algorithm 1 converge to their optimal ones. Let us denote the residual of the power balance in a power grid at iteration number $k \geq 0$ during market trading by $r^k := s^k + \sigma^k + \bar{B}\theta^k - d^k$. We have the following theorem regarding the convergence property of the prices updated according to Algorithm 1.

Theorem 2 *Let Assumptions 1–6 be held. If the step size γ in (9.36) satisfies*

$$0 < \gamma < \frac{2(\lambda^0 - \lambda^*)^{\mathrm{T}} r^{k+1}}{\left(r^{k+1} + 2\sum_{l=1}^{k} r^l\right)^{\mathrm{T}} r^{k+1}}, \quad \forall k = 0, 1, \ldots, \tag{9.37}$$

then the electricity price λ^k updated according to Algorithm 1 converges to λ^* as $k \to \infty$.

Proof We set a candidate of the Lyapunov function to be

$$V(\lambda^k) = \left\| \lambda^k - \lambda^* \right\|^2. \tag{9.38}$$

From (9.36) and the definition of r^k, the updating equation of the price λ^k is given by

$$\lambda^{k+1} = \lambda^k - \gamma r^{k+1}. \tag{9.39}$$

By using this equation, $V(\lambda^{k+1})$ becomes

Algorithm 1 Distributed regional price decision algorithm with alternating decision making of market players

---

1: The ISO sets the initial electricity price $\lambda^0 = [\lambda_1^0, \dots, \lambda_L^0]^T$ arbitrarily, and transmits them to consumers and generators in each area.

2: Based on the electricity price λ_i^k, consumers and generators in each area determine their power demand d_i^{k+1} and supply s_i^{k+1}:

$$d_i^{k+1} = \arg \max_{d_i^{\min} \leq d_i \leq d_i^{\max}} \mathcal{W}_{d_i}\left(d_i, \lambda_i^k\right), \quad i \in \mathcal{A}, \tag{9.31}$$

$$s_i^{k+1} = \arg \max_{s_i^{\min} \leq s_i \leq s_i^{\max}} \mathcal{W}_{s_i}\left(s_i, \lambda_i^k\right), \quad i \in \mathcal{A}, \tag{9.32}$$

and report them to the ISO.

3: By using the power demand and supply reported from consumers and generators in each area, the ISO determines the power supply σ^{k+1} from large scale power-generating facilities and voltage phase angles θ^{k+1}:

$$(\sigma^{k+1}, \theta^{k+1}) = \arg \min_{\sigma, \theta} \tilde{\mathcal{L}}(\sigma, \theta), \quad \text{subject to } (9.20)-(9.22), \tag{9.33}$$

where

$$\tilde{\mathcal{L}}(\sigma, \theta) := \sum_{i=1}^{L} \left\{c_{L_i}(\sigma_i) + f_i(\theta_i)\right\} - \lambda^{k^T}\left(\sigma + \bar{B}\theta\right) + \frac{\gamma}{2}\left\|\sigma + \bar{B}\theta + s^{k+1} - d^{k+1}\right\|^2, \tag{9.34}$$

and $d^{k+1} = \left[d_1^{k+1}, \dots, d_L^{k+1}\right]^T$, $s^{k+1} = \left[s_1^{k+1}, \dots, s_L^{k+1}\right]^T$, and $\gamma > 0$ is a constant parameter.

4: If the power balance in a power grid given by

$$s^{k+1} + \sigma^{k+1} + \bar{B}\theta^{k+1} = d^{k+1} \tag{9.35}$$

is satisfied, the ISO terminates this market trading and sets λ_i^k to the actual electricity price. If (9.35) is not satisfied, the ISO updates the electricity prices in each area according to

$$\lambda_i^{k+1} = \lambda_i^k - \gamma\left(s_i^{k+1} + \sigma_i^{k+1} + \sum_{j \in \mathcal{A}_i} \bar{B}_{ij}\theta_j^{k+1} - d_i^{k+1}\right), \quad i \in \mathcal{A}, \tag{9.36}$$

and transmits the updated prices λ_i^{k+1} to consumers and generators in each area.

5: Change the iteration number k to $k+1$ and go back to 2.

$$\begin{aligned} V(\lambda^{k+1}) &= \|\lambda^{k+1} - \lambda^*\|^2 \\ &= \|\lambda^k - \gamma r^{k+1} - \lambda^*\|^2 \\ &= \|\lambda^k - \lambda^*\|^2 + \gamma^2 \|r^{k+1}\|^2 - 2\gamma(\lambda^k - \lambda^*)^T r^{k+1} \\ &= V(\lambda^k) + \gamma^2 \|r^{k+1}\|^2 - 2\gamma(\lambda^k - \lambda^*)^T r^{k+1}. \end{aligned} \tag{9.40}$$

Therefore, the time difference of the Lyapunov function defined in (9.38) is given by

$$\Delta V := V(\lambda^{k+1}) - V(\lambda^k)$$
$$= \gamma^2 \left\| r^{k+1} \right\|^2 - 2\gamma(\lambda^k - \lambda^*)^{\mathrm{T}} r^{k+1}. \tag{9.41}$$

On the other hand, the price λ^k at the iteration number k (≥ 1) is given by

$$\lambda^k = \lambda^0 - \gamma \sum_{l=1}^{k} r^l \tag{9.42}$$

where λ^0 is the initial values of the prices.

Substituting this λ^k into (9.41), ΔV becomes

$$\Delta V = \gamma^2 \|r^{k+1}\|^2 - 2\gamma \left(\lambda^0 - \gamma \sum_{l=1}^{k} r^l - \lambda^* \right)^{\mathrm{T}} r^{k+1}$$

$$= \gamma^2 \left(r^{k+1} + 2 \sum_{l=1}^{k} r^l \right)^{\mathrm{T}} r^{k+1} - 2\gamma(\lambda^0 - \lambda^*)^{\mathrm{T}} r^{k+1}. \tag{9.43}$$

From this equation, the time difference ΔV becomes negative ($\Delta V < 0$), if the step size γ satisfies (9.37) in each iteration step, and thus, the price λ^k converges to λ^* as $k \to \infty$.

Theorem 2 shows that the electricity prices converge to the optimal ones with the information exchange between the ISO and the other market players through market trading with the alternating decision making based on Algorithm 1, if its step size is set appropriately.

9.2.3 Simulation Verification

We verify the validity and effectiveness of the proposed price decision algorithm through numerical simulation.

9.2.3.1 Simulation Conditions

Figure 9.3 shows the power grid model used in this simulation. We use the IEEJ EAST 30-machine system model [24] and divide it into four areas. In addition, we set the peak loads of each area to be 185.5, 349.0, 93.0, and 98.5 MW, respectively, which are also determined based on the load data given in [24]. Furthermore, in this verification, we assume that the utility of consumers is different depending on time. Specifically, we set the utility function $v_{i,t}$ of consumers in area i between time t and $t + 1$ [h] to be

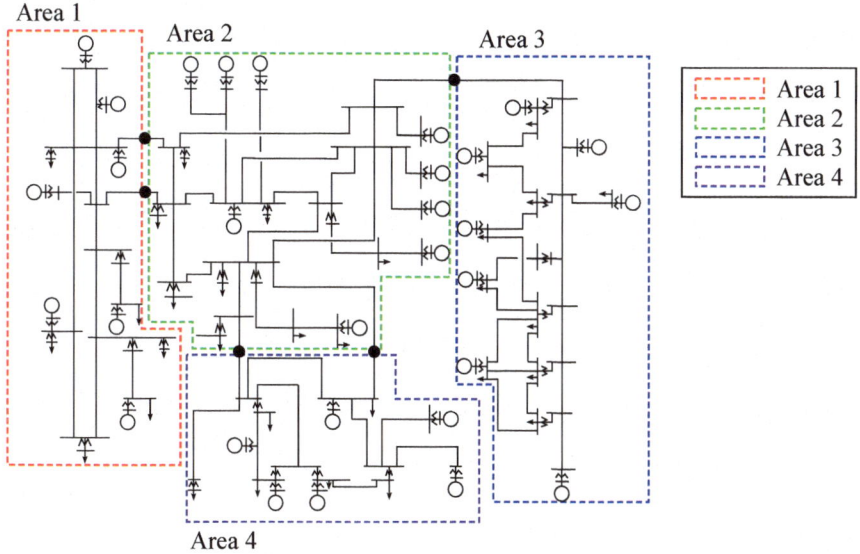

Fig. 9.3 IEEJ EAST 30-machine system divided into four areas

$$v_{i,t}(d_i) = a_{i,t}\mu_2 \log\left(\frac{d_i - \mu_1 d_{i,t}^{\text{peak}}}{\mu_2} + 1\right), \tag{9.44}$$

where $d_{i,t}^{\text{peak}} \in \mathbb{R}$ is a peak load in area i between t and $t+1$ calculated with the peak loads in each area described above. On the other hand, we set the cost function c_i of generators in area i, and the cost functions c_{L_i} and f_i to operate large-scale power-generating facilities and the voltage phase angles in area i, respectively, to be

$$c_i(s_i) = 5b_i s_i^2, \quad c_{L_i}(\sigma_i) = b_i \sigma_i^2, \tag{9.45}$$

$$f_i(\boldsymbol{\theta}_i) = \sum_{k=1}^{n_i} \zeta_{i_k} \theta_{i_k}^2. \tag{9.46}$$

Note that the utility and cost functions and the power grid model described above satisfy Assumptions 1–6, respectively. The other parameters of these utility and cost functions are determined as follows. We first set $\mu_1 d_{i,t}^{\text{peak}}$ and μ_2 in (9.44) to represent the elasticity of the power demand for a price based on the above-mentioned peak loads of each area. Next, we set coefficients $a_{i,t}$ and b_i of the utility function and cost functions by using the fixed electricity price $\lambda_f = 25.19$ [Yen/kWh]. In addition, we set $\mu_1 = 0.8$, $\mu_2 = 0.2, \zeta_{i_k} = 1.0 \times 10^9, \forall k \in \{1, 2, \ldots, n_i\}, \forall i \in \mathcal{A}$. Furthermore, the lower and upper bounds of each variable are listed in Table 9.1.

Table 9.1 Simulation parameters: lower and upper bounds of each variable

Parameter [unit]	Symbol	Value
Lower and upper bounds of demand of consumers [MWh]	$d_i^{\min}, d_i^{\max}$	$\mu_1 d_i^{\text{peak}}, \infty$
Lower and upper bounds of supply from generators [MWh]	$s_i^{\min}, s_i^{\max}$	$0, \dfrac{\mu_1 d_i^{\text{peak}}}{5}$
Lower and upper bounds of supply from the ISO [MWh]	$\sigma_i^{\min}, \sigma_i^{\max}$	$0, \infty$
Lower and upper bound of angle [degree]	$\theta_i^{\max}$	1
Lower and upper bound of transmission capacity [MW]	$P_{i_k j_l}^{\max}$	5

9.2.3.2 Simulation Results

Figure 9.4 shows the results of the convergence with respect to the electricity prices in areas 1–4 between 13:00 and 14:00 [h] through market trading. In this figure, 'Step' on the horizontal axis represents the number of iterations, which is denoted by k in Algorithm 1. The solid lines in this figure show the results obtained by using our proposed method with a constant step size $\gamma = 2.5 \times 10^2$, while the dashed lines show those obtained by using the conventional gradient method with a constant step size $\gamma = 1.0 \times 10^2$. This figure shows that the electricity prices of each area converge to their optimal ones $\boldsymbol{\lambda}^* = [25.21, \ 28.87, \ 27.67, \ 25.66]^{\mathsf{T}}$ [Yen/kWh] by using our proposed method, where those optimal values are derived by numerically solving the primal optimization problem (9.16)–(9.22). Also, it is shown that our proposed method improves its convergence speed with less fluctuation, compared with the results by using the conventional one in dashed lines. On the other hand, Fig. 9.5 shows the step size used in our proposed method described above and its upper bounds given in the right-hand side of (9.37) at each iteration. As shown in this figure, the step size is lower than its upper bound in every iteration, and thus, the condition (9.37) is satisfied. Consequently, we verify the validity of Theorem 2 from these two figures that the electricity prices in each area converge to their optimal ones with our proposed market trading algorithm if we set its step size appropriately.

Next, Figs. 9.6 and 9.7 show the hourly electricity prices of each area and the hourly power balance in area 2, respectively, which are the results obtained by using our price decision method. From Fig. 9.6, it is confirmed that the prices are determined depending on time and area. Furthermore, Fig. 9.7 shows that the power demand is consistent with the sum of the power supply from generators and that from large-scale power-generating facilities with power flow, which are denoted by s and $\sigma + \mathrm{PF}(\theta)$, respectively, in every hour. This result concludes that the regional power balances including power flow among areas are maintained based on hourly and regional prices determined with our proposed price decision method.

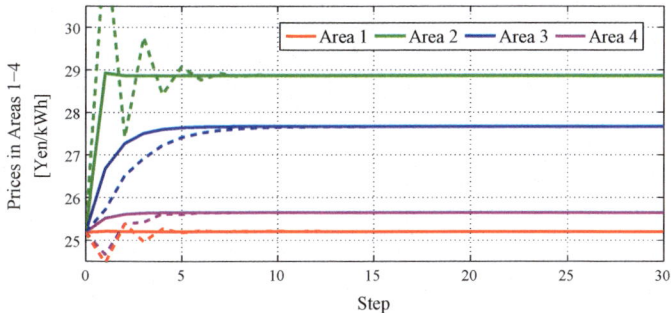

Fig. 9.4 Results of convergence of regional electricity prices in market trading

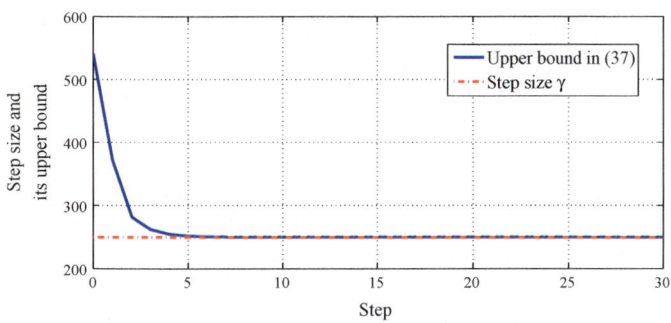

Fig. 9.5 Step size γ and its upper bound given in the right-hand side of (9.37)

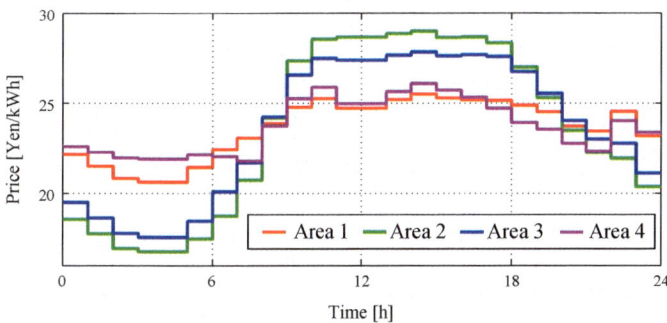

Fig. 9.6 Hourly regional electricity prices

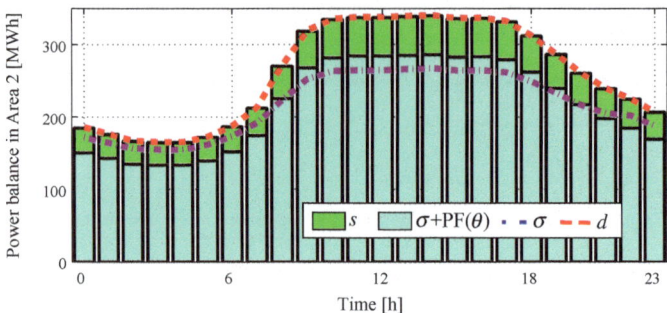

Fig. 9.7 Power balance in area 2

9.3 Optimal Demand Adjustment of Consumers with Various Electric Appliances Using Dynamic Pricing by Aggregator

Throughout Sect. 9.2, we showed the distributed price decision procedure through market trading to maximize social welfare in a power grid. This proposed price decision method is based on the representative behavior models of consumers and generators in each area. However, consumers obtain different utility depending on their lifestyles even if they live in the same regional area and consume the same amount of demand at the same time. For this issue, in this section, we design the behavior models of consumers with the utility depending on the types of electric appliances (EAs) to represent their power-consuming behavior more precisely. Then, we show a price decision procedure to solve the demand adjustment problem of consumers with the participation of an aggregator based on the alternating decision making of those market players in day-ahead market trading. We also show the convergence property of the pricing algorithm proposed in this section and illustrate its validity and effectiveness through numerical simulation.

9.3.1 Problem Formulations of Electricity Market with Aggregator

The market model including the participation of an aggregator considered in this section is shown in Fig. 9.8. Note that this model can be regarded as the market model of one specific regional area shown in Fig. 9.1 without power flow. Besides the market players introduced in Sect. 9.2, an aggregator who manages some consumers participates in market trading. Specifically, the aggregator purchases power supply from generators and the ISO with the price ρ and sells them to consumers with the price μ. As a result, the aggregator obtains its profit from the difference between

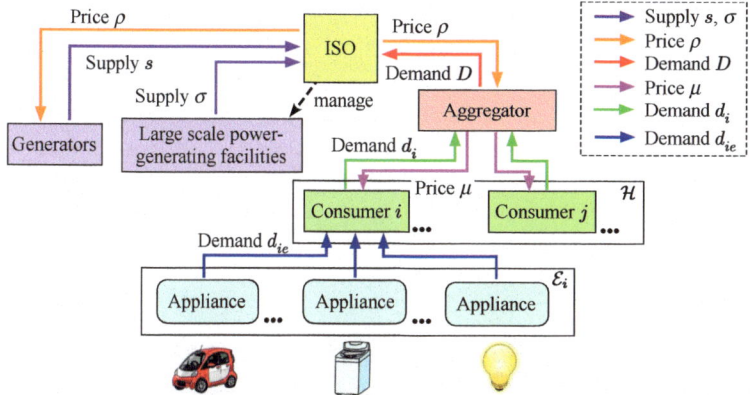

Fig. 9.8 Market model with the participation of an aggregator

the sale of electricity and its operation cost. On the other hand, we assume that each consumer has some different types of EAs in this section. To simplify the notations in the rest of this section, let $\mathcal{H}$ be the set of consumers managed by the aggregator, $\mathcal{E}_i$ be the set of EAs owned by the consumer $i \in \mathcal{H}$, as shown in Fig. 9.8. In addition, we divide one day into $N - 1$ time zones and determine the power demand, supply, and prices for each time $t \in \mathcal{T} := \{0, 1, \ldots N - 1\}$ at one time in a day-ahead market.

9.3.1.1 Behavior Models of Consumers Having Different Types of EAs

In general, electric appliances (EAs) owned by consumers can be classified into the following three types depending on their properties:

1. Electric appliances that can be controlled freely and generate utility for consumers depending on its electricity consumption, e.g., lights and TVs
2. Electric appliances that require a certain amount of electric power consumption in a day and generate utility for consumers depending on the total amount of electricity consumption, e.g., washers, electric vehicles (EVs), and dryers
3. Electric appliances that generate utility for consumers depending on not only electricity consumption but also temperature, e.g., air conditioners (ACs)

In the following, we individually introduce the utility obtained by consumers in detail according to the types of EAs described above, as shown in [12, 14, 23].

Type 1 EAs (e.g., lights, TVs)

Let $\mathcal{E}_{i,1} \subset \mathcal{E}_i$ be the set of EAs belonging to Type 1 owned by consumer i, and $\mathcal{T}_{ie,1} \subset \mathcal{T}$ be the set of times when this type of EA is used. The total utility that the consumer i obtains by using an EA $e \in \mathcal{E}_{i,1}$ in a day is given by

$$U_{ie,1}\left(\boldsymbol{d}_{ie,1}\right) = \sum_{t \in \mathcal{T}_{ie,1}} v_{ie,1}(d_{ie,1}(t)), \tag{9.47}$$

where $v_{ie,1} : \mathbb{R} \to \mathbb{R}$ is a utility function to represent monetary satisfaction that the consumer i obtains by using the EA $e \in \mathcal{E}_{i,1}$, while $d_{ie,1}(t) \in \mathbb{R}$ is a power demand consumed because of this use and $\boldsymbol{d}_{ie,1} := [d_{ie,1}(0), \ldots, d_{ie,1}(N-1)]^{\mathrm{T}}$. We set the lower and upper bounds of the demand $d_{ie,1}(t)$ at each time to be

$$d_{ie,1}^{\min}(t) \le d_{ie,1}(t) \le d_{ie,1}^{\max}(t), \tag{9.48}$$

where $d_{ie,1}^{\min}(t) \in [0, \infty)$ and $d_{ie,1}^{\max}(t) \in [0, \infty)$ are the lower and upper bounds of the power demand $d_{ie,1}(t)$.

Type 2 EAs (e.g., washers, EVs)

As with Type 1 EAs, let $\mathcal{E}_{i,2} \subset \mathcal{E}_i$ be the set of EAs belonging to Type 2 owned by consumer i, and $\mathcal{T}_{ie,2} \subset \mathcal{T}$ be the set of times when this type of EA is used. The utility obtained by using this type of EA depends on the total amount of power consumption throughout the day. That is, the utility is given by

$$U_{ie,2}(\boldsymbol{d}_{ie,2}) = v_{ie,2}\left(\sum_{t \in \mathcal{T}_{ie,2}} d_{ie,2}(t)\right), \tag{9.49}$$

where $v_{ie,2} : \mathbb{R} \to \mathbb{R}$ is a utility function regarding the usage of EAs $e \in \mathcal{E}_{i,2}$ at time $t \in \mathcal{T}_{ie,2}$, $d_{ie,2}(t) \in \mathbb{R}$ is a demand consumed at that time, and $\boldsymbol{d}_{ie,2} := [d_{ie,2}(0), \ldots, d_{ie,2}(N-1)]^{\mathrm{T}}$. We also set the lower and upper bounds of the demand $d_{ie,2}(t)$ at each time and its total amount, respectively, to be

$$d_{ie,2}^{\min}(t) \le d_{ie,2}(t) \le d_{ie,2}^{\max}(t), \tag{9.50}$$

$$d_{i,e}^{\text{day,min}} \le \sum_{t \in \mathcal{T}_{ie,2}} d_{ie,2}(t) \le d_{i,e}^{\text{day,max}}, \tag{9.51}$$

where $d_{ie,2}^{\min}(t) \in [0, \infty)$ and $d_{ie,2}^{\max}(t) \in [0, \infty)$ are the lower and upper bounds of power demand $d_{ie,2}(t)$, and $d_{i,e}^{\text{day min}} \in [0, \infty)$ and $d_{i,e}^{\text{day max}} \in [0, \infty)$ are those regarding total amount of power demand consumed by using the EA $e \in \mathcal{E}_{i,2}$ throughout the day.

Type 3 EAs (e.g., ACs)

As with the other types, let $\mathcal{E}_{i,3} \subset \mathcal{E}_i$ be the set of EAs belonging to Type 3 owned by consumer i, and $\mathcal{T}_{ie,3} \subset \mathcal{T}$ be the set of times when this type of EA is used.

The utility of this type of EA depends on the temperature. Let us denote the room temperature at time $t \in \mathcal{T}_{ie,3}$ by $T_{ie}^{\text{in}}(t) \in \mathbb{R}$ and also denote the outdoor temperature by $T_{ie}^{\text{out}}(t) \in \mathbb{R}$. According to [23], $T_{ie}^{\text{in}}(t)$ is given by

$$T_{ie}^{\text{in}}(t) = (1 - \alpha)^t T_{ie}^{\text{in}}(-1) + \sum_{\tau=0}^{t}(1 - \alpha)^{t-\tau}\alpha T_{ie}^{\text{out}}(\tau) + \sum_{\tau=0}^{t}(1 - \alpha)^{t-\tau}\beta d_{ie,3}(\tau),$$

(9.52)

where $d_{ie,3}(t) \in \mathbb{R}$ is a power demand consumed by using the EA $e \in \mathcal{E}_{i,3}$ at time $t \in \mathcal{T}_{ie,3}$, and α and β are parameters that represent the thermal characteristics. According to the temperature $T_{ie}^{\text{in}}(t)$ given in (9.52), the utility obtained by using Type 3 EAs is given by

$$U_{ie,3}(\boldsymbol{d}_{ie,3}) = \sum_{t \in \mathcal{T}_{ie,3}} v_{ie,3}\left(T_{ie}^{\text{in}}(t)\right),$$

(9.53)

where $\boldsymbol{d}_{ie,3} := [d_{ie,3}(0), \ldots, d_{ie,3}(N-1)]^{\text{T}}$. We set the lower and upper bounds of the demand $d_{ie,3}(t)$ and those of the temperature $T_{ie}^{\text{in}}(t)$ with which the consumer feels comfortable, respectively, to be

$$d_{ie,3}^{\text{min}}(t) \le d_{ie,3}(t) \le d_{ie,3}^{\text{max}}(t),$$

(9.54)

$$T_{ie}^{\text{min}}(t) \le T_{ie}^{\text{in}}(t) \le T_{ie}^{\text{max}}(t),$$

(9.55)

where $d_{ie,3}^{\text{min}}(t) \in [0, \infty), d_{ie,3}^{\text{max}}(t) \in [0, \infty), T_{ie}^{\text{min}}(t) \in \mathbb{R}$ and $T_{ie}^{\text{max}}(t) \in \mathbb{R}$ are bounds of each variable at time $t \in \mathcal{T}_{ie,3}$.

With the utility regarding each type of EA described so far, the total utility $U_i(\boldsymbol{d}_i)$ that consumer i obtains throughout the day is given by

$$U_i(\boldsymbol{d}_i) = \sum_{e \in \mathcal{E}_{i,1}} U_{ie,1}(\boldsymbol{d}_{ie,1}) + \sum_{e \in \mathcal{E}_{i,2}} U_{ie,2}(\boldsymbol{d}_{ie,2}) + \sum_{e \in \mathcal{E}_{i,3}} U_{ie,3}(\boldsymbol{d}_{ie,3}),$$

(9.56)

where $\boldsymbol{d}_i$ is a vector that consists of power demand $d_{ie,j}(t)$, $j \in \{1, 2, 3\}$, $t \in \mathcal{T}$, consumed by using EAs $e \in \mathcal{E}_i = \{\mathcal{E}_{i,1}, \mathcal{E}_{i,2}, \mathcal{E}_{i,3}\}$. With this total utility, we define the welfare function regarding the consumer i as

$$\mathcal{W}_{\boldsymbol{d}_i}^{+}(\boldsymbol{d}_i, \boldsymbol{\mu}) := U_i(\boldsymbol{d}_i) - \sum_{t=0}^{N-1} \mu(t)\bar{d}_i(t),$$

(9.57)

where $\mu(t)$ is the price for consumers determined by the aggregator and $\boldsymbol{\mu} = [\mu(0), \ldots, \mu(N-1)]^{\text{T}}$. In addition, $\bar{d}_i(t) := \sum_{j \in \{1,2,3\}} \sum_{e \in \mathcal{E}_{i,j}} d_{ie,j}(t)$ is a total amount of the power demand consumed at time t. Based on this welfare function, the optimization problem that each consumer solves to maximize their own profits according to the price μ is given by

$$\max_{d_i} \mathcal{W}_{d_i}^{+}(d_i, \mu), \tag{9.58}$$

$$\text{subject to } \bar{d}_i^{\min}(t) \le \bar{d}_i(t) \le \bar{d}_i^{\max}(t), \ \forall t \in \mathcal{T}, \tag{9.59}$$

$$(9.48), \ \forall t \in \mathcal{T}_{ie,1}, \ \forall e \in \mathcal{E}_{i,1}, \tag{9.60}$$

$$(9.50) \text{ and } (9.51), \ \forall t \in \mathcal{T}_{ie,2}, \ \forall e \in \mathcal{E}_{i,2}, \tag{9.61}$$

$$(9.54) \text{ and } (9.55), \ \forall t \in \mathcal{T}_{ie,3}, \ \forall e \in \mathcal{E}_{i,3}, \tag{9.62}$$

where $\bar{d}_i^{\min}(t) \in [0, \infty)$ and $\bar{d}_i^{\max}(t) \in [0, \infty)$ are the lower and upper bounds of the total amount of power demand consumed at time $t \in \mathcal{T}$, respectively.

We introduce the following assumption regarding the utility functions described in this subsection.

Assumption 7 The utility functions $v_{ie,1}$, $v_{ie,2}$, and $v_{ie,3}$ are in $C^2[0, \infty)$ and strictly concave.

9.3.1.2 Behavior Model of Aggregator

The aggregator is a market player obtaining its profit from the difference between the sale of electricity and its operation cost. To represent this behavior mathematically, we define the welfare function regarding the aggregator as

$$\mathcal{W}_a^{+}(D, \mu, \rho) := \sum_{t=0}^{N-1} \{(\mu(t) - \rho(t)) D(t) - f_D(D(t))\}, \tag{9.63}$$

where $\rho(t) \in \mathbb{R}$ and $D(t) \in \mathbb{R}$ are the price determined by the ISO and an amount of the power purchased from generators and the ISO at time $t \in \mathcal{T}$, respectively. In addition, $\rho := [\rho(0), \ldots, \rho(N-1)]^{\mathrm{T}}$, $D := [D(0), \ldots, D(N-1)]^{\mathrm{T}}$, and $f_D : \mathbb{R} \to \mathbb{R}$ is a cost function regarding the operation cost on power trading. Based on this welfare function, the optimization problem that the aggregator solves to maximize its own welfare becomes

$$\max_{D} \mathcal{W}_a^{+}(D, \mu, \rho), \tag{9.64}$$

$$\text{subject to } D^{\min}(t) \le D(t) \le D^{\max}(t), \ \forall t \in \mathcal{T}, \tag{9.65}$$

where $D^{\min}(t) \in [0, \infty)$ and $D^{\max}(t) \in [0, \infty)$ are the lower and upper bounds of $D(t)$.

We introduce the following assumption regarding this cost function.

Assumption 8 The cost function f_D is in $C^2[0, \infty)$, strictly increasing and strictly convex.

9.3.1.3 Behavior Model of Generator

Based on the behavior model of generators given in (9.4) described in Sect. 9.2, we define the welfare function regarding the generator i throughout the day as

$$\mathcal{W}_s^+(s, \rho) := \sum_{t=0}^{N-1} \{\rho(t)s(t) - c(s(t))\}, \tag{9.66}$$

where $s := [s(0), \ldots, s(N-1)]^{\mathrm{T}}$ and $c : \mathbb{R} \to \mathbb{R}$ is a cost function satisfying Assumption 2. The optimization problem that each generator solves to maximize their own welfare becomes

$$\max_s \mathcal{W}_s^+(s, \rho), \tag{9.67}$$

$$\text{subject to } s^{\min}(t) \leq s(t) \leq s^{\max}(t), \ \forall t \in \mathcal{T}, \tag{9.68}$$

where $s^{\min}(t) \in [0, \infty)$ and $s^{\max}(t) \in [0, \infty)$ are the lower and upper bounds of $s(t)$.

9.3.2 Distributed Maximization of Social Welfare by Adjusting Power Demand of Consumers in a Day-Ahead Market

9.3.2.1 Design of Social Welfare Function Considering the Participation of Aggregator

Similar to the social welfare maximization problem described in Sect. 9.2, we consider maximizing social welfare including the profit of the aggregator as well as the other market players through market trading. For this purpose, let $d_{\mathcal{H}}$ be the vector that consists of power demand $d_i, i \in \mathcal{H}$. Then, we define the social welfare function with the utility and cost of each market player including the aggregator as

$$\mathcal{W}^+(d_{\mathcal{H}}, D, s, \sigma) := \sum_{i \in \mathcal{H}} U_i(d_i) - \sum_{t=0}^{N-1} \{f_D(D(t)) - c(s(t)) - c_L(\sigma(t))\}, \tag{9.69}$$

where $\sigma(t) \in \mathbb{R}$ and $c_L : \mathbb{R} \to \mathbb{R}$ are power supply from large-scale power-generating facilities and its operating cost function satisfying Assumption 5, respectively, as defined in Sect. 9.2, and $\sigma := [\sigma(0), \ldots, \sigma(N-1)]^{\mathrm{T}}$. According to this welfare function, the optimization problem to be solved in a day-ahead market is given by

$$\max_{d_{\mathcal{H}}, D, s, \sigma} \mathcal{W}^+ (d_{\mathcal{H}}, D, s, \sigma), \tag{9.70}$$

subject to (9.59)–(9.62), (9.65), (9.68),

$$s(t) + \sigma(t) - D(t) = 0, \ \forall t \in \mathcal{T}, \tag{9.71}$$

$$D(t) - \sum_{i \in \mathcal{H}} \bar{d}_i(t) = 0, \ \forall t \in \mathcal{T}, \tag{9.72}$$

where the constraint (9.71) represents the power balance between the aggregator and generators, and also, (9.72) represents that between the aggregator and consumers.

9.3.2.2 Distributed Optimal Pricing Algorithm and Convergence Analysis

To solve the optimization problem (9.70)–(9.72) in a distributed manner through market trading, we define the partially augmented Lagrange function $\hat{\mathcal{L}}^+$ as

$$\hat{\mathcal{L}}^+ (d_{\mathcal{H}}, D, s, \sigma, \rho_0, \mu_0)$$

$$= \mathcal{W}^+ (d_{\mathcal{H}}, D, s, \sigma) + \rho_0^{\mathrm{T}} (s + \sigma - D) + \mu_0^{\mathrm{T}} \left(D - \sum_{i \in \mathcal{H}} d_i \right)$$

$$- \frac{\iota}{2} \sum_{t=0}^{N-1} \| s(t) + \sigma(t) - D(t) \|^2 - \frac{\varepsilon}{2} \sum_{t=0}^{N-1} \left\| D(t) - \sum_{i \in \mathcal{H}} \bar{d}_i(t) \right\|^2, \tag{9.73}$$

where $\rho_0 = [\rho_0(0), \dots, \rho_0(N-1)]^{\mathrm{T}}$ and $\mu_0 = [\mu_0(0), \dots, \mu_0(N-1)]^{\mathrm{T}}$ are the Lagrange multipliers associated with the constraints (9.71) and (9.72), respectively. In addition, ι and ε are positive constant parameters. As we have shown in Sect. 9.2, the Lagrange multipliers can be regarded as the prices in market trading, and thus, we use the notations ρ and μ instead of ρ_0 and μ_0 in the rest of this section.

Similar to the market trading algorithm for the dynamic electricity pricing given in Algorithm 1 proposed in the previous section, we show a price decision procedure to derive the optimal prices ρ^* and μ^* that induce market players including an aggregator to behave optimally and maximize the social welfare $\mathcal{W}^+$ given in (9.69) through market trading. The proposed algorithm is given in Algorithm 2. In this algorithm, two kinds of prices: μ and ρ, are updated by the aggregator and the ISO, respectively, while each market player determines their decision variables alternately. Because of this alternating decision making of market players, the ISO and the aggregator update their prices based on Algorithm 2 without knowing any information of the utility and cost functions regarding the other market players.

We also show the convergence property of the proposed algorithm. Let us denote the residual of the power balance between generators and the aggregator by $\delta^k := s^k + \sigma^k - D^k$, and that between the aggregator and consumers by

Algorithm 2 Distributed price decision algorithm for the optimal demand adjustment of consumers in a day-ahead market

1: The ISO sets the initial value of the electricity price $\boldsymbol{\rho}^0 = [\rho^0(0), \dots, \rho^0(N-1)]^{\mathrm{T}}$ arbitrarily and transmits it to the aggregator and generators. Also, the aggregator sets the initial value of the electricity price $\boldsymbol{\mu}^0 = [\mu^0(0), \dots, \mu^0(N-1)]^{\mathrm{T}}$ arbitrarily and transmits it to consumers, while receiving the initial value of power supply $\boldsymbol{s}^0 = [s^0(0), \dots, s^0(N-1)]^{\mathrm{T}}$ from generators. Furthermore, the ISO set the initial value of power supply $\boldsymbol{\sigma}^0 = [\sigma^0(0), \dots, \sigma^0(N-1)]^{\mathrm{T}}$ of large scale power-generating facilities and transmits it to generators and the ISO.

2: Consumers determine the power demand $\boldsymbol{d}_i^{k+1}$:

$$\boldsymbol{d}_i^{k+1} = \arg\max_{\boldsymbol{d}_i} \; \mathcal{W}_{d_i}^+ \left(\boldsymbol{d}_i, \boldsymbol{\mu}^k \right), \; \text{subject to (9.59)--(9.62)}, \; i \in \mathcal{H}, \tag{9.74}$$

and report it to the aggregator.

3: The aggregator determines an amount of the power $\boldsymbol{D}^{k+1}$ purchased from the generators:

$$\boldsymbol{D}^{k+1} = \arg\max_{\boldsymbol{D}} \; \mathcal{W}_a^+ \left(\boldsymbol{D}, \boldsymbol{\mu}^k, \boldsymbol{\rho}^k \right) - \frac{\iota}{2} \sum_{t=0}^{N-1} \left\| s^k(t) + \sigma^k(t) - D(t) \right\|^2 \tag{9.75}$$

$$- \frac{\varepsilon}{2} \sum_{t=0}^{N-1} \left\| D(t) - \sum_{i \in \mathcal{H}} \bar{d}_i^{k+1}(t) \right\|^2, \; \text{subject to (9.65)},$$

where $\bar{d}_i^{k+1}(t) := \sum_{j \in \{1,2,3\}} \sum_{e \in \mathcal{E}_{i,j}} d_{ie,j}^{k+1}(t)$, and report it to generators through the ISO.

4: Generators determine the power supply $\boldsymbol{s}^{k+1}$:

$$\boldsymbol{s}^{k+1} = \arg\max_{\boldsymbol{s}} \; \mathcal{W}_s^+ \left(\boldsymbol{s}, \boldsymbol{\rho}^k \right) - \frac{\iota}{2} \sum_{t=0}^{N-1} \left\| s(t) + \sigma^k(t) - D^{k+1}(t) \right\|^2, \; \text{subject to (9.68)},$$

$$\tag{9.76}$$

and report it to the ISO.

5: The ISO determines the power supply of large scale power-generating facilities $\boldsymbol{\sigma}^{k+1}$:

$$\boldsymbol{\sigma}^{k+1} = \arg\max_{\boldsymbol{\sigma}} \sum_{t=0}^{N-1} \left\{ \rho^k(t)\sigma(t) - c_L(\sigma(t)) - \frac{\iota}{2} \left\| s^{k+1}(t) + \sigma(t) - D^{k+1}(t) \right\|^2 \right\}. \tag{9.77}$$

6: If the power balance between the power purchased by the aggregator and power supply from generators and large scale power-generating facilities and that between the power purchased by the aggregator and the sum of demand of consumers, which are, respectively, given by

$$s^{k+1} + \sigma^{k+1} - D^{k+1} = 0, \quad D^{k+1} - \sum_{i \in \mathcal{H}} d_i^{k+1} = 0, \tag{9.78}$$

are satisfied, this market trading algorithm is terminated, and the ISO and the aggregator set the prices $\boldsymbol{\rho}^k$ and $\boldsymbol{\mu}^k$ to their actual prices, respectively. If the power balances given in (9.78) are not satisfied, they update their prices $\boldsymbol{\rho}^k$ and $\boldsymbol{\mu}^k$, respectively, according to

$$\boldsymbol{\rho}^{k+1} = \boldsymbol{\rho}^k - \iota \left(s^{k+1} + \sigma^{k+1} - D^{k+1} \right), \quad \boldsymbol{\mu}^{k+1} = \boldsymbol{\mu}^k - \varepsilon \left(D^{k+1} - \sum_{i \in \mathcal{H}} d_i^{k+1} \right), \tag{9.79}$$

and transmit the updated price $\boldsymbol{\rho}^{k+1}$ to the aggregator and generators, and the updated price $\boldsymbol{\mu}^{k+1}$ to consumers.

7: Change the iteration number k to $k+1$ and go back to 2.

$\xi^k := D^k - \sum_{i \in \mathcal{H}} d_i^k$, where k is the number of iterations in Algorithm 2. We obtain the following proposition with respect to the convergence property of Algorithm 2:

Proposition 1 *Let Assumptions 2, 5, 7, and 8 be held. If the step sizes ι and ε in (9.79), respectively, satisfy*

$$0 < \iota < \frac{2(\rho^0 - \rho^*)^{\mathrm{T}} \delta^{k+1}}{(\delta^{k+1} + 2\sum_{l=1}^{k} \delta^l)^{\mathrm{T}} \delta^{k+1}}, \quad \forall k = 0, 1, \ldots, \tag{9.80}$$

$$0 < \varepsilon < \frac{2(\mu^0 - \mu^*)^{\mathrm{T}} \xi^{k+1}}{(\xi^{k+1} + 2\sum_{l=1}^{k} \xi^l)^{\mathrm{T}} \xi^{k+1}}, \quad \forall k = 0, 1, \ldots, \tag{9.81}$$

then the electricity prices ρ^k and μ^k updated according to Algorithm 2 converge to ρ^* and μ^*, respectively, as $k \to \infty$.

The proof of this proposition is omitted since it is easily proven in the same way as Theorem 2.

9.3.3 Simulation Verification

We show the validity and effectiveness of the proposed market trading algorithm in a day-ahead market through numerical simulation.

9.3.3.1 Simulation Conditions

In this verification, we use a day-ahead market model in which eight households of consumers, generators, an aggregator and the ISO participate in market trading. Specifically, we divide those eight households into two groups according to their lifestyles: four households staying at home all day are Group A and the rest of four households going out from 8:00 to 18:00 are Group B. Furthermore, we assume that every household has five kinds of EAs (lighting, TV, EV, washer, and AC), and we number each of them as $e = 1$ is lighting, $e = 2$ is TV, $e = 3$ is EV, $e = 4$ is a washer, and $e = 5$ is AC.

Design of utility functions

We set the utility function regarding the Type 1 EAs to be

$$v_{ie,1}(d_{ie,1}(t)) = c_{ie} - b_{ie} \left(d_{ie,1}(t) - d_{ie}^{\mathrm{pref}}(t) \right)^2, \tag{9.82}$$

where $d_{ie}^{\mathrm{pref}}(t)$ is the preferable power demand that consumers obtain the maximum utility at time t, and b_{ie} and c_{ie} are positive constants. In addition, all households use lightings for $\mathcal{T}_{i1,1} = \{18, \ldots, 23\}$, and those in Group A use TVs for $\mathcal{T}_{i2,1} = \{6, \ldots, 23\}$, while those in Group B use TVs for $\mathcal{T}_{i2,1} = \{18, \ldots, 24\}$.

Also, we set the utility function regarding the Type 2 EAs to be

$$v_{ie,2} \left(\sum_{t \in \mathcal{T}_{ie,2}} d_{ie,2}(t) \right) = b_{ie} \log \left(\sum_{t \in \mathcal{T}_{ie,2}} d_{ie,2}(t) + 1 \right). \tag{9.83}$$

All households charge EVs for $\mathcal{T}_{i3,2} = \{18, \ldots, 24, 0, \ldots, 7\}$. In addition, we assume that households in Group A can use washers throughout the day, while those in Group B can only use them for $\mathcal{T}_{i4,2} = \{18, \ldots, 24, 0, \ldots, 7\}$.

As with the other types, we set the utility function regarding the Type 3 EAs to be

$$v_{ie,3} \left(T_{ie}^{\text{in}}(t) \right) = c_{ie} - b_{ie} \left(T_{ie}^{\text{in}}(t) - T_{ie}^{\text{comf}}(t) \right)^2. \tag{9.84}$$

In this verification, we use the temperature data in Tokyo, Japan on July 3, 2016, given in [8] as the outdoor temperatures. We assume that households in Group A use ACs throughout the day, while those in Group B only use them for $\mathcal{T}_{i4,3} = \{18, \ldots, 24, 0, \ldots, 7\}$ according to their lifestyles. The details of each parameter are listed in Table 9.2.

Design of cost function

On the other hand, we set the cost functions regarding an aggregator, generators, and large-scale power-generating facilities managed by the ISO, respectively, to be

$$\text{Aggregator's cost function: } f_D \left(D(t) \right) = b_D D^2(t), \tag{9.85}$$

$$\text{Generators' cost function: } c \left(s(t) \right) = b_S s^2(t), \tag{9.86}$$

$$\text{ISO's cost function: } c_L \left(\sigma(t) \right) = b_L \sigma^2(t), \tag{9.87}$$

where the coefficients b_D, b_S, and b_L are positive constants.

9.3.3.2 Simulation Results

Figures 9.9a, b and 9.10 show the simulation results of the hourly power demand consumed by four households in Group A, those in Group B, and the sum of their demand, respectively. In these figures, the horizontal axes represent the time and each line shows the power demand consumed by using each EA at each time and its total amount in four or eight households, respectively. These figures show that consumers adjust their demand while satisfying the constraints on the use of EAs listed in Table 9.2. Also, Fig. 9.11 shows the results of two kinds of prices obtained through market trading with our proposed price decision method. From this figure, it is shown that both kinds of prices are changed depending on time.

Table 9.2 Simulation parameters

Parameter [unit]		Symbol	Value
Number of time zones		N	25
Initial electricity price [yen/kWh]		ρ_f	25, $\forall t \in \mathcal{T}$
Lower and upper bounds of supply [kW]		$[s^{\min}, s^{\max}]$	$[0, \infty]$
Lighting	Lower and upper bounds of demand [kWh]	$[d_{i1,1}^{\min}, d_{i1,1}^{\max}]$	$[0.2, 0.8]$
(Type 1)	Preferable demand [kW]	d_{i1}^{pref}	0.6
TV	Lower and upper bounds of demand [kWh]	$[d_{i2,1}^{\min}, d_{i2,1}^{\max}]$	$[0.1, 0.5]$
(Type 1)	Preferable demand [kW]	d_{i2}^{pref}	0.3
EV	Lower and upper bounds of demand [kWh]	$[d_{i3,2}^{\min}, d_{i3,2}^{\max}]$	$[0, 1.5]$
(Type 2)	Lower and upper bounds of total demand [kWh]	$[d_{i3}^{\text{day},\min}, d_{i3}^{\text{day},\max}]$	$[5.1, 6.0]$
Washer	Lower and upper bounds of demand [kWh]	$[d_{i4,2}^{\min}, d_{i4,2}^{\max}]$	$[0, 1.0]$
(Type 2)	Lower and upper bounds of total demand [kWh]	$[d_{i4}^{\text{day},\min}, d_{i4}^{\text{day},\max}]$	$[1.6, 2.5]$
Lower and upper bounds of temperature [°C]		$[T_{i5}^{\min}, T_{i5}^{\max}]$	$[25, 28]$
Initial indoor temperature [°C]		$T_{i5}^{\text{in}}(-1)$	27
AC	Lower and upper bounds of demand [kWh]	$[d_{i5,3}^{\min}, d_{i5,3}^{\max}]$	$[0, 4.0]$
(Type 3)	Coefficient of heat transfer	α	0.9
	Coefficient of thermal efficiency	β	-1.2×10^{-3}
Step size		ι	7.0×10^{-7}
Step size		ε	9.0×10^{-7}

Next, Fig. 9.12a, b shows the results of the convergence with respect to the residuals δ^k and ξ^k of the power balances during market trading with our proposed algorithm, respectively. Note that the horizontal axes in these figures represent the number of iterations, which is denoted by k in Algorithm 2. These figures show that both residuals converge to 0 after a sufficient number of iterations have passed.

Finally, Fig. 9.13a, b show the results of the convergence with respect to the price $\rho^k(t)$ at $t = 14$ by using the proposed method given in Algorithm 2 and that by using a conventional gradient-based one, respectively, where we use a grater step size $\iota = 3.0 \times 10^{-6}$ in both methods. From these two results, it is shown that the price converges with our proposed method even if we use a grater step size that cannot achieve the convergence with a conventional gradient-based one. This result concludes that the proposed pricing algorithm given in this section also enables us to improve its convergence speed as well as to maximize the social welfare in a distributed manner with information privacy of market players through market trading.

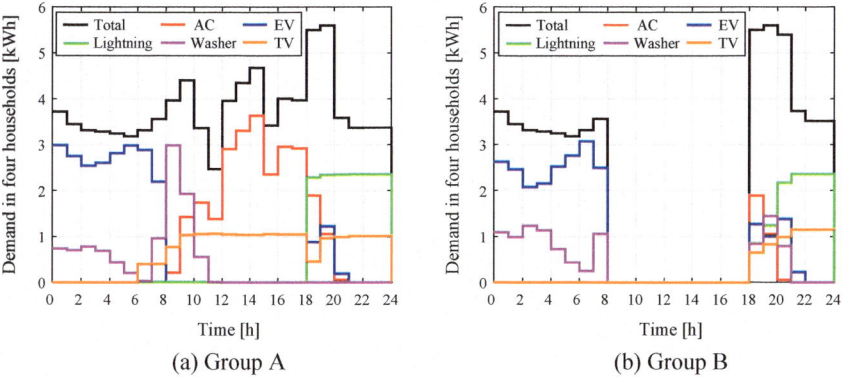

Fig. 9.9 Hourly power demand consumed by using each electric appliance

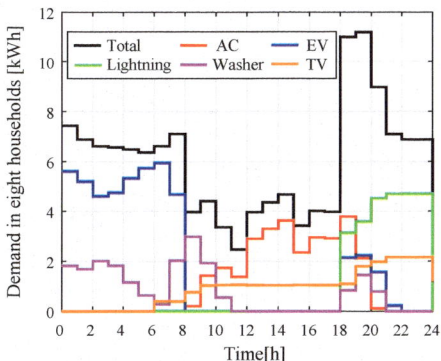

Fig. 9.10 Total power demand

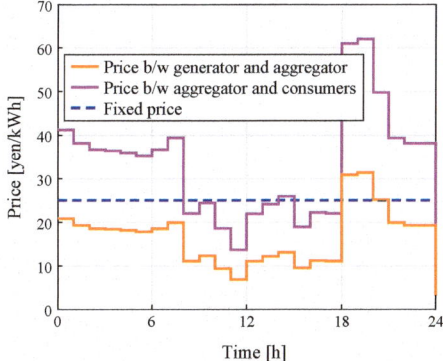

Fig. 9.11 Hourly electricity prices

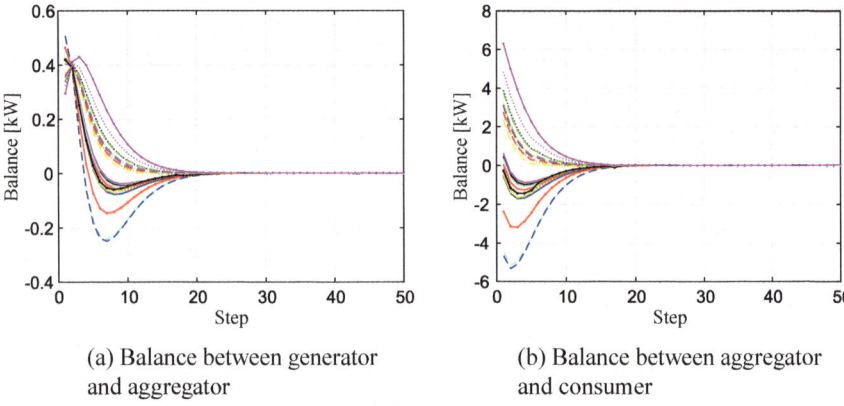

(a) Balance between generator   (b) Balance between aggregator
and aggregator   and consumer

Fig. 9.12 Results of convergence of power balance in market trading

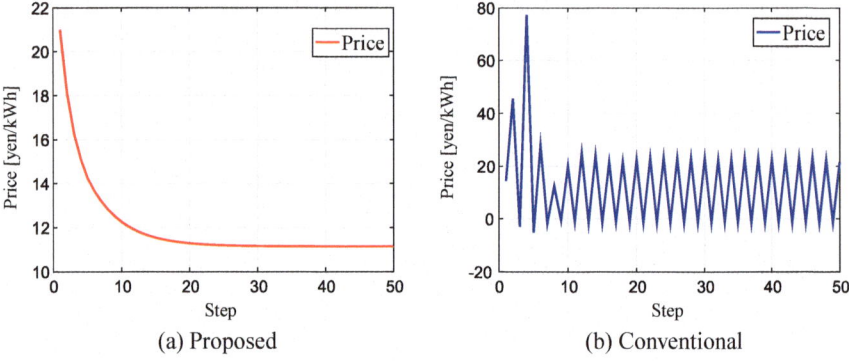

(a) Proposed   (b) Conventional

Fig. 9.13 Results of convergence of the price in market trading

9.4 Conclusion

In this chapter, we have introduced distributed price decision procedures regarding the dynamic pricing through market trading to derive the optimal prices that maximize social welfare with information privacy of power consumers and generators. Specifically, we have first dealt with an electricity market that covers some regional areas in a power grid and proposed a market trading algorithm for the dynamic electricity pricing based on the alternating decision making of market players. This price decision method enables us to maximize social welfare of the entire power grid in a distributed manner with the satisfaction of the physical constraints regarding a power grid including power flow among areas. Subsequently, we have dealt with an electricity market in one specific regional area where an aggregator participates in its trading as a mediator between the market operator and consumers. For this market model, we have designed the behavior model of consumers with the utility depending on the types of their electric appliances to represent their lifestyles more precisely and proposed a market trading algorithm to adjust their demand in a distributed manner including the participation of an aggregator in a day-ahead market.

This chapter has also shown the convergence properties of our proposed market trading algorithms for each market model, respectively. Furthermore, we have verified the validity and effectiveness of these proposed algorithms through numerical simulation by showing that they enable us not only to derive the optimal electricity prices to maximize social welfare regarding each electricity market in a distributed manner but also to improve their convergence speed with less fluctuation, compared with those by using conventional gradient-based ones.

References

1. Arrow KJ, Hurwicz L, Uzawa H (1958) Studies in linear and nonlinear programming. Stanford University Press, Palo Alto
2. Boyd S, Parikh N, Chu E, Peleato B, Eckstein J (2011) Distributed optimization and statistical learning via the alternating direction method of multipliers. Found Trends® Mach Learn 3(1):1–122
3. Chasparis GC, Rantzer A, Jörnsten K (2013) A decomposition approach to multi-region optimal power flow in electricity networks. In: Proceedings of 2013 European control conference, pp 3018–3024
4. Conejo AJ, Aguado JA (1998) Multi-area coordinated decentralized DC optimal power flow. IEEE Trans Power Syst 13(4):1272–1278
5. Dall'Anese E, Zhu H, Giannakis GB (2013) Distributed optimal power flow for smart microgrids. IEEE Trans Smart Grid 4(3):1464–1475
6. Erseghe T (2014) Distributed optimal power flow using ADMM. IEEE Trans Power Syst 29(5):2370–2380
7. Gkatzikis L, Koutsopoulos I, Salonidis T (2013) The role of aggregators in smart grid demand response markets. IEEE J Sel Areas Commun 31(7):1247–1257
8. Japan Meteorological Agency (2016) Temperature data of Tokyo. http://www.jma.go.jp/jma/indexe.html. Accessed 17 July 2019
9. Jokic A, Lazar M, van den Bosch PPJ (2009) Real-time control of power systems using nodal prices. Int J Electr Power Energy Syst 31(9):522–530
10. Joskow PL, Wolfram CD (2012) Dynamic pricing of electricity. Am Econ Rev 102(3):381–385
11. Kundur P (1994) Power system stability and control. McGraw-Hill Professional, New York
12. Li N, Chen L, Low SH (2011) Optimal demand response based on utility maximization in power networks. In: Proceedings of 2011 IEEE power and energy society general meeting, pp 1–8
13. Mas-Colell A, Whinston MD, Green JR (1995) Microeconomic theory. Oxford University Press, Oxford
14. Muto K, Okawa Y, Namerikawa T (2018) Optimal demand adjustment of consumers with various appliances using dynamic pricing. SICE J Control Meas Syst Integr 11(3):146–153
15. Namerikawa T, Okubo N, Sato R, Okawa Y, Ono M (2015) Real-time pricing mechanism for electricity market with built-in incentive for participation. IEEE Trans Smart Grid 6(6):2714–2724
16. Okawa Y, Muto K, Namerikawa T (2018) Passivity-based stability analysis of electricity market trading with dynamic pricing. SICE J Control Meas Syst Integr 11(5):390–398
17. Okawa Y, Namerikawa T (2014) Dynamic pricing considering constraints of power grids. In: Proceedings of SICE annual conference 2014, pp 1484–1489

18. Okawa Y, Namerikawa T (2016) Multi-period regional energy management based on dynamic pricing with non-deficit real-time market trading. SICE J Control Meas Syst Integr 9(5):207–215
19. Okawa Y, Namerikawa T (2016) Optimal power demand management among consumers with aggregator considering state and control constraints. In: Proceedings of IEEE 55th conference on decision and control, pp 801–806
20. Okawa Y, Namerikawa T (2017) Distributed optimal power management via negawatt trading in real-time electricity market. IEEE Trans Smart Grid 8(6):3009–3019
21. Roozbehani M, Dahleh M, Mitter S (2010) Dynamic pricing and stabilization of supply and demand in modern electric power grids. In: Proceedings of 2010 First IEEE international conference on smart grid communications, pp 543–548
22. Samadi P, Mohsenian-Rad AH, Schober R, Wong VWS, Jatskevich J (2010) Optimal real-time pricing algorithm based on utility maximization for smart grid. In: Proceedings of 2010 First IEEE international conference on smart grid communications, pp 415–420
23. Shi W, Li N, Xie X, Chu CC, Gadh R (2014) Optimal residential demand response in distribution networks. IEEE J Sel Areas Commun 32(7):1441–1450
24. The Institute of Electrical Engineers of Japan (2007) Japanese Power System Model. http://www2.iee.or.jp/ver2/pes/23-st_model/english/index.html. Accessed 17 July 2019
25. Warrington J, Goulart P, Mariéthoz S, Morari M (2012) A market mechanism for solving multi-period optimal power flow exactly on AC networks with mixed participants. In: Proceedings of 2012 American control conference, pp 3101–3107
26. Wei E, Ozdaglar A (2012) Distributed alternating direction method of multipliers. In: Proceedings of 51st IEEE conference on decision and control, pp 5445–5450

Chapter 10
Real-Time Pricing for Electric Power Systems by Nonlinear Model Predictive Control

Toshiyuki Ohtsuka, Yu Kawano, Tomoaki Hashimoto, Yusuke Okajima
and Kenji Kashima

Abstract This chapter discusses real-time pricing for electric power systems by using nonlinear model predictive control (NMPC). In NMPC, the price of electricity is regarded as a control input, and it is optimized in real time to achieve supply–demand balance and load frequency regulation. The chapter describes the formulation of several problems on the basis of different models of consumers, suppliers, and renewable energy resources. Moreover, it shows that when NMPC is used in combination with nonlinear estimation and/or stochastic optimization, it can take into account inherent uncertainties in consumer characteristics and renewable energy source. Numerical simulations show that NMPC provides a flexible and effective framework for real-time pricing with realistic computational costs.

T. Ohtsuka (✉) · K. Kashima
Graduate School of Informatics, Kyoto University, Yoshida-honmachi, Sakyo-ku,
Kyoto 606-8501, Japan
e-mail: ohtsuka@i.kyoto-u.ac.jp

K. Kashima
e-mail: kashima@amp.i.kyoto-u.ac.jp

Y. Kawano
Graduate School of Engineering, Hiroshima University, 1-4-2 Kagamiyama,
Higashi-Hiroshima-shi, Hiroshima 739-8527, Japan
e-mail: ykawano@hiroshima-u.ac.jp

T. Hashimoto
Faculty of Engineering, Osaka Institute of Technology, 5-16-1 Omiya, Asahi-ku,
Osaka 535-8585, Japan
e-mail: tomoaki.hashimoto@oit.ac.jp

Y. Okajima
Graduate School of Education, Joetsu University of Education, 1 Yamayashiki-machi,
Joetsu-shi, Niigata 943-8512, Japan
e-mail: okajima@juen.ac.jp

© Springer Nature Singapore Pte Ltd. 2020
T. Hatanaka et al. (eds.), *Economically Enabled Energy Management*,
https://doi.org/10.1007/978-981-15-3576-5_10

10.1 Introduction

To make effective use of limited energy resources, the trend has been toward greater deregulation of the power industry and liberalization of the power market. This trend has made itself felt in Japan, where momentum is growing for large-scale introduction of renewable energy [22]. Of the various renewable energy sources, wind power and solar power can be introduced relatively easily on a large scale, and they actually have accounted for the majority of renewable energy generation in recent years [13]. However, they have the disadvantage that the amount of power they generate depends on the weather conditions. In an AC system, it is necessary to maintain the balance of supply and demand of electric power and maintain the frequency within a reference range. If the amount of renewable energy on the grid is small, it is possible to cope with the output fluctuation by adjusting the output of the existing generators. However, as the amount of renewable energy increases, it becomes more difficult to cope with the output fluctuation with only the generators, and the impact of an accident on the grid also increases [4].

With the above background in mind, various studies have been conducted on the operation and management of new power grid systems. One line of research is on real-time pricing. Real-time pricing is a variable pricing strategy, and it is a generic term for a scheme with a relatively short time scale in which the electricity price fluctuates from moment to moment. Other variable pricing strategies include critical peak pricing and time-of-use pricing. It has been reported that real-time pricing is most effective in terms of smoothing the load curve of electricity [6, 7]. Therefore, in this chapter, we will consider optimal control of power systems by real-time pricing.

The question of what kind of policy should be used to set the price of electricity in consideration of social welfare has been studied on the basis of the principle that participants in a market change their behaviors selfishly in response to dynamic changes in electricity prices [3, 26]. However, if we treat this problem only on relatively long time scales, any solution would not account for frequency fluctuations in the power system. Nor would it consider nonlinearities or constraints affecting the frequency fluctuations.

On the other hand, the demand response has been considered as a way of dealing with fluctuations of renewable energy besides adjusting the supply by using generators and batteries. Demand response means that not only the supply side but also the demand side adjusts the supply–demand balance. There are two approaches [5]. One is direct control of the power consumed by the consumers' equipment; the other is real-time pricing to indirectly reduce the power consumption by changing the electricity price presented to the consumer. Real-time pricing is more preferable for consumers than direct control because consumers can determine their own power consumption according to the electricity price and demand situation. In particular, the literature describes real-time pricing methods that maximize the utility of customers and suppliers under the constraint of supply–demand balance [15, 21].

This chapter formulates the real-time pricing problem as nonlinear model predictive control (NMPC), which is a kind of optimal control, to determine the electricity price as a control input to a power system, and it discusses a method for dealing with nonlinearities and constraints. In Sect. 10.2, for a power system consisting of consumers, suppliers, generators, and an independent system operator (ISO), an objective function is defined taking into account social welfare and load frequency fluctuation. Then, the basic concept of NMPC-based real-time pricing is verified through a numerical simulation from the viewpoint of stabilizing the load frequency and shortening the computation time enough for real-time optimization. In Sect. 10.3, we consider online estimation of the consumers' model and integrate the online estimation into real-time pricing by NMPC. The consumers' response to the electricity price depends on the weather, time, and season, and it is necessary to update the consumers' model online and changes to the electricity price. We describe a method to optimize the electricity price by using NMPC; a particle filter is used to estimate the parameters of the demand function online. In Sect. 10.4, real-time pricing is formulated as a stochastic model predictive control (SMPC) that considers the fluctuation of renewable energy explicitly as a stochastic disturbance and deals with the constraint on frequency fluctuation as a chance constraint. It is shown that both constraint satisfaction and improved control performance can be achieved by explicitly considering stochastic disturbances and chance constraints.

10.2 Maximization of Social Welfare

10.2.1 Objective

We will consider the problem of controlling the load frequency of an electric power system that consists of consumers, suppliers, generators, and an ISO. In [3], real-time pricing was used to assist in the control of the load frequency of such a system. Since real-time pricing is considered to be especially useful for reducing the peak load and for flattening the load curve, we will focus on using it for stabilizing the load frequency deviation.

Reference [26] describes a framework for modeling and analyzing electric power systems composed of consumers, suppliers, and an ISO. The ISO in this case is a nonprofit entity whose primary role is to optimally determine electricity prices subject to network and operational constraints. The authors analyzed real-time price optimization for maximizing the benefits of both consumers and suppliers subject to a constraint on the balance between supply and demand. However, they did not model the generator dynamics with frequency fluctuations, which would be needed for real-time price optimization.

The generator dynamics in power networks can be described by nonlinear coupled swing equations. The problem of controlling such dynamics by using real-time pricing has been studied in [3]. However, in that study, the swing equations were linearized. Although linearized swing equations are useful for considering coherent swing instabilities, they cannot be used to account for non-coherent instabilities, for which the nonlinearity of the swing equations cannot be neglected. Hence, in this section, we model the generator dynamics as nonlinear coupled swing equations.

In particular, we will describe an optimization-based real-time pricing method for stabilizing the nonlinear coupled swing equations of the load frequency deviations and for maximizing social welfare (which means maximizing the benefits of both consumers and suppliers). For this purpose, we will apply the model predictive control approach.

Model predictive control (MPC), also called receding horizon control, is a well-established control method in which the current control input is obtained by solving a finite-horizon open-loop optimal control problem using the current state of the system as the initial state; the procedure is repeated at each sampling instant. Thus, MPC is a kind of optimal feedback control in which the control performance over a finite future is optimized with an objective function that has a moving initial time and terminal time. Although NMPC, i.e., MPC for nonlinear systems, requires a numerical solution of a nonlinear optimal control problem to be computed in real time, a fast numerical algorithm tailored for NMPC has been established in [24]. Below, we describe an optimal electric price operation based on NMPC for achieving a steady state by taking into account the benefits of both consumers and suppliers. Interested readers may consult [16], which describes the method in detail.

10.2.2 Power System Model

We will consider an electric power system (EPS) composed of n areas connected by AC networks without loops. Each area has a unique consumer, supplier, and generator. The EPS is exclusively operated by a nonprofit and neutral organization, the ISO. The ISO decides electricity prices to maximize social welfare, as well as to balance supply and demand while stabilizing the frequencies of the generators. We will consider a real-time price optimization problem in which consumers determine their consumption and the suppliers determine the supply amounts to maximize their benefits based on the electricity prices provided by the ISO.

10.2.2.1 Consumer and Supplier Models

Suppose that the consumer (or supplier) in area i has a utility function $v_i(z)$ (or cost function $c_i(z)$) which represents the profit from using z units of electricity (or the cost required to generate z units of electricity). All consumers (or suppliers) decide on the amount they will consume (or the amount of power they will generate) in response to the electricity prices given by the ISO so as to maximize their benefits. Thus, the electricity demand D_i and supply S_i in area i are, respectively, determined by:

$$D_i = \arg\max_{z}\{v_i(z) - \lambda_i z\}, \tag{10.1}$$

$$S_i = \arg\max_{z}\{\lambda_i z - c_i(z)\}, \tag{10.2}$$

where λ_i is the electricity price of area i. Here, we assume that the utility (or cost) functions are $C^2(0, \infty)$ functions, strictly increasing, and strictly concave (or convex) for all i. Accordingly, D_i and S_i are uniquely determined as functions of the price:

$$D_i(\lambda_i) = (dv_i/d\lambda_i)^{-1}(\lambda_i), \tag{10.3}$$

$$S_i(\lambda_i) = (dc_i/d\lambda_i)^{-1}(\lambda_i). \tag{10.4}$$

10.2.2.2   Generator Model

The swing equations of the generators in the power system composed of consumers and suppliers are given as follows:

$$\frac{d\theta_i(t)}{dt} = \tilde{\omega}_i(t), \tag{10.5}$$

$$M_i \frac{d\tilde{\omega}_i(t)}{dt} = -D_i(t) + S_i(t) + \sum_{j \in \mathcal{N}_i} V_i V_j B_{ij} \sin(\theta_j(t) - \theta_i(t)), \tag{10.6}$$

where the system parameters are defined in Table 10.1.

To simplify the controller design, we compute nondimensionalized equations of (10.5) and (10.6) under the assumption that $V_i = V$, $M_i = M$, and $B_{ij} = B$ for all i, j. After changing the variable $\tau = \omega_0 t$, from (10.5) and (10.6), we have

$$M\omega_0^2 \frac{d^2\theta_i(\tau)}{d\tau^2} = -D(t) + S(t) + \sum_{j \in \mathcal{N}_i} V^2 B \sin(\theta_j(\tau) - \theta_i(\tau)). \tag{10.7}$$

Table 10.1 System parameters (Reprinted from [16] Copyright 2014 ISCIE)

θ_i	Phase difference angle in area i
$\tilde{\omega}_i$	Phase difference angular velocity in area i
D_i	Power demand by consumers in area i
S_i	Power supply by suppliers in area i
R_i	Power supply by renewable energy in area i
M_i	Inertia constant in area i
V_i	Voltage in area i
B_{ij}	Susceptance of transmission line between areas i and j
$\mathcal{N}_i$	Set of areas connecting area i

Hereafter, $d\theta_i(\tau)/d\tau$ will also be expressed as $\dot{\theta}_i(\tau)$. We chose ω_0 as follows: $\omega_0 := \sqrt{V^2 B/M}$, which implies that $V^2 B/M\omega_0^2 = 1$.

Next, let d_i and s_i be defined by

$$d_i(\tau) := \frac{D_i(t)}{M\omega_0^2} = \frac{D_i(t)}{V^2 B}, \tag{10.8}$$

$$s_i(\tau) := \frac{S_i(t)}{M\omega_0^2} = \frac{S_i(t)}{V^2 B}, \tag{10.9}$$

where d_i and s_i are the dimensionless demand and supply in area i, respectively. By using this new notation, (10.7) can be rewritten as

$$\ddot{\theta}_i(\tau) = s_i(\tau) - d_i(\tau) + \sum_{j \in \mathcal{N}_i} \sin(\theta_j(\tau) - \theta_i(\tau)). \tag{10.10}$$

Accordingly, we obtain dimensionless equations as follows.

$$\dot{\theta}_i(\tau) = \omega_i(\tau), \tag{10.11}$$

$$\dot{\omega}_i(\tau) = s_i(\tau) - d_i(\tau) + \sum_{j \in \mathcal{N}_i} \sin(\theta_j(\tau) - \theta_i(\tau)), \tag{10.12}$$

where ω_i is dimensionless angular velocity. Note that $\omega_i(\tau) = \tilde{\omega}_i(t)/\omega_0 = \sqrt{M/V^2 B}\tilde{\omega}_i(t)$. We will model the generator dynamics with nonlinear coupled swing equations (10.11) and (10.12).

10.2.3 Real-Time Price Optimization

We formulate the real-time pricing problem of the ISO as an NMPC problem in which there are three objectives.

(1) To maximize the benefits of the consumer and supplier $(v_i(d_i) - \lambda_i d_i) + (\lambda_i s_i - c_i(s_i))$ in each area.
(2) To minimize both the phase difference angular velocity ω_i and phase difference angle θ_i of each generator.
(3) To minimize the absolute value of the difference between power demand and supply $d_i - s_i$ in each area.

Thus, the optimization problem is to minimize the following objective function.

$$J = x^{\mathsf{T}}(t+T)Px(t+T) + \int_t^{t+T} \left[x^{\mathsf{T}}(\tau)Qx(\tau) + \sum_{i=1}^{n} \Phi_i(\lambda_i) \right] d\tau \tag{10.13}$$

$$\Phi_i(\lambda_i) := R_i(c_i(s_i) - \lambda_i s_i + \lambda_i d_i - v_i(d_i)) + W_i(s_i - d_i)^2, \tag{10.14}$$

where P, Q, R_i, and W_i are weighting coefficients and T is the length of the horizon. The state vector x is defined by $x = [\theta_1, \omega_1, \ldots, \theta_n, \omega_n]^T$. In accordance with (10.3) and (10.4), both s_i and d_i are functions of λ_i. Thus, Φ_i is a function of λ_i. Here, negative social welfare Φ_i is assumed to be a convex function. Accordingly, the real-time pricing problem can be reduced to one of minimizing the objective function J subject to the swing equations (10.11) and (10.12). The electricity price is determined by the ISO so as to minimize J in (10.13) subject to (10.11) and (10.12). This online optimization procedure is repeated every sampling time. To solve this NMPC problem, we apply the C/GMRES [24] algorithm which combines a continuation approach with a Krylov solver and leads to a fast numerical solver for NMPC.

10.2.4 Numerical Simulations

Simulations were conducted with the aim of verifying the effectiveness of the optimization-based real-time pricing method described above.

Here, we define the utility and cost functions as $v_i = 2\sqrt{d_i}$ and $c_i = s_i^2/2$, respectively. According to (10.3) and (10.4), d_i and s_i are

$$d_i = \frac{1}{\lambda_i^2}, \tag{10.15}$$

$$s_i = \lambda_i. \tag{10.16}$$

Note that $\lambda_i = 1$ when $(s_i - d_i) = 0$ for all i. Namely, $\lambda_i = 1$ is the equilibrium price that balances the demand and supply of electricity. Here, we see that Φ_i can be expressed as

$$\begin{aligned}
\Phi_i(\lambda_i) &= R_i \left(-\frac{\lambda_i^2}{2} - \frac{1}{\lambda_i} \right) + W_i \left(\lambda_i - \frac{1}{\lambda_i^2} \right)^2 \\
&= R_i \left(-\frac{\lambda_i^2}{2} - \frac{1}{\lambda_i} \right) + W_i \left(\lambda_i^2 - \frac{2}{\lambda_i} + \frac{1}{\lambda_i^4} \right) \\
&= \left(-\frac{R_i}{2} + W_i \right) \lambda_i^2 + (-R_i - 2W_i) \frac{1}{\lambda_i} + \frac{W_i}{\lambda_i^4}.
\end{aligned} \tag{10.17}$$

The above equation indicates that Φ_i is a convex function if $2W_i > R_i$. Thus, not only the benefits of both consumers and suppliers but also the balance between demand and supply need to be considered to properly formulate the optimization problem.

Here, let us consider a power network composed of ten areas, as shown in Fig. 10.1. We will use the C/GMRES [24] algorithm to solve the NMPC problem. The weighting coefficients P and Q are set as I and $10I$, respectively, where I is the identity matrix. Furthermore, R_i and W_i are set to 1 for all i so as to satisfy the convexity condition $2W_i > R_i$.

Fig. 10.1 Power network
model (Reprinted from [16]
Copyright 2014 ISCIE)

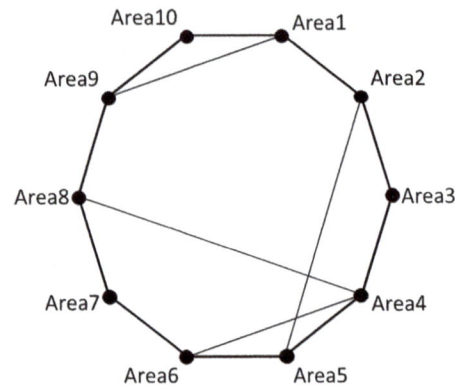

Fig. 10.2 Time responses of
θ_i ($i = 1, \ldots, 10$)
(Reprinted from [16]
Copyright 2014 ISCIE)

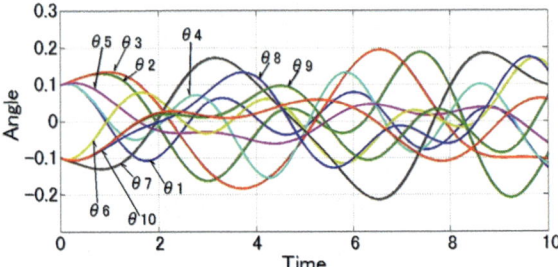

First, consider a situation where λ_i has a fixed value of 1, which means $(s_i - d_i) = 0$. The time responses of θ_i and ω_i for the case without the electricity pricing operation are shown in Figs. 10.2 and 10.3, respectively. We see that the phase difference angles and phase difference angular velocities do not converge to zero. Next, the time responses of θ_i and ω_i for the case with optimal electricity pricing operation based on the proposed method are shown in Figs. 10.4 and 10.5. Moreover, the time responses of λ_i are shown in Fig. 10.6. Here, we see that both the phase difference angles and phase difference angular velocities converge to zero and the electric prices converge to one. Thus, these results verify the effectiveness of the method described above.

The simulation was performed on a computer having the following specs—CPU: Intel Core i7 3.20 (GHz), RAM: 3.00 (GB), OS: Windows 7, Software: Visual C++. The simulation parameters of the C/GMRES method were the length of the horizon $T = 1 - e^{-0.5\tau}$, the number of discretization steps of the horizon $N = 10$, the discretization step of the dimensionless time τ: $\Delta\tau = 0.002$, and the number of iterations in the GMRES method: 5. The average computational time per update (one control cycle) was 0.33 (ms), which is sufficiently short for a real-time implementation.

Fig. 10.3 Time responses of ω_i $(i = 1, \ldots, 10)$ (Reprinted from [16] Copyright 2014 ISCIE)

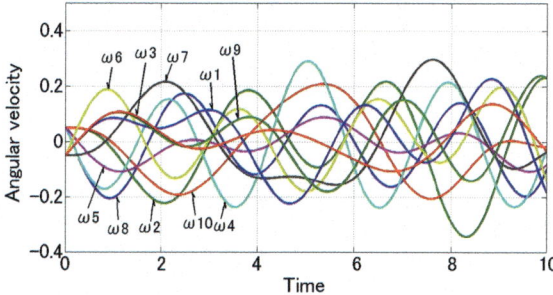

Fig. 10.4 Time responses of θ_i $(i = 1, \ldots, 10)$ (Reprinted from [16] Copyright 2014 ISCIE)

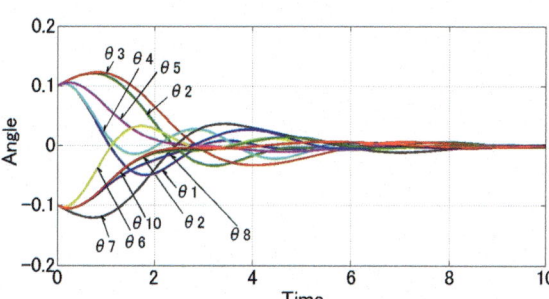

Fig. 10.5 Time responses of ω_i $(i = 1, \ldots, 10)$ (Reprinted from [16] Copyright 2014 ISCIE)

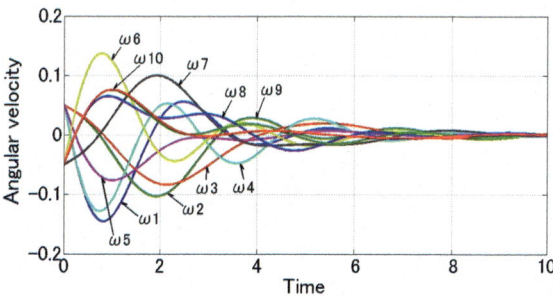

Fig. 10.6 Time responses of λ_i $(i = 1, \ldots, 10)$ (Reprinted from [16] Copyright 2014 ISCIE)

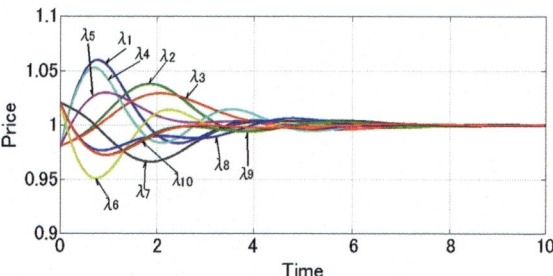

10.3 Online Estimation of Consumers' Characteristics

10.3.1 Objective

In this section, we present a real-time pricing mechanism for the EPS model proposed in [14]. Our objective is to stabilize the load frequency fluctuation of the generator caused by supply and demand imbalances of the thermal power plant, battery energy storage system (BESS), heat pump water heater (HPWH), and disturbances caused by the wind power generator and load power. Irie et al. [14] described a method to stabilize the load frequency by directly adjusting the electricity generation and/or consumption of the thermal power plant, BESS, and HPWH. However, the owner of the HPWH is a consumer, and directly controlling the electricity consumption of the HPWH is not realistic. To make it work with real-time pricing, we have to modify the formulation of this problem by modeling the demand function of the consumer owning the HPWH as a sigmoid function. The demand function depends upon the weather, time, and season, so we can design the algorithm to make online estimates of its parameters by using a particle filter [17]. The pricing mechanism also includes a particle filter. Interested readers may consult [28], which describes the method in detail.

10.3.2 Power System Model

Let us aggregate sets of identical components, such as the set of generators, into one. In the EPS model, the operating state is chosen to be the one where the load frequency is stabilized and supply and demand are balanced. We will thus model the deviation from the operating state. In a similar manner as described in [14], we can treat wind power generation as a disturbance and assume that its amount is around 2000 (MW), a value corresponding to the target set by the Japanese government [1]. We will also assume that the base load power remains constant.

10.3.2.1 Load Frequency of Generator

Suppose that all generators in the grid are synchronized. Then, the deviation in load frequency of the (aggregated) generator satisfies the following swing equation [18],

$$M_{eq}\Delta\dot{\omega} + D\Delta\omega = P_m - P_L, \tag{10.18}$$

where M_{eq} and D denote the sum of the inertia constants of all generators in the grid and the damping coefficient, respectively, and P_m and P_L are the sum generated powers and sum consumed power in the grid, respectively.

10.3.2.2 Thermal Power Plant Model and BESS

The thermal power plant illustrated in Fig. 10.7 is modeled with the parameters in Table 10.2. BESS is modeled as in Fig. 10.8. To determine the charge (discharge) and storage capacity, BESS tracks the ideal charge (discharge) signal within the upper and lower bounds, and its time constant is $T_b = 1$.

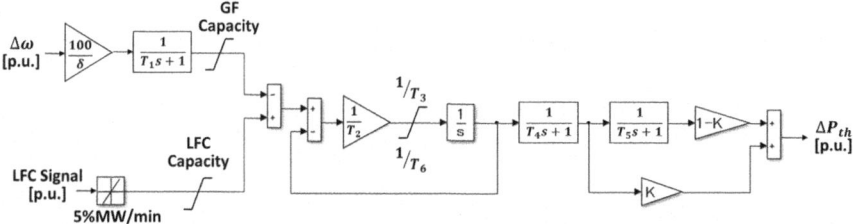

Fig. 10.7 Thermal power plant model (Reprinted from [28] Copyright 2017 SICE)

Table 10.2 Parameters of thermal power plant model (Reprinted from [28] Copyright 2017 SICE)

δ	Permanent speed variation (%)	5
T_1	Speed relay time constant (s)	0.2
T_2	CV servo time constant (s)	0.2
T_3	CV servo open time (s)	5
T_4	High-pressure turbine time constant (s)	0.25
T_5	Low-pressure turbine time constant (s)	9.0
T_6	CV servo close time (s)	−0.001
K	High-pressure output dispatching rate (p.u.)	0.3

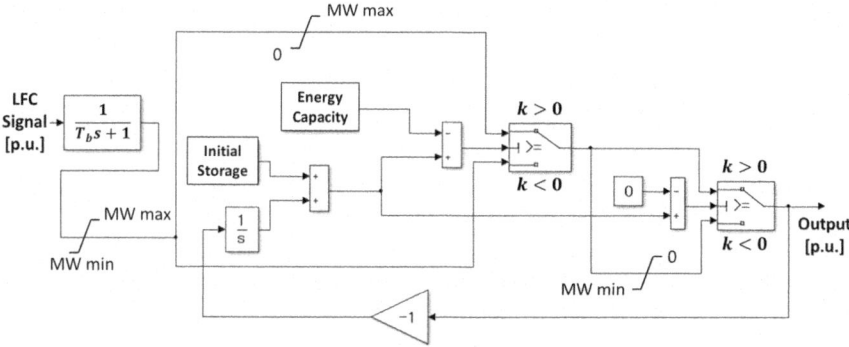

Fig. 10.8 BESS model (Reprinted from [28] Copyright 2017 SICE)

10.3.2.3 HPWH

The HPWH boils water using a refrigeration cycle, so it cannot be switched on and off frequently. According to [14], the HPWH consumes 90% of its rated value when uncontrolled, and the consumption capacity can be adjusted by ± 10 from 90%. We will assume that there are 1.26 million HPWHs in the grid, and their total capacity is 1520 (MW). As in [14], we will also assume that there is sufficient storage, i.e., the stored energy does not exceed the capacity. Figure 10.9 shows the model of the HPWH for the parameters in Table 10.3.

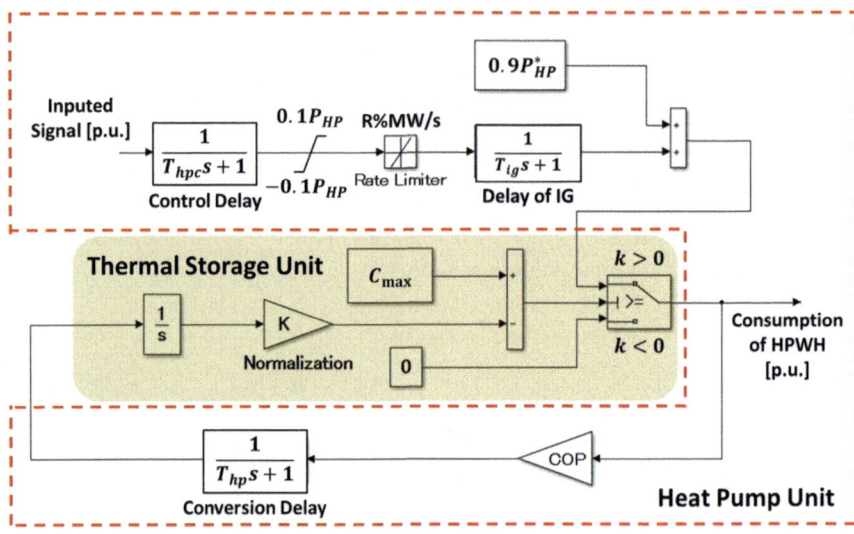

$* P_{HP}$: Rated Power Consumption of HPWH [p.u.]

Fig. 10.9 HPWH model (Reprinted from [28] Copyright 2017 SICE)

Table 10.3 Parameters of HPWH model (Reprinted from [28] Copyright 2017 SICE)

T_{hpc}	Control delay (s)	30
R	Rate limiter characteristic (%MW/s)	1
T_{ig}	Delay of induction motor (s)	1
COP	Coefficient of performance	4
T_{hp}	Conversion delay (s)	300
C_{max}	Capacity of thermal storage unit (MWh)	7560

10.3.2.4   Relationships Between Prices and HPWH's Consumptions

Suppose that the relationship between the HPWH's power consumption and the price of electricity is described with a sigmoid function, whose value saturates to the left and to the right. We denote the adjustable capacity of the HPWH's power consumption and the deviation in price as C_{HP} and λ_{HP}, respectively. By multiplying and shifting a standard sigmoid function, we can generate another sigmoid function as follows (see Fig. 10.10),

$$\Delta P_{HP} = -C_{HP}\left(\frac{1}{1 + e^{-a_{HP}\lambda_{HP}}} - \frac{1}{2}\right), \tag{10.19}$$

where a_{HP} is a factor that determines the slope of the function and can be regarded as the consumer's reaction to the price change.

10.3.3   Control Schemes

Load frequency control (LFC) in the EPS is conducted by calculating the area requirement (AR) from the load frequency deviation and sending it to the generator as an LFC signal. The frequency range of the HPWH is much lower than those of the thermal power plant and BESS. Therefore, it is reasonable to determine the control scheme depending on the frequency. That is, we can directly control the two pieces of equipment having high frequencies, whereas we can control each HPWH by presenting an electricity price to its owner. The slope of the sigmoid function can be made to depend on situational factors such as the weather, time, and season, and this can be estimated online. We designed a particle filter to make this estimation.

Fig. 10.10 Demand function (Reprinted from [28] Copyright 2017 SICE)

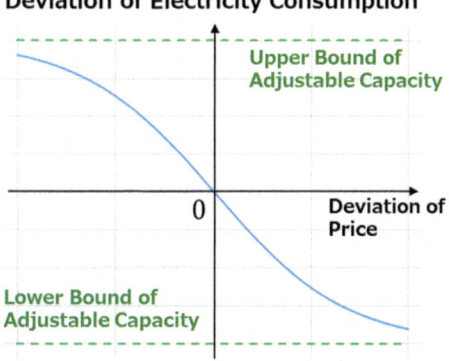

10.3.3.1 Nonlinear Model Predictive Control for Price Presentation

The NMPC method for controlling the HPWH works as follows. A low-pass filter
with time constant T_{AR} extracts the low-frequency components from the load frequency deviation. In addition, the integrated value of the low-frequency range of
the AR is calculated in order to cope with the wind generation output increases and
decreases that occur over a long period of time. In short, we solve the following MPC
problem with objective function,

$$J = \varphi(x(t+T), t+T) + \int_t^{t+T} L(x(\tau), u(\tau), \tau) d\tau, \qquad (10.20)$$

$$\varphi = \frac{1}{2} x^T S_f x, \quad L = \frac{1}{2} x^T Q x + \frac{1}{2} r u^2, \qquad (10.21)$$

where S_f, Q, and r are weighting coefficients and state equation

$$x = [x_1, x_2, x_3, x_4]^T, \qquad (10.22)$$

$$\dot{x}_1 = \frac{1}{T_{hpc}} \left[-x_1 + C_{HP} \left(\frac{1}{1 + e^{-a_{HP}u}} - \frac{1}{2} \right) \right], \qquad (10.23)$$

$$\dot{x}_2 = \frac{1}{M} (-Dx_2 - x_1), \qquad (10.24)$$

$$\dot{x}_3 = -\frac{1}{T_{AR}} \left[x_3 + \frac{K_{sys} x_2}{2} \right], \qquad (10.25)$$

$$\dot{x}_4 = x_3, \qquad (10.26)$$

where x_1, x_2, x_3, and x_4 are the deviations in the HPWH's consumption, frequency
deviation in the generator, AR, and integrated value of AR, respectively, and u is the
price presented to the HPWH's owner.

10.3.3.2 PI Controller for Thermal Power Plant and BESS

Let us compare two different PI controllers that can be used to control the thermal
power plant and BESS directly. The first one, shown in Fig. 10.11, is from [14], and
it uses the anti-windup PI control shown in Fig. 10.12 with $T_{AR} = 10$ (s), $K_p = 5$,
and $K_i = 0.1$. Note that, because the heat pump is controlled by presenting the price
to the consumer, we do not send signals to the HPWH in Fig. 10.11.

The second PI control uses HP (Heat Pump) shortage compensation as shown in
Fig. 10.13. Note that only the bottom right part of the first PI controller in Fig. 10.11 is
changed; thus, the other sections of Fig. 10.11 have been omitted from Fig. 10.13. In
the "HP Estimation" block in Fig. 10.13, the measured heat pump power consumption

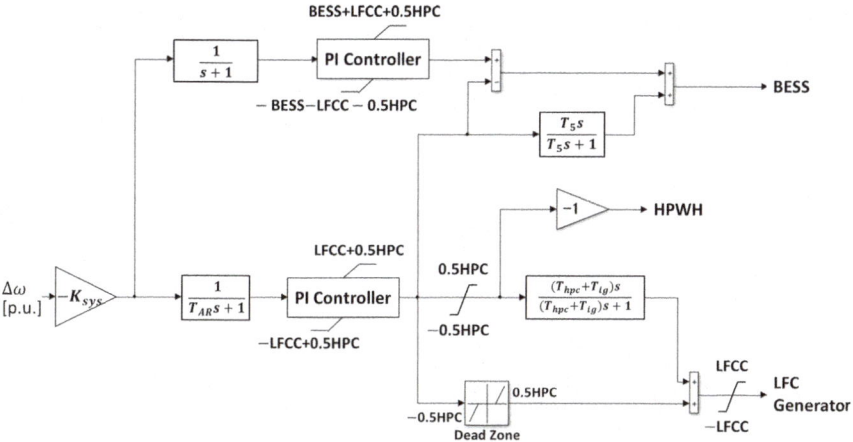

Fig. 10.11 PI controller from [14] (Reprinted from [28] Copyright 2017 SICE)

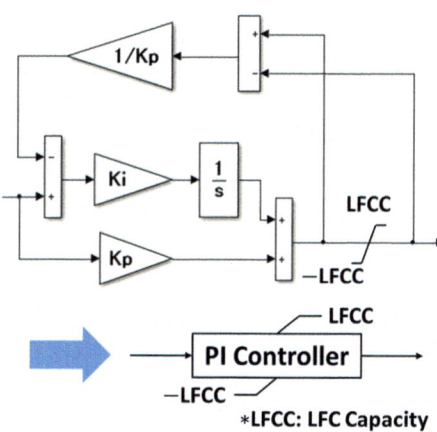

Fig. 10.12 Anti-windup PI controller (Reprinted from [28] Copyright 2017 SICE)

x_{HP} and electricity price u as well as the estimated parameter a_{HP} are updated every 10 s. x_{HP} for the next 10 s is estimated by using a Runge–Kutta method to solve (10.23).

10.3.3.3   Estimation of Consumer Parameter

Suppose that the controller administrator obtains the variation of the power consumption of a HPWH $x_{HP} = x_1$ as an observation y at a fixed sampling time T. Then, using y and the presented price u, it estimates the parameter $\hat{\theta} = a_{HP}$ in the demand function (10.19) in real time. The discrete-time model for estimating the consumer parameter is as follows.

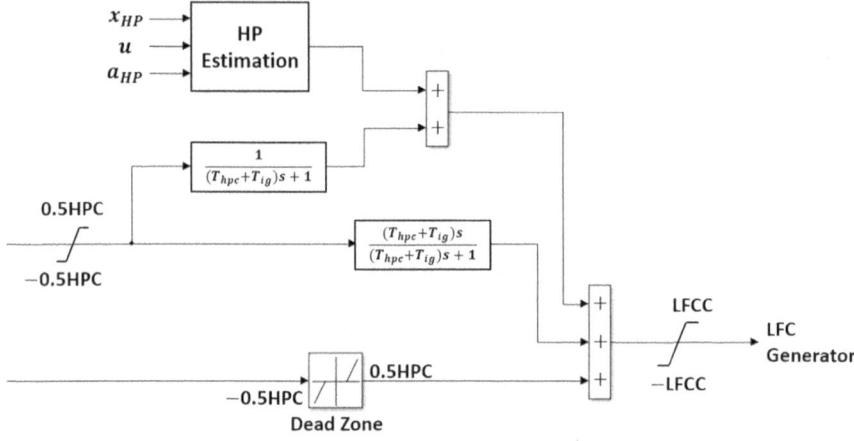

Fig. 10.13 PI controller with HP shortage compensation mechanism (Reprinted from [28] Copyright 2017 SICE)

$$x_{\text{HP}}[t+1] = \left(1 - \frac{T}{T_{\text{hpc}}}\right) x_{\text{HP}}[t] - \frac{T}{T_{\text{hpc}}} C_{\text{HP}} \left(\frac{1}{1 + e^{-\hat{\theta}[t]u[t]}} - \frac{1}{2}\right) + w^{(\text{HP})}[t],$$

$$(10.27)$$

$$\hat{\theta}[t+1] = \hat{\theta}[t] + w^{(p)}[t],$$ 	(10.28)

$$y[t] = x_{\text{HP}}[t] + v[t],$$ 	(10.29)

where $u[t]$ is the control input at the tth sampling time, $w^{(\text{HP})}[t]$ and $w^{(p)}[t]$ are system noise, and $v[t]$ is observation noise.

10.3.3.4 Model Predictive Control with Parameter Estimation Using a Particle Filter

The method is summarized as follows (see also Fig. 10.14).

1. Obtain the state $x(t)$ of the power system at time $t = n \times \Delta s$, $(n = 0, 1, \dots)$.
2. Let $y[t] = x_1(t)$. Then, update the particle filter by using a predictive ensemble.
3. By using the estimated parameter values, solve the optimal control problem in (10.20)–(10.26).
4. By using the obtained control input, update the time step of the particle filter. Then, save the obtained predictive ensemble.
5. Propose control input, i.e., electricity price, to consumers, and then keep this price until the next sampling time.

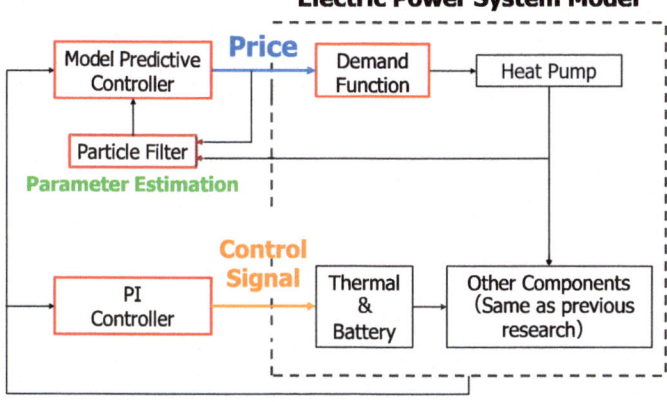

Fig. 10.14 Concept of the control scheme (Reprinted from [28] Copyright 2017 SICE)

10.3.4 Numerical Simulations

10.3.4.1 Simulation Conditions

The simulation considers small-scale systems in two areas joined by an interconnection, following the Institute of Electrical Engineers of Japan (IEEJ) East 30 Machine
system model [12], which is a power system model based on the work in [14]. For
the wind power generation and load fluctuations, the simulation uses the values in
Figs. 10.15 and 10.16. Table 10.4 shows the parameters of the power systems.

The length and time step of the horizon for NMPC are 90 (s) and 1 (s), respectively.
Due to large fluctuations in wind power, the weight of integration of AR is chosen to
be larger than the others; $S_f = Q = \mathrm{diag}(10^{-6}, 10^{-6}, 10^{-6}, 1)$ and $r = 0.25$. Note
that the `fsolve` function in MATLAB/Simulink was used as the solver.

The number of ensemble particles in the particle filter is 200. The initial ensemble
for $[x_{\mathrm{HP}}, \hat{\theta}]^{\mathrm{T}}$ is generated from a two-dimensional normal distribution with mean
value $[0, 4.5]^{\mathrm{T}}$ and covariance matrix $\mathrm{diag}(5, 0.5)$. In addition, after some trial and
error, the system noise and observation noise were set to be $(w^{(\mathrm{HP})}[t], w^{(p)}[t]) \sim$
$(N(0, 2), N(0, 0.05))$ and $v[t] \sim N(0, 2)$, respectively.

Numerical simulation uses a fourth-order Runge–Kutta method with a time step of
0.01 (s). To verify the efficacy of estimating the consumer parameter with the particle
filter, the following four scenarios are considered, and the control performances in
the four scenarios are compared with that in the previous research [14].

Scenario 1 (consumer parameter known/conventional PI control): The consumer
parameter is $a_{\mathrm{HP}} = 5$. The PI control [14] is used for direct control.

Scenario 2 (consumer parameter changed/conventional PI control): The same
controller as in Scenario 1 is used. However, a_{HP} changes from $a_{\mathrm{HP}} = 5$ to $a_{\mathrm{HP}} = 4$
after 900 s.

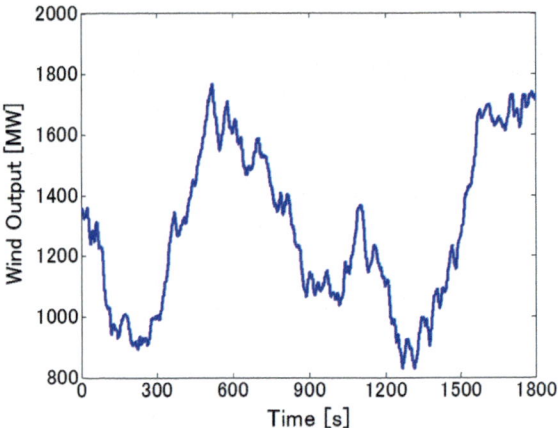

Fig. 10.15 Wind power generated output (Reprinted from [28] Copyright 2017 SICE)

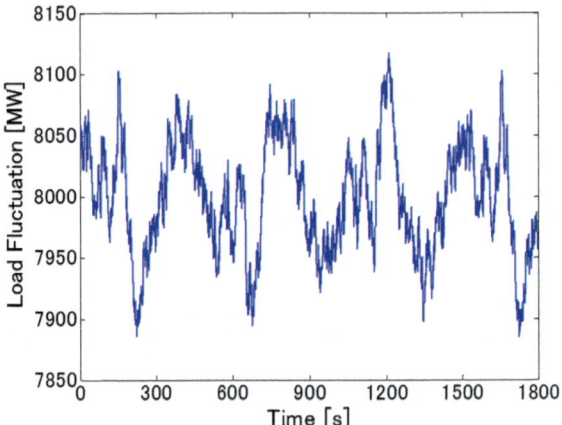

Fig. 10.16 Load fluctuation (Reprinted from [28] Copyright 2017 SICE)

Scenario 3 (with particle filter/conventional PI control): The consumer parameter changes as in Scenario 2, and the particle filter is used to estimate a_{HP}.

Scenario 4 (with particle filter/control with HP shortage compensation): The consumer parameter changes as in Scenario 2, and the particle filter is used to estimate a_{HP}. Here, we use the PI controller with HP shortage compensation for direct control.

10.3.4.2 Simulation Results

To quantitatively evaluate the frequency fluctuations, we can use the maximum frequency deviation and root mean square (RMS) value [23], defined as

Table 10.4 Simulation condition (Reprinted from [28] Copyright 2017 SICE)

Reference frequency (Hz)		50
Power system constant (%MW/Hz)		9
Load-damping constant D (p.u.)		2
LFC operation mode		FFC
GF capacity	(%)	±5
	[MW]	±278
LFC capacity	(%)	±1.5
	(MW)	±125
Simulation time (s)		1800

Table 10.5 Comparison of RMS and maximum deviation (Reprinted from [28] Copyright 2017 SICE)

	RMS (Hz)	max Δf (Hz)
Previous research [14]: direct commands	0.0220	0.0587
Scenario 1: known parameter and conventional PI controller	0.0251	0.0702
Scenario 2: parameter changes and conventional PI controller	0.0255	0.0721
Scenario 3: particle filter and conventional PI controller	0.0253	0.0705
Scenario 4: particle filter and PI controller with HP shortage compensation	0.0248	0.0671

$$x_{\mathrm{rms}} = \sqrt{\frac{1}{N} \sum_{i=1}^{N} \Delta f_i^2}, \qquad (10.30)$$

where N is the number of samples within an evaluation period and Δf_i is the frequency deviation.

Table 10.5 shows the results of from the previous research and Scenarios 1–4. In Scenario 1 (parameter known) and Scenario 2 (parameter changed), the frequency fluctuation suppression deteriorates when an incorrect parameter is used for price presentation. In Scenario 3 (with particle filter/conventional PI control), when an unknown parameter is estimated online, the performance of the control is almost the same as in Scenario 1, where the parameter is known.

Figures 10.17, 10.18, 10.19, 10.20, 10.21, and 10.22 show the results of Scenario 4 (with particle filter/PI control with HP shortage compensation). In Fig. 10.17, the frequency fluctuations tend to be larger near 500 and 1600 s when the wind power generation greatly fluctuates (Fig. 10.15). In Fig. 10.18, even when the supply is insufficient, the power consumption of the heat pump is not largely adjusted by the price presentation. This shortage is compensated by the thermal power generator and BESS (Figs. 10.19 and 10.20). Because the demand function in Fig. 10.10 saturates, the consumption does not change dramatically even if the price does.

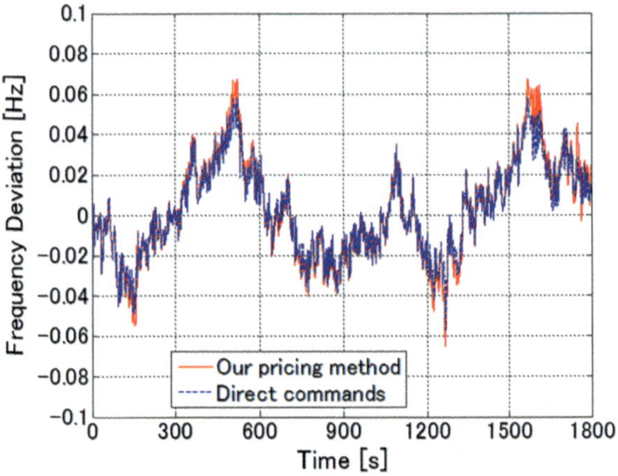

Fig. 10.17 Frequency
fluctuation (Scenario 4)
(Reprinted from [28]
Copyright 2017 SICE)

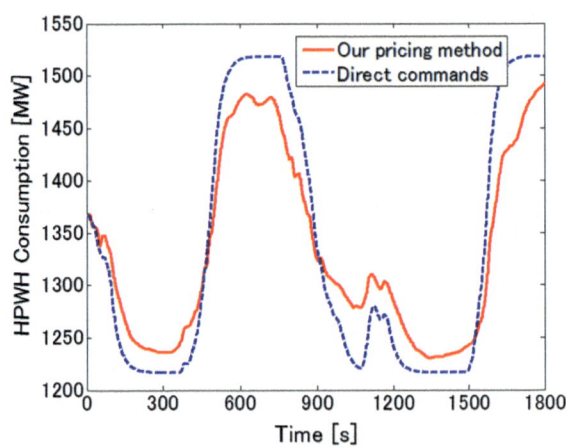

Fig. 10.18 HPWH
(Scenario 4) (Reprinted
from [28] Copyright 2017
SICE)

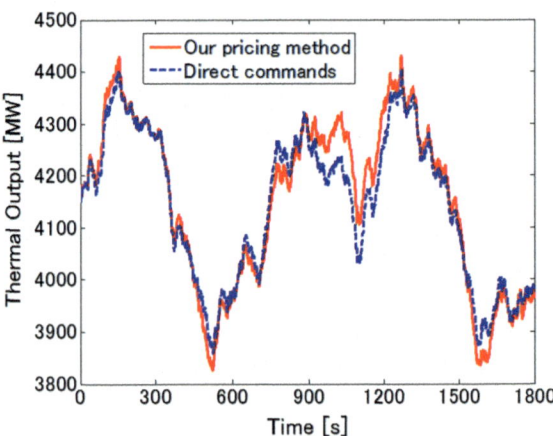

Fig. 10.19 Thermal
generation (Scenario 4)
(Reprinted from [28]
Copyright 2017 SICE)

Fig. 10.20 BESS (Scenario 4) (Reprinted from [28] Copyright 2017 SICE)

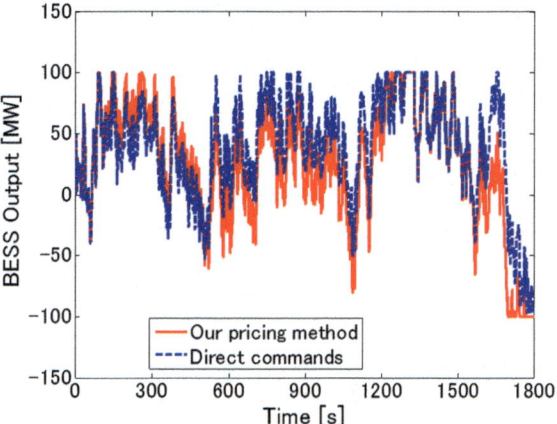

Fig. 10.21 Electricity price (Scenario 4) (Reprinted from [28] Copyright 2017 SICE)

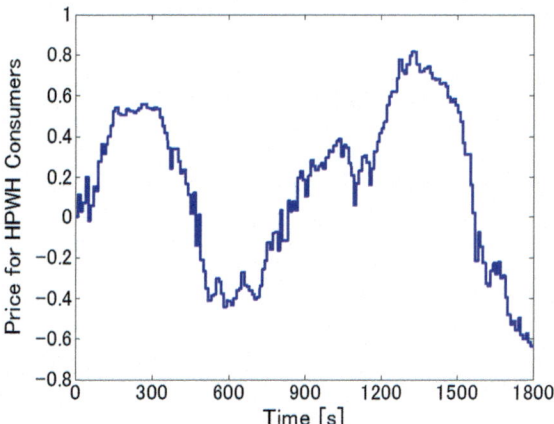

Fig. 10.22 Value of parameter a_{HP} (Reprinted from [28] Copyright 2017 SICE)

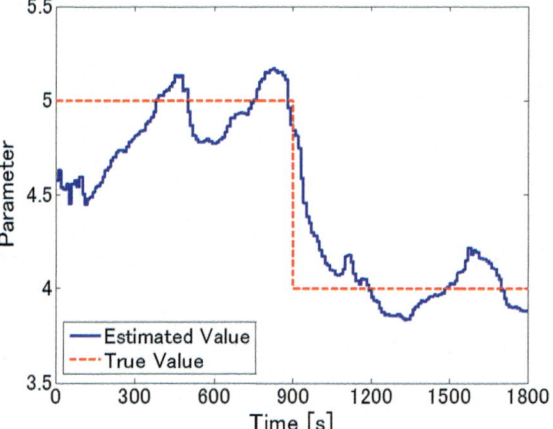

10.4   Stochastic Optimization for High Penetration of Renewable Energy

10.4.1   Objective

This section describes an integrated frequency control system with SMPC on the basis of the work of Satouchi et al. [27, 28]. SMPC explicitly incorporates disturbances in the model. We assume that the HPWH can adjust its own power consumption in response to the price. A demand function can be used to estimate the HPWH's power consumption, but it can vary with the weather, time, and season. We introduce the particle filter proposed by Satouchi et al. [28] in order to estimate the demand function in real time.

Since SMPC can deal with disturbances by using their probability distributions, it can evaluate the objective function and constraints in a stochastic manner. In SMPC, chance constraints are often considered instead of deterministic constraints. A chance constraint requires that the probability of satisfying an inequality is above a designated bound. In this section, we impose chance constraints on the frequency and suppress the frequency deviation to within a bound with a designated probability.

Various approaches to SMPC have been developed in the literature [19]. Most of them are for linear systems, while much less work has been done on nonlinear systems [19]. The SMPC approaches for linear systems can be roughly classified into two categories [8]. The first is a class of analytic approximation methods [25]. These methods reformulate the chance constraints and the objective function into deterministic forms in order for them to be included in the MPC formulation. The second class is based on randomized methods [29]. A sufficient number of samples of disturbances are generated, and deterministic constraints instead of chance constraints have to be satisfied for every sample. For nonlinear systems, Gaussian-mixture models are used to describe the evolution of the distributions of states [20]. However, the calculation costs of these methods are generally high [19].

The model derived in this section is a so-called Hammerstein model [9], in which nonlinearities are included only in the input term. If the cumulative distribution function (CDF) of disturbances is analytically defined, the SMPC problem for linear systems can be transformed into an equivalent deterministic optimization problem [11]. Here, we extend this transformation to Hammerstein models and transform the SMPC problems for Hammerstein models into equivalent deterministic optimizations. Moreover, open-loop SMPC, which optimizes only the input sequence, is conservative in that it does not consider any feedback on the prediction horizon. Affine disturbance feedback parametrization is often used in linear SMPC in order to relax this sort of conservativeness [10]. SMPC for Hammerstein models with affine disturbance feedback parametrization can also be approximately transformed into a deterministic problem, which can be solved with a common optimization solver.

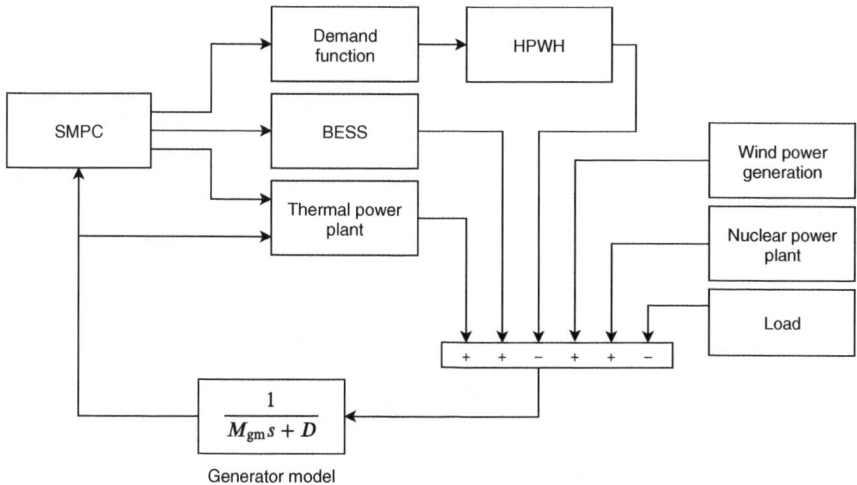

Fig. 10.23 Electric power system

10.4.2   Power System Model

The EPS considered here is shown in Fig. 10.23. This system consists of thermal
power plants, nuclear power plants, a BESS, wind power plants, and HPWHs. The
controller controls the thermal power plants and BESS directly and the HPWHs indirectly with real-time pricing. We assume that the HPWHs are installed in households
and can regulate the power consumption in response to the electricity price. The
controller observes the gross power generation/consumption of the power system.
Nuclear power plants are normally controlled over a long time scale; thus, they are
not controlled in this system. All generators are assumed to be synchronized, and
each kind of generator and piece of equipment is represented by its own model. Power
generation or consumption is represented by the deviation from the initial value. The
models of the components in Fig. 10.23 are the same as those in the previous section.

Wind power and demand fluctuation are considered disturbances. We create
patterns of wind power generation data, considering the wind power capacity as
described in 10.4.4.1. The power spectrum of wind power can be represented by a
function proportional to $1/\|f\|^{\frac{5}{3}}$, where f is frequency [2]. In the simulation, we
used MATLAB's colored noise generator to generate data having this property. Similarly, we generated patterns of demand fluctuation data by referring to the work of
Irie et al. [14] and using the same tool.

10.4.3 *Controller*

In 10.4.3.1, the models are simplified and discretized so that they can be used in the SMPC. In 10.4.3.2, two types of SMPC are formulated, i.e., open-loop SMPC and closed-loop SMPC. These SMPC problems can be transformed into deterministic optimization problems, which can be solved with a common optimization solver. Here, we will use the particle filter proposed by Satouchi et al. [28] to estimate the demand characteristic in real time.

10.4.3.1 Models for SMPC

First, we simplify the models described in Sect. 10.4.2 by ignoring terms whose time constant is sufficiently small. We define i to be discrete time and Δt to be the control cycle in what follows.

A simplified model of a thermal power generator is obtained as follows:

$$p_t(i) = (1 - K_{tp})x_{ts}(i) + K_{tp}\left(-\frac{100}{\delta}\omega(i) + u_t(i)\right), \tag{10.31}$$

$$x_{ts}(i+1) = x_{ts}(i) + \frac{1}{T_s}(-x_{ts}(i) - \frac{100}{\delta}\omega(i) + u_t(i))\Delta t, \tag{10.32}$$

where u_t represents the command input to the thermal power generator, x_{ts} represents the output of slow dynamics, and p_t is the output of the generator.

Since the BESS model has a sufficiently fast response, its output ideally follows the input:

$$p_b(i) = u_b(i), \tag{10.33}$$

where p_b is positive if the BESS is discharging and p_b is negative otherwise.

The HPWH is simplified as follows:

$$p_h(i+1) = p_h(i) + \frac{1}{T_{hpc}}(-p_h(i) + u_h(i))\Delta t, \tag{10.34}$$

where u_h is the HPWH input (electric power signal) and p_h represents the HPWH output (electric power).

We treat the wind power generation and demand fluctuation as one variable p_w because their power spectra are similar, and we assume that the fluctuation follows the model,

$$p_w(i+1) = p_w(i) + w(i), \tag{10.35}$$

$$w(i) \sim \mathcal{N}(0, \sigma). \tag{10.36}$$

Finally, we define the state vector x, input vector u, and output vector y as follows:

$$x = \begin{bmatrix} \omega & p_h & u_t & u_b & p_w & x_{ts} \end{bmatrix}^T, \tag{10.37}$$

$$u = \begin{bmatrix} u_p & \Delta u_t & \Delta u_b \end{bmatrix}^T, \tag{10.38}$$

$$y = \begin{bmatrix} \omega & p_h & p_t & p_b, \end{bmatrix}^T, \tag{10.39}$$

where Δu_t and Δu_b denote the variations of u_t and u_b per control cycle, respectively. We use Δu_t and Δu_b as the input variables in order to constrain the variation of the inputs per control cycle. The actual inputs u_t and u_b to the system are obtained by integrating Δu_t and Δu_b with respect to time.

As a whole, the state equation and output equation are represented as follows:

$$x(i+1) = Ax(i) + Bg(u(i)) + Cw(i), \tag{10.40}$$

$$y(i) = Dx(i), \tag{10.41}$$

where

$$A = \begin{bmatrix} 1 - \left(D_{gm} - \frac{100}{\delta} K_{tp}\right)\frac{\Delta t}{M_{gm}} & -\frac{\Delta t}{M_{gm}} & \frac{K_{tp}}{M_{gm}}\Delta t & \frac{\Delta t}{M_{gm}} & \frac{\Delta t}{M_{gm}} & \frac{1-K_{tp}}{M_{gm}}\Delta t \\ 0 & 1 - \frac{1}{T_{hpc}}\Delta t & 0 & 0 & 0 & 0 \\ 0 & 0 & 1 & 0 & 0 & 0 \\ 0 & 0 & 0 & 1 & 0 & 0 \\ 0 & 0 & 0 & 0 & 1 & 0 \\ -\frac{100}{\delta}\cdot\frac{\Delta t}{T_5} & 0 & \frac{\Delta t}{T_5} & 0 & 0 & 1 - \frac{\Delta t}{T_5} \end{bmatrix},$$

$$B = \begin{bmatrix} 0 & 0 & 0 \\ \frac{\Delta t}{T_{hpc}} & 0 & 0 \\ 0 & \Delta t & 0 \\ 0 & 0 & \Delta t \\ 0 & 0 & 0 \\ 0 & 0 & 0 \end{bmatrix}, \quad C = \begin{bmatrix} 0 \\ 0 \\ 0 \\ 0 \\ 1 \\ 0 \end{bmatrix}, \quad D = \begin{bmatrix} 1 & 0 & 0 & 0 & 0 & 0 \\ 0 & 1 & 0 & 0 & 0 & 0 \\ -\frac{100}{\delta}K_{tp} & 0 & K_{tp} & 0 & 0 & 1 - K_{tp} \\ 0 & 0 & 0 & 1 & 0 & 0 \end{bmatrix},$$

$$g(u) = \begin{bmatrix} \Delta P_{HP}(u_p) \\ \Delta u_t \\ \Delta u_b \end{bmatrix}.$$

The state equation (10.40) is a Hammerstein model [9] because the nonlinearity exists only in the input term. Note that $\Delta P_{HP}(u_p)$ is the demand function defined in (10.19) with λ_{HP} replaced by u_p; thus, the nonlinear function $g(\cdot)$ is one-to-one and its inverse exists. This property is necessary for transforming it into a deterministic problem.

10.4.3.2   Formulation of SMPC

Open-Loop SMPC

First, we formulate open-loop SMPC, in which only one sequence of inputs over the horizon is optimized for all possible sequences of disturbances. At each time step i, open-loop SMPC solves the following problem, setting the present state $\bar{x}(i)$ as the initial state $x(0)$. Note that open-loop SMPC is a kind of feedback control, in spite of its name.

Problem 1 *(Open-loop SMPC problem)*

$$\min_{u(0),\dots,u(N-1)} \mathbb{E}[J], \tag{10.42}$$

subject to

$$\mathbb{E}[J] = \mathbb{E}\left[\sum_{j=1}^{N} \frac{1}{2} y(j)^{\mathsf{T}} Q y(j) + \sum_{j=0}^{N-1} \frac{1}{2} u(j)^{\mathsf{T}} R u(j)\right], \tag{10.43}$$

$$x(j+1) = Ax(j) + Bg(u(j)) + Cw(j) \quad (j = 0, \dots, N-1), \tag{10.44}$$

$$y(j) = Dx(j) \quad (j = 1, \dots, N), \tag{10.45}$$

$$x(0) = \bar{x}(i), \tag{10.46}$$

$$\mathbb{P}[E_x x(j) \le h_x] \ge p_x \quad (j = 1, \dots, N), \tag{10.47}$$

$$E_u u(j) \le h_u \quad (j = 0, \dots, N-1). \tag{10.48}$$

Equation (10.43) represents the objective function, which is a quadratic form of the output and the input. Equations (10.44) and (10.45) are the state equation and output equation defined in 10.4.3.1. Inequality (10.47) denotes chance constraints, which require that the probability of satisfying the state inequalities be above a designated bound p_x. Let $\bar{u}(i)$ denote the actual input applied to the plant at time i in what follows. The controller optimizes the control inputs $\{u(0), \dots, u(N-1)\}$ at each time step, and the first one $u(0)$ is used as the actual input $\bar{u}(i)$. The procedure above accomplishes feedback control in that it derives an input $\bar{u}(i)$ depending on the present state $\bar{x}(i)$. We call Problem 1 "open-loop SMPC," because it simply optimizes the input sequence, i.e., open-loop control, over the horizon.

Closed-Loop SMPC

The state sequence $x(1), \dots, x(N)$ on the prediction horizon is affected by the past disturbances and thus has uncertainty. Therefore, the optimal input on the prediction horizon should change depending on the state trajectory earlier than the input. This implies that it is preferable to find an optimal control policy rather than only one sequence of inputs. A control policy can be generally written as a function of states that are observed when the input is actually applied. In particular, for linear systems, it can be parameterized as an affine function of the past states:

$$u(j) = \sum_{k=1}^{j} M'_{j,k} x(k) + \gamma'_j. \tag{10.49}$$

On the other hand, the future states $x(k)$ are uniquely determined by the state equation and future disturbances. Thus, an input on the prediction horizon can be also written as an affine function of the past disturbances. This sort of parametrization is called affine disturbance feedback parametrization [10, 25] and is represented as follows:

$$u(j) = \sum_{k=0}^{j-1} M_{j,k} w(k) + \gamma_j. \tag{10.50}$$

Both parametrizations (10.49) and (10.50) are equivalent, while the latter leads to a convex problem that is easy to deal with [10]. In this section, we call SMPC with affine disturbance feedback parametrization "closed-loop SMPC," because it optimizes the feedback policy, i.e., closed-loop control, over the horizon. However, the model we will consider is a Hammerstein model with nonlinearities in the input; thus, the parametrization cannot be used directly.

In this case, we parametrize the term $g(u)$ instead of input u, as follows:

$$g(u(j)) = \sum_{k=0}^{j-1} M_{j,k} w(k) + g(\gamma_j). \tag{10.51}$$

The nonlinear function $g(\cdot)$ is one-to-one; thus, the input is represented by the inverse function.

$$u(j) = g^{-1}\left(\sum_{k=0}^{j-1} M_{j,k} w(k) + g(\gamma_j)\right). \tag{10.52}$$

Consequently, we can formulate a closed-loop SMPC problem as follows. At each time step i, the closed-loop SMPC solves the following problem, setting the present state $\bar{x}(i)$ as the initial state $x(0)$.

Problem 2 (*Closed-loop SMPC problem*)

$$\min_{\{M_{j,k}\},\{\gamma_j\}} \mathbb{E}[J], \tag{10.53}$$

subject to

$$\mathbb{E}[J] = \mathbb{E}\left[\sum_{j=1}^{N} \frac{1}{2} y(j)^{\mathsf{T}} Q y(j) + \sum_{j=0}^{N-1} \frac{1}{2} u(j)^{\mathsf{T}} R u(j)\right], \tag{10.54}$$

$$x(j+1) = Ax(j) + Bg(u(j)) + Cw(j) \quad (j = 0, \ldots, N-1), \tag{10.55}$$

$$y(j) = Dx(j) \quad (j = 1, \ldots, N), \tag{10.56}$$

$$g(u(j)) = \sum_{k=0}^{j-1} M_{j,k} w(k) + g(\gamma_j) \quad (j = 1, \ldots, N), \tag{10.57}$$

$$x(0) = \bar{x}(i), \tag{10.58}$$

$$\mathbb{P}[E_x x(j) \le h_x] \ge p_x \quad (j = 1, \ldots, N), \tag{10.59}$$

$$\mathbb{P}[E_u u(j) \le h_u] \ge p_u \quad (j = 0, \ldots, N-1). \tag{10.60}$$

The controller optimizes the sequences $\{M_{j,k}\}$ and $\{\gamma_j\}$ at each time step i and sets the first one of the optimal input sequence $u(0) = \gamma_0$ as $\bar{u}(i) = u(0)$. The above procedure accomplishes feedback control in that it finds an input $\bar{u}(i)$ depending on the present state $\bar{x}(i)$.

The SMPC problems defined in this section are in general hard to deal with directly because of their stochastic formulations. However, the chance constraints on Hammerstein models can be converted into deterministic constraints, as in the case of linear models [11]. The converted problems are deterministic nonlinear programming problems that are solvable with a common optimization solver [30, 31].

10.4.4 Numerical Simulations

10.4.4.1 Simulation Settings

The EAST 30 system model [12] is a model of an electrical power system, and the simulation uses the smaller grid of the two interconnected areas in it. The capacity and initial output of the generators are shown in Table 10.6. The model consists of six thermal power plants and two nuclear power plants. However, nuclear power generators are not controlled in the simulation since the controller is designed for a fast time scale. Although this power system model was originally interconnected to the other grid, the simulation treats it as a single grid in order to simply examine the effect of a supply–demand imbalance on the frequency.

The parameters of the power system are shown in Table 10.7. The GF and LFC capacities have been chosen so that they represent a typical composition of generators

Table 10.6 Parameters of the generators

	Rated capacity (MW)	Initial output (MW)
Nuclear	3000	2800
Thermal	5560	4150
Wind	2000	1050

Table 10.7 Parameters of the power system

Reference frequency	(Hz)	50
Power system constant	(%MW/Hz)	9
GF capacity	(%)	±5
	(MW)	±278
LFC capacity	(%)	±1.5
	(MW)	±125

Table 10.8 Parameters of the controller

Time interval	Δt	1 (s)
Prediction horizon	N	7
Weight matrix	Q	diag(10000, 10, 100, 0.1)
	R	diag(100, 10, 0.01)
Constraints	ω_{max}	0.05 (Hz)
	$u_{t,max}$	125 (MW)
	$u_{b,max}$	100 (MW)
	$\Delta u_{t,max}$	4.63 (MW/s)
		(5%MW/h)
	$\Delta u_{b,max}$	100 (MW/s)
Standard deviation of W	σ	9.9406 (MW/s)

at night, when the effects of wind power fluctuations become apparent and LFC capacity is small. We will assume 2000 (MW) of wind power on the basis of Japan's target in 2030 [1].

The BESS capacity is set to 100 (MW). Irie et al. [14] show that if BESS has a 200 (MW) capacity, the HPWH is not needed for frequency control. However, in terms of cost, it is preferable to reduce the BESS capacity. Thus, our goal here will be to show that we can adjust the HPWH output to deal with the fluctuation when the BESS capacity is 100 (MW). In particular, we suppose that there are 1.26 million HPWHs, so the capacity is 1512 (MW) [14]. Irie et al. [14] investigated the number of households in the actual area depicted in the simulation and calculated the capacity assuming that 30% of them have HPWHs. Here, we will assume that 20% of the rated capacity of HPWH is controlled; thus, the controllable capacity is 302.4 (MW).

The parameters of the controller, the upper and lower bounds of the constraints, and the variance of the disturbance are shown in Table 10.8. The probability bounds for chance constraints are all set to 90%.

The simulation was carried out in MATLAB/Simulink for 1800 (s). The control methods were open-loop SMPC (Problem 1), closed-loop SMPC (Problem 2), and deterministic MPC, which does not consider disturbances. We prepared ten patterns

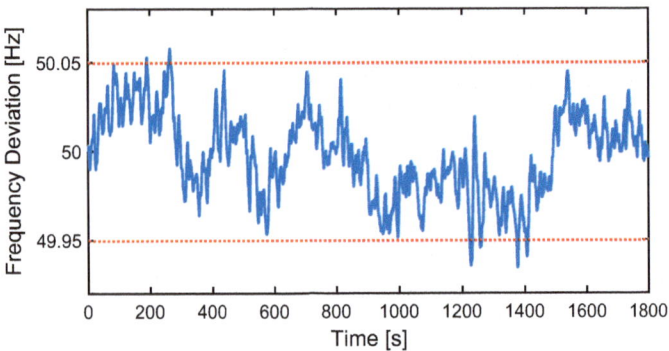

Fig. 10.24 Frequency deviation of closed-loop SMPC

of disturbance data, as mentioned in Sect. 10.4.2. We carried out simulations with each disturbance pattern.

As for the estimation of the demand characteristic, we changed the parameter a_{HP} in the middle of the simulation in order to confirm that the uncertainty in it is properly coped with and does not lead to failure of the frequency control. The initial value was set to $a_{HP} = 5$ and was changed to $a_{HP} = 4$ at 900 (s) in the simulation. The initial estimate of the estimator was $a_{HP} = 4.5$.

10.4.4.2 Performance Comparison

Figure 10.24 shows the simulated frequency deviation in closed-loop SMPC; it is clear that the deviation does not exceed the bounds for most of the simulation time. Let us compare the responses of closed-loop SMPC, open-loop SMPC, and deterministic MPC in terms of two performance measures: (a) the percentage of time that the frequency exceeds the limit ω_{max}, and (b) the average and variance of the stage cost $\frac{1}{2}(y(i)^{\mathsf{T}} Q y(i) + u(i)^{\mathsf{T}} R u(i))$ calculated from the simulation output. Figure 10.25 plots measure (a) for the three control methods for ten different disturbance patterns. The performance values of the three control methods for the same disturbance pattern are connected with a line. It can be seen that the closed-loop SMPC has the smallest percentage, followed in order by open-loop SMPC and deterministic MPC, for all disturbance patterns. SMPC can predict future state distributions affected by disturbances; thus, it handles the constraints more properly than deterministic MPC does. Additionally, closed-loop SMPC can consider future feedback; thus, it can deal with constraints even better than open-loop SMPC can.

Figure 10.26 shows the average stage cost. It can be seen that the smallest cost is achieved by closed-loop SMPC, followed in order by open-loop SMPC and deterministic MPC. One of the reasons for this result is that closed-loop SMPC reduces conservativeness against constraints.

Fig. 10.25 Percentage of time in which the frequency exceeds the limit value ω_{max}

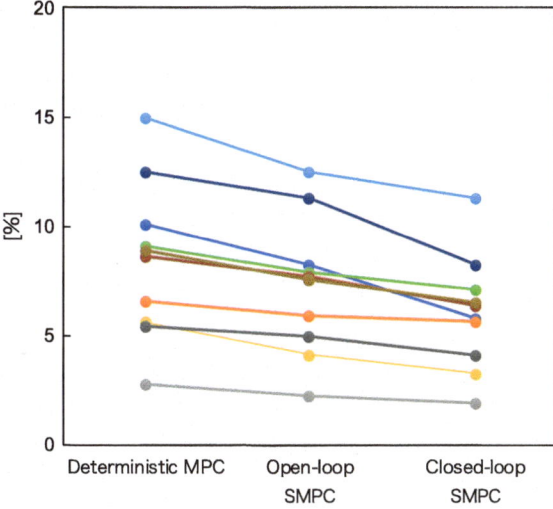

Fig. 10.26 Average stage cost

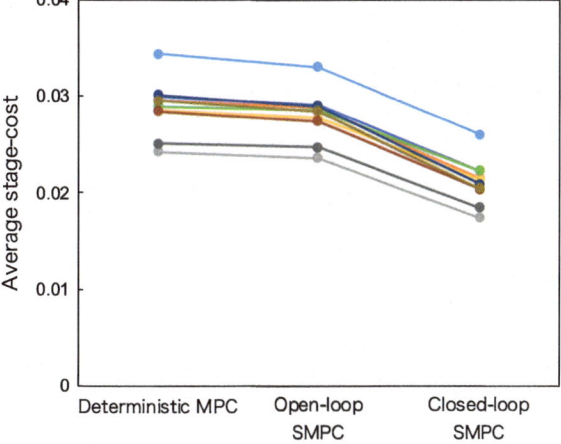

In summary, closed-loop SMPC can avoid violations of constraints more effectively and at the same time, reduce the stage cost. Thus, we conclude that it is the best control method of the three. However, a disadvantage is its high calculation cost due to the large number of decision variables. As for the other two methods, their calculation times were less than the control cycle of 1 (s) for the simulation's setup (CPU: Intel(R) Xeon(R) E5-2650 v4 2.20 GHz). In contrast, the calculation time for closed-loop SMPC was about 40 (s). Note that we used MATLAB's `fmincon`, which is an off-the-shelf nonlinear optimization solver; thus, the calculation time may be further reduced if a dedicated algorithm for closed-loop SMPC is developed.

10.5   Conclusions

In this chapter, we formulated load frequency control using real-time pricing as an NMPC for power systems where the power market is liberalized and a large amount of renewable energy is used. NMPC is a general framework that can explicitly take nonlinear models and constraints into account, and it was shown in this chapter that it can deal with various problem settings of real-time pricing flexibly. In NMPC, a nonlinear optimization problem has to be solved at each time step to determine the control input, and the computation cost can be an obstacle to its implementation. However, it was shown that the computation time for NMPC is sufficiently short when maximizing the social welfare and minimizing the load frequency fluctuation. Furthermore, it was shown that SMPC that explicitly considers the fluctuation of renewable energy as a stochastic disturbance can improve control performance while satisfying constraints on the load frequency fluctuation as much as possible. The combination of online estimation and NMPC should also be effective in various problem settings other than those described in this chapter.

Acknowledgements Sections 10.2 and 10.3 have been shortened and translated from their Japanese versions [16, 28] under permission of the Institute of Systems, Control and Information Engineers and the Society of Instrument and Control Engineers, respectively.

References

1. Agency for Natural Resources and Energy (2009) Long-term energy supply and demand outlook (in Japanese
2. Apt J (2007) The spectrum of power from wind turbines. J Power Sources 169(2):369–374
3. Berger AW, Schweppe FC (1989) Real time pricing to assist in load frequency control. IEEE Trans Power Syst 4(3):920–926. https://doi.org/10.1109/59.32580
4. Bevrani H, Ghosh A, Ledwich G (2010) Renewable energy sources and frequency regulation: survey and new perspectives. IET Renew Power Gener 4(5):438–457
5. Borenstein S, Jaske M, Rosenfeld A (2002) Dynamic pricing, advanced metering, and demand response in electricity markets. Center for the Study of Energy Markets, UC Berkeley
6. Borenstein S, Jaske M, Rosenfeld A (2002) Dynamic pricing, advanced metering, and demand response in electricity markets. J Am Chem Soc 128(12):4136–4145
7. Crossley D (2008) The role of advanced metering and load control in supporting electricity networks. In: Research report no. 5 task XV of the international energy agency demand side management programme. Energy Futures Australia PTY LTD
8. Farina M, Giulioni L, Scattolini R (2016) Stochastic linear model predictive control with chance constraints—a review. J Process Control 44:53–67
9. Fruzzetti KP, Palazoğlu A, McDonald KA (1997) Nonlinear model predictive control using Hammerstein models. J Process Control 7(1):31–41
10. Goulart PJ, Kerrigan EC, Maciejowski JM (2006) Optimization over state feedback policies for robust control with constraints. Automatica 42:523–533
11. Hashimoto T (2017) Transformation of a chance constraint in stochastic model predictive control problems (in Japanese). System/Control/Information 61(2):63–68
12. IEE Japan (1999) Standard model in the power system (in Japanese). IEEJ Tech Rep 754
13. International Energy Agency (2016) Medium-term renewable energy market report 2016—market analysis and forecasts to 2021—executive summary

14. Irie H, Yokoyama A, Tada Y (2010) System frequency control by coordination of batteries and heat pump based water heaters on customer side in power system with a large penetration of wind power generation (in Japanese). Trans IEEJ B 130(3):338–346
15. Jokic A, Lazar M, van den Bosch PPJ (2009) Real-time control of power systems using nodal prices. Int J Electr Power Energy Syst 31(9):522–530
16. Kamemoto H, Hashimoto T, Kashima K, Ohtsuka T (2014) Model predictive control based real-time pricing for load frequency control in electric power systems (in Japanese). Trans Inst Syst Control Inf Eng 27(10):405–411
17. Katayama T (2011) Nonlinear Kalman filter (in Japanese). Asakura Publishing
18. Kunder P (1994) Power system stability and control. McGraw-Hill
19. Mesbah A (2016) Stochastic model predictive control: an overview and perspectives for future research. IEEE Control Syst Mag 36(6):30–44
20. Mesbah A, Streif S, Findeisen R, Braatz RD (2014) Stochastic nonlinear model predictive control with probabilistic constraints. In: Proceedings of the 2014 American control conference, pp 2413–2419
21. Namerikawa T, Okubo N, Sato R, Okawa Y, Ono M (2015) Real-time pricing mechanism for electricity market with built-in incentive for participation. IEEE Trans Smart Grid 6(6):2714–2724
22. New Energy and Industrial Technology Development Organization (2014) Renewable energy technology white paper, 2nd edn
23. Nishizaki Y, Irie H, Yokoyama A, Tada Y (2009) Blade pitch angle control and its capacity reduction effect on battery for load frequency control in power system with a large capacity of wind power generation (in Japanese). IEEJ Trans Power Energy 129(1):50–56. https://doi.org/10.1541/ieejpes.129.50
24. Ohtsuka T (2004) A continuation/GMRES method for fast computation of nonlinear receding horizon control. Automatica 40(4):563–574. https://doi.org/10.1016/j.automatica.2003.11.005
25. Oldewurtel F, Jones CN, Morari M (2008) A tractable approximation of chance constrained stochastic MPC based on affine disturbance feedback. In: Proceedings of the 47th IEEE conference on decision and control, pp 4731–4736
26. Roozbehani M (2010) On the stability of wholesale electricity markets under real-time pricing. In: Proceedings of the 49th IEEE conference on decision and control, pp 1911–1918
27. Satouchi R, Kawano Y, Ohtsuka T (2016) Load frequency control by integrating real-time price presentations for consumers and direct commands issued to generators and batteries. In: IEEE international conference on sustainable energy technologies, pp 396–400
28. Satouchi R, Kawano Y, Ohtsuka T (2017) Real-time pricing with consumers estimation by a particle filter (in Japanese). Trans Soc Instrum Control Eng 53(8):463–472
29. Schildbach G, Fagiano L, Frei C, Morari M (2014) The scenario approach for stochastic model predictive control with bounds on closed-loop constraint violations. Automatica 50(12):3009–3018
30. Yanagiya T (2019) Load frequency control by stochastic model predictive control utilizing real-time pricing (in Japanese). In: Master's thesis, Department of Systems Science, Graduate School of Informatics, Kyoto University
31. Yanagiya T, Okajima Y, Hashimoto T, Ohtsuka T (2018) Real-time pricing using stochastic model predictive control with high penetration of wind power generation (in Japanese). In: Proceedings of the 61st joint automatic control conference, pp 978–985

Chapter 11
Distributed Multi-Agent Optimization Protocol over Energy Management Networks

Izumi Masubuchi, Takayuki Wada, Yasumasa Fujisaki and Fabrizio Dabbene

Abstract Distributed multi-agent optimization is a key methodology to solve problems arising in large-scale networks, including recent energy management systems that consist of such many entities as suppliers, consumers, and aggregators, who behave as independent agents. In this chapter, we focus on distributed multi-agent optimization based on the linear consensus protocol and exact penalty methods, which can deal with a wide variety of convex problems with equality and inequality constraints. The agents in the network do not need to disclose their objective and constraint functions to the other agents, and the optimization is executed only by sending and receiving decision variables that are common and are needed to coincide among the agents in the network. The protocol shown in this chapter can work over unbalanced networks to obtain a Pareto optimal solution, which is applied to solve minimax problems via distributed computation. After describing the protocol, we provide a concrete proof of the convergence of the decision variables to an optimal consensus point and show numerical examples including an application to a direct-current optimal power flow problem.

I. Masubuchi (✉)
Graduate School of System Informatics, Kobe University, 1-1 Rokkodai,
Nada 657-8501, Japan
e-mail: msb@harbor.kobe-u.ac.jp

T. Wada · Y. Fujisaki
Graduate School of Information Science and Technology, Osaka University,
1-5 Yamadaoka Suita, Osaka 565-0871, Japan
e-mail: t-wada@ist.osaka-u.ac.jp

Y. Fujisaki
e-mail: fujisaki@ist.osaka-u.ac.jp

F. Dabbene
Institute of Electronics Computer and Telecommunication Engineering,
National Research Council of Italy (CNR-IEIIT), Torino, Italy
e-mail: fabrizio.dabbene@polito.it

© Springer Nature Singapore Pte Ltd. 2020
T. Hatanaka et al. (eds.), *Economically Enabled Energy Management*,
https://doi.org/10.1007/978-981-15-3576-5_11

11.1 Introduction

Distributed multi-agent optimization has been receiving a great deal of attention as a key methodology for decision making over large-scale networks to attain a common purpose over a network by independent agents in the network. Network-wide decision making is now an important issue in energy management systems, consisting of a complicated network and many entities such as thermal plants, wind farms, photovoltaic farms, consumers, and aggregators, whose goal is to maximize their own profit. In the same time, energy management systems are physical systems whose stable operation is of the first importance to maintain the quality of energy service. A typical problem of large-scale networks of energy management systems is the optimal power flow (OPF) problem, the minimization of the total generation cost subject to limitations of power outputs, transmission capacities, and balance between supply and demand [26, 28]. Such optimal problems can be solved by gathering all information on objective and constraint functions of the agents in the network. However, because of the privacy reasons, agents often cannot disclose their objective and constraint functions to other agents; these functions contain information on power network parameters, generation costs, load demands, limits on generators and transmission lines, and so on. Then, the optimization with hiding the objective and constraint functions of the agents can only be realized via distributed optimization, executed by the agents with limited communication exchanging, e.g., only decision variables that are common over the network.

One of the approaches to realize distributed optimization is the multi-agent optimization, where agents modify the decision variable to decrease their objective function under their own constraints as well as make their decision variable closer to those of other agents available in each iteration. The latter, called consensus of the decision variables, is a foundation of various methods of distributed optimization and there are a considerable number of methods proposed in the literature based on linear protocols for consensus [12, 23] and subgradient methods for convex optimization [3]. A protocol for unconstrained distributed multi-agent optimization has been proposed in [21] with time-varying symmetric information exchange, where agents do both of sending and receiving of information to other agents in one iteration. This leads to an undirected graph that represents the structure of the information exchange in the network. The results are extended to optimization with asymmetric communication in [22], which employs only inequality constraints that are common among all agents in the network, while balanced directed graphs are admitted. An optimization problem with inequality and equality constraints is solved in [30] via a primal–dual subgradient method with the linear protocol, which can deal with individual inequality constraints of agents if no equality constraints are posed, while equality and inequality constraints need to be common among the agents to include an equality constraint. Another important approach to distributed optimization is alternating direction methods of multipliers (ADMM) [5]. Applications of ADMM to OPF problems can be found in, e.g., [13] and [1], where symmetric communication is assumed. Notice that in ADMM-based methods and also in primal–dual methods such as [6], each agent solves an optimization problem in every iteration and the network structure is often assumed to be symmetric, i.e., the connections in the network are symmetric, which is represented with an undirected (and hence balanced) graph.

Subgradient-based methods have been further extended to be able to deal with more general formulations of distributed optimization problems. In particular, exact penalty methods [3, 7] have been exploited to multi-agent optimization in [15–18, 20], where the protocols can work over directed graphs and handle individual equality and inequality constraints. At each iteration, the agents only compute the subgradients of the objective and constraint functions in addition to averaging of his/her own decision variable and decision variables received from the neighboring agents. These protocols are applicable to unbalanced networks [17, 27, 29] to obtain a Pareto optimal solution, by which we can solve constrained distributed minimax optimization problems over unbalanced networks. In this chapter, we provide a most updated protocol in this line of our study based on [17, 19, 20] that achieves partial consensus of decision variables of agents at a Pareto optimal solution over an unbalanced network, where we mean by partial consensus that each agent can posses a common decision variable and an individual decision variable where only the common variable needs to converge to a common value among the agents while the individual variable does not. Then, the protocol enjoys more freedom in the formulation of optimization problems than full-consensus protocols, by introducing individual auxiliary variables. We provide a concrete proof of the convergence of the decision variable generated by the proposed protocol and numerical examples of the applications of the proposed protocol to an academic example of the constrained minimax optimization problem and a direct-current optimal power flow (DC-OPF) problem with thirty buses.

The rest of this chapter is organized as follows. A multi-agent distributed optimization problem is formulated in Sect. 11.2. Section 11.3 provides a distributed multi-agent optimization protocol, and the proof of the convergence of the decision variable is shown in Sect. 11.4. Section 11.5 presents applications of the protocol to a minimax problem and a DC-OPF problem. Finally, Sect. 11.6 concludes this chapter.

Notation

Let $\mathbb{R}^n$ and $\mathbb{R}^{m \times n}$ denote the sets of real n-vectors and $m \times n$-matrices, respectively. For vectors $x, y \in \mathbb{R}^n$, $\|x\|$ is the Euclidean norm of x and $\langle x, y \rangle$ is the standard inner product of x, y, while $\|x\|_q$ denotes the q-norm of x. Inequalities on real vectors are elementwise. Let $\mathbf{1}_N = [1 \ 1 \ \cdots \ 1]^\top \in \mathbb{R}^N$ and I_N be the identity matrix of size N. The diagonal matrix whose diagonal entries are $d_1, d_2, \ldots, d_n$ is represented as $\mathrm{diag}\{d_1, d_2, \ldots, d_n\}$. Let $\overline{\mathrm{B}}(r; x)$ be the closed ball whose center is x and whose radius is r. For matrices A and B, $A \otimes B$ stands for the Kronecker product of A and B. The gradient of a continuously differentiable function $f : \mathbb{R}^n \to \mathbb{R}$ at x is denoted by $\nabla f(x)$. We write $[x_1^\top x_2^\top \cdots x_k^\top]^\top = (x_1, x_2, \ldots, x_k)$ for column vectors $x_1, x_2, \ldots, x_k$. For a set S, S^c stands for the complement of S. The interior of $S \subset \mathbb{R}^n$ is denote by S°.

11.2 Problem Formulation

11.2.1 Multi-agent Pareto Optimization Problem

We consider a multi-agent optimization problem on a network of agents. Let $\mathcal{A} = \{1, 2, \ldots, N\}$ be the set of agents. Suppose that each agent i has functions $f^i : \mathbb{R}^n \times \mathbb{R}^{l^i} \to \mathbb{R}$, $g^i : \mathbb{R}^n \times \mathbb{R}^{l^i} \to \mathbb{R}$, $h^i : \mathbb{R}^n \times \mathbb{R}^{l^i} \to \mathbb{R}^{m^i}$, and that the benefit of agent i is to minimize $f^i(x^i, y^i)$ subject to inequality constraint $g^i(x^i, y^i) \leq 0$ and equality constraint $h^i(x^i, y^i) = 0$. Variable x^i is common to the network and hence the consensus among agents, i.e., the coincidence of variables x^i is needed, while variable y^i is used only by agent i to express its objective and constraint functions. Accordingly, we formulate the following problem:

Problem 1 Find a Pareto optimal solution $(x^i, y^i) \in \mathbb{R}^{n+l^i}, i = 1, 2, \ldots, N$ for

$$\min_{(x^i, y^i) \in \mathbb{R}^{n+l^i}, i=1,2,\ldots,N} \{f^1(x^1, y^1), f^2(x^2, y^2), \ldots, f^N(x^N, y^N)\} \qquad (11.1)$$

subject to constraints:

$$g^i(x^i, y^i) \leq 0, \quad h^i(x^i, y^i) = 0, \quad i = 1, 2, \ldots, N \qquad (11.2)$$

and the partial consensus of decision variables, namely

$$x^1 = x^2 = \cdots = x^N. \qquad (11.3)$$

Here, we say that a feasible solution $\{(x^i, y^i), i = 1, 2, \ldots, N\}$ is *Pareto optimal* if it is not possible to move from that point and decrease at least one objective function without increasing any other objective functions. See e.g. [14].

To realize the consensus, the agents can communicate with each other to exchange variables x^i. Let $\mathcal{J}^i$ be a subset of $\mathcal{A} \setminus \{i\}$ and suppose that agent i can receive variables x^j from agents j that belong to $\mathcal{J}^i$. By this, we define a directed graph $\mathcal{G} = (\mathcal{A}, \mathcal{E})$, where $\mathcal{E} = \{(j, i) \in \mathcal{A} \times \mathcal{A} : j \in \mathcal{J}^i\}$.

Assumption 1 Graph $\mathcal{G}$ is strongly connected.

Definition 1 A convex function $f : \mathbb{R}^n \to \mathbb{R}$ is said to be proper if its domain D is not empty. A vector $v \in \mathbb{R}^n$ is a subgradient of f at $x_0 \in D$ if $f(x) - f(x_0) \geq \langle v, x - x_0 \rangle$ holds for all $x \in D$. Let $\partial f(x_0)$ be the set of all subgradients of f at x_0, which set is called subdifferential of f at x_0. Then, f is said to be subdifferentiable at $x_0 \in D$ if $\partial f(x_0)$ is not empty.

We assume the following on functions f^i, g^i, h^i.

Assumption 2 Functions $f^i, g^i, i = 1, 2, \ldots, N$ are proper subdifferentiable convex on $\mathbb{R}^{n+l^i}$ and $h^i, i = 1, 2, \ldots, N$ are affine functions on $\mathbb{R}^{n+l^i}$.

Let $\partial f^i(x^i, y^i)$ denote the subdifferential of f^i at $(x^i, y^i) \in \mathbb{R}^{n+l^i}$. From Assumption 2, subdifferentials $\partial f^i(x^i, y^i)$, $\partial g^i(x^i, y^i)$, $\partial h^i(x^i, y^i)$ are non-empty closed sets [24].

Assumption 3 Functions $f^i, i = 1, 2, \ldots, N$ are globally Lipschitz continuous on $\mathbb{R}^{n+l^i}$ and functions $g^i, i = 1, 2, \ldots, N$ are locally Lipschitz continuous at every point of $\mathbb{R}^{n+l^i}$.

Let $\mathbb{P}^N$ be the subset of $\mathbb{R}^N$ such that $\pi = [\pi^1 \pi^2 \cdots \pi^N]^\top \in \mathbb{P}^N$ satisfies $\pi^i > 0$, $i = 1, 2, \ldots, N$ and $\sum_{i=1}^{N} \pi^i = 1$. A Pareto optimal solution is obtained by solving the following problem that minimizes a weighted sum of the objective functions with any vector $\pi \in \mathbb{P}^N$, and vice versa if the problem is convex [14].

Problem 2 Find $(x^i, y^i) \in \mathbb{R}^{n+l^i}, i = 1, 2, \ldots, N$ that minimizes

$$\sum_{i=1}^{N} \pi^i f^i(x^i, y^i)$$

subject to (11.2) and (11.3).

Problem 1 can be applied to several standard objective functions without depending on π when the objective functions are common over the network and consequently depend only on the common variable x, i.e., $f^1(x, y^1) = f^2(x, y^2) = \cdots = f^N(x, y^N) = f(x)$ for some $f : \mathbb{R}^n \to \mathbb{R}$. An important application of this is the distributed minimax optimization:

$$\min_{(\tilde{x}^i, y^i) \in \mathbb{R}^{\tilde{n}+l^i}, i=1,2,\ldots,N} \quad \max_{1 \leq i \leq N} \tilde{f}^i(\tilde{x}^i, y^i) \tag{11.4}$$

subject to $\tilde{g}^i(\tilde{x}^i, y^i) \leq 0$, $\tilde{h}^i(\tilde{x}^i, y^i) = 0$, and $\tilde{x}^1 = \tilde{x}^2 = \cdots = \tilde{x}^N$. This problem is subsumed in Problem 1 by defining $x^i = (\tilde{x}^i, \gamma^i)$, $f^i(x^i, y^i) = \gamma^i$, $g^i(x^i, y^i) = \max\{\tilde{f}^i(\tilde{x}^i, y^i) - \gamma^i, \tilde{g}^i(\tilde{x}^i, y^i)\}$, $h^i(x^i, y^i) = \tilde{h}^i(\tilde{x}^i, y^i)$, where $\gamma^i \in \mathbb{R}$ is a slack variable. Notice that a distributed minimax optimization over a balanced network with only inequality constraints has been considered in [25].

Also, we can handle the minimization problem of sum of objective functions:

$$\min_{(\tilde{x}^i, y^i) \in \mathbb{R}^{\tilde{n}+l^i}, i=1,2,\ldots,N} \quad \sum_{i=1}^{N} \tilde{f}^i(\tilde{x}^i, y^i) \tag{11.5}$$

subject to $\tilde{g}^i(\tilde{x}^i, y^i) \leq 0$, $\tilde{h}^i(\tilde{x}^i, y^i) = 0$ and $\tilde{x}^1 = \tilde{x}^2 = \cdots = \tilde{x}^N$, where the sum in (11.5) does not involve π. A technique is provided in [29] to solve the min-sum problem in unbalanced networks with slack variable $\gamma^i \in \mathbb{R}^N, i = 1, 2, \ldots, N$ to be made consensus. The settings are as follows: $x^i = (\tilde{x}^i, \gamma^i)$, $f^i(x^i, y^i) = \sum_{j=1}^{N} \gamma^{ij}$, $g^i(x^i, y^i) = \max\{\tilde{f}^i(\tilde{x}^i, y^i) - \gamma^{ii}, \tilde{g}^i(\tilde{x}^i, y^i)\}$, $h^i(x^i, y^i) = \tilde{h}^i(\tilde{x}^i, y^i)$, where γ^{ij} denotes the jth element of vector γ^i. Note that this rephrasing of the problem requires slack variable γ^i of size N, the number of agents in the network, and the optimal values of $\tilde{f}^i(\tilde{x}^i, y^i)$ are shared by all the agents through γ^i, which coincide with each other under consensus. This method of additional variable (γ^i above) can be applied to inequality and equality conditions on the summation of functions of agents over the network, that is, inequality $\sum_{i=1}^{N} \tilde{g}_s^i(\tilde{x}^i, y^i) \leq 0$ is decomposed with additional variable $\gamma_s^i \in \mathbb{R}^N$ as $\tilde{g}_s^i(\tilde{x}^i, y^i) \leq \gamma^{ii}$, $\sum_{j=1}^{N} \gamma^{ij} \leq 0$ with $x^i = (\tilde{x}^i, \gamma^i)$. Equality conditions can be handled similarly.

In addition to the purpose of describing objective and constraint functions that depend on variables other than x which are not common among agents, local variables y^i can be utilized to manipulate objective functions that are not globally Lipschitz continuous. Represent Problem 1 as

$$\min_{(x^i, \tilde{y}^i) \in \mathbb{R}^{n+\tilde{l}^i}, i=1,2,\dots,N} \{\tilde{f}^1(x^1, \tilde{y}^1), \tilde{f}^2(x^2, \tilde{y}^2), \dots, \tilde{f}^N(x^N, \tilde{y}^N)\}$$

subject to $\tilde{g}^i(x^i, \tilde{y}^i) \leq 0, \tilde{h}^i(x^i, \tilde{y}^i) = 0, i = 1, 2, \dots, N$, and $x^1 = x^2 = \dots = x^N$. If objective functions $\tilde{f}^i$ are not globally Lipschitz, we can rewrite the problem in the form of Problem 1 with setting $y^i = (\tilde{y}^i, \phi^i)$, $f^i(x^i, y^i) = \phi^i$, $g^i(x^i, y^i) = \max\{\tilde{f}^i(x^i, \tilde{y}^i) - \phi^i, \tilde{g}^i(x^i, \tilde{y}^i)\}$, $h^i(x^i, y^i) = \tilde{h}^i(x^i, \tilde{y}^i)$, where ϕ^i is a local slack variable. Apparently the new objective functions f^i satisfy Assumption 3.

11.2.2 Penalized Objective Function and Its Exactness

The protocol of this paper will be based on exact penalty methods [3, 7] for the following problem. Let $y = (y^1, y^2, \dots, y^N)$, $l = \sum_{i=1}^{N} l^i$.

Problem 3 Find $(x, y) \in \mathbb{R}^{n+l}$ that minimizes

$$F(x, y) = \sum_{i=1}^{N} \pi^i f^i(x, y^i)$$

subject to

$$g^i(x, y^i) \leq 0, \quad h^i(x, y^i) = 0, \quad i = 1, 2, \dots, N. \tag{11.6}$$

Define the feasible set of Problem 3 as

$$X_F = \{(x, y) \in \mathbb{R}^{n+l} : g^i(x, y^i) \leq 0, \quad h^i(x, y^i) = 0, \quad i = 1, 2, \dots, N\}.$$

Problem 3 is a version of Problem 2 with the consensus being previously assumed: $x = x^1 = x^2 = \dots = x^N$. Note that Problem 3 is invoked only for the problem formulation and the proof of the convergence of our protocol. Let $X_{*\pi}$ be the set of optimal solutions $(x_*, y_*) \in \mathbb{R}^{n+l}$ to Problem 3 for vector π, where $y_* = (y_*^1, y_*^2, \ldots, y_*^N)$.

Assumption 4 For all $\pi \in \mathbb{P}^N$, $X_{*\pi}$ is a non-empty compact subset of $\mathbb{R}^{n+l}$.

For $(x_*, y_*) \in X_{*\pi}$, let $f_* = \sum_{i=1}^{N} \pi^i f^i(x_*, y_*^i)$. By definition, $g^i(x_*, y_*^i) \leq 0$ and $h^i(x_*, y_*^i) = 0, i = 1, 2, \ldots, N$.

Next, let $g^i(x, y^i)$ be given as

$$g^i(x, y^i) = \max_{1 \leq j \leq N^i} g^{ij}(x, y^i),$$

where $g^{ij}(x, y^i)$ satisfies Assumptions 2 and 3, and set

$$g_{col}^i(x, y^i) = \left[\max\{g^{i1}(x, y^i), 0\} \quad \cdots \quad \max\{g^{i N^i}(x, y^i), 0\} \right]^\top.$$

Then define

$$G^i(x, y^i) = \left\| \begin{bmatrix} g_{col}^i(x, y^i) \\ h^i(x, y^i) \end{bmatrix} \right\|_q, \qquad i = 1, 2, \ldots, N, \qquad (11.7)$$

where $1 \leq q \leq \infty$. Clearly $G^i(x, y^i) = 0, i = 1, 2, \ldots, N$ if and only if (x, y) is feasible. For each constant $p > 0$, define the following *penalized objective function*

$$F(x, y; p) = \sum_{i=1}^{N} \pi^i (f^i(x, y^i) + p G^i(x, y^i)). \qquad (11.8)$$

Since $G^i(x, y^i) \geq 0$, we have a monotonicity in p as

$$F(x, y; p_1) \leq F(x, y; p_2) \quad \text{if} \quad p_1 \leq p_2. \qquad (11.9)$$

Let $K \subset \mathbb{R}^{n+l}$ be a compact set that satisfies

$$X_{*\pi} \subset K, \qquad X_F \cap K^\circ \neq \emptyset \qquad (11.10)$$

and define $X_{*\pi}^K(p)$ be the set of optimal solutions $(x, y) \in \mathbb{R}^{n+l}$ to the following penalized optimization problem:

Problem 4 Find $(x, y) \in K, i = 1, 2, \ldots, N$ that minimizes $F(x, y; p)$.

Since $F(x, y; p)$ is continuous and K is compact, $X_{*\pi}^K(p) \neq \emptyset$.

Assumption 5 For each $\pi \in \mathbb{P}^N$ and compact set $K \subset \mathbb{R}^{n+l}$ that satisfies (11.10), there exists a $p_{*\pi}^K > 0$ such that $X_{*\pi}^K(p) = X_{*\pi}$ holds for all $p \geq p_{*\pi}^K$.

We say that penalized objective function $F(x, y; p)$ is *exact* if Assumption 5 holds.

The exactness of a penalized function in the penalized optimization problem over a compact set can be satisfied under mild constraint qualifications such as *Mangasarian–Fromovitz constraint qualification* (MFCQ) investigated in [10] and [8, 9]. See e.g. [7] for a survey on exact penalty methods and constraint qualifications. For notational simplicity, here we refer to MFCQ for constrained optimization problems in a standard form, namely $\min_{x \in \mathbb{R}^n} f(x)$ such that $g^i(x) \leq 0$, $i = 1, 2, \ldots, m$ and $h^j(x) = 0$, $j = 1, 2, \ldots, l$, where g^i and h^j are scalar-valued continuously differentiable functions. Let $x_1 \in \mathbb{R}^n$ be an optimal solution to this problem and denote by I_1 the set of indexes for which the inequality constraints are active, i.e., $I_1 = \{1 \leq i \leq m : g^i(x_1) = 0\}$. Then, MFCQ is said to hold at x_1 if $\nabla h^j(x_1), j = 1, 2, \ldots, l$ are linearly independent and there exists a $z \in \mathbb{R}^n$ such that $\nabla g^i(x_1)^\top z < 0$, $i \in I_1$ and $\nabla h^j(x_1)^\top z = 0$, $j = 1, 2, \ldots, l$ are satisfied, whereas $z \neq 0$. The exactness of penalized objective functions in the sense of Assumption 5 is proved in [9] under MFCQ, based on [10] and [8]. Note that the objective function is not restricted by MFCQ.

If in addition the constraints are convex as in our formulation, i.e., functions g^i are convex and functions h^j are affine, then MFCQ is guaranteed by the *Slater condition* (see e.g. [4]), where the Slater condition requires the existence of an $x_0 \in \mathbb{R}^n$ such that $g^i(x_0) < 0$, $i = 1, 2, \ldots, m$ and $h^j(x_0) = 0$, $j = 1, 2, \ldots, l$. In fact, from the convexity, we have $g^i(x_0) \geq \nabla g^i(x_1)^\top (x_0 - x_1) + g^i(x_1)$ for x_0, x_1 as above and for $i \in I_1$. Since $g^i(x_0) < 0$ and $g^i(x_1) = 0$, we obtain $0 > \nabla g^i(x_0)^\top (x_0 - x_1)$. Moreover, since we can write $h^j(x) = (a^j)^\top x + b^j$ for constant $a^j \in \mathbb{R}^n$ and $b^j \in \mathbb{R}$, we see $\nabla h^j(x_1)^\top (x_0 - x_1) = (a^j)^\top (x_0 - x_1) = b^j - b^j = 0$. Therefore $z = x_0 - x_1$ satisfies MFCQ. Note that the linear independence condition on $\nabla h^j(x)$ in MFCQ is in fact not needed if the equality conditions are given with affine functions, for which $\nabla h^j(x)$ is constant, in view of how the linear independence is exploited in the proof of Theorem 2.2 in [10].

11.2.3 Radial Unboundedness of Constraint Functions

For the inequality constraint functions, we assume the following conditions, which can be seen as a certain radial unboundedness of a function involving its subgradient.

Assumption 6 For $i = 1, 2, \ldots, N$, there exist positive numbers ρ^i, λ^i, and σ^i such that

$$G^i(x, y^i) \geq \lambda^i \|(x, y^i)\|, \qquad G^i(x, y^i) \geq \sigma^i \|v\| \, \|(x, y^i)\| \qquad (11.11)$$

hold whenever $\|(x, y^i)\| > \rho^i$ and $v \in \partial G(x, y^i)$.

Notice that the second inequality in (11.11) is not needed if $G^i(x, y^i)$ is globally Lipschitz continuous, whereas subgradients of G^i is bounded.

Here, we show a function $g^i(x, y^i)$ satisfying Assumption 6. For notational simplicity, let $z^i = (x, y^i)$ and drop index i, i.e., we represent $g^i(x, y^i) = g(z)$. Let $g_j(z)$, $j = 1, 2, \ldots, m$ be polynomials of order at most $M (\geq 1)$ and let $g(z) = \max_{1 \leq j \leq m} g_j(z)$. Assume that the agents share a positive number $r_{\max}$ such that the norm of any optimal solution is no more than $r_{\max}$, which is in practice not hard to assume. Then, without changing $X_{*\pi}$, we can redefine $g(z)$ as

$$g(z) = \max\{g_1(z), \ldots, g_m(z), \hat{g}(z)\}, \quad \hat{g}(z) = \|z\|^{M+1} - r_{\max}^{M+1}. \quad (11.12)$$

For some ρ large enough with satisfying $\rho^{M+1} \geq 2r_{\max}^{M+1}$, the maximum in (11.12) is attained only by $\hat{g}(z)$ for $\|x\| > \rho$, whereas

$$g(z) = \|z\|^{M+1} - r_{\max}^{M+1}, \quad \partial g(z) = \{v\}, \quad v = (M+1)\|z\|^{M-1}z,$$

where $\|v\| = (M+1)\|z\|^M$. One can deduce for $\|z\| > \rho$ that

$$g(z) \geq \frac{1}{2}\rho^M \|z\|, \quad g(z) \geq \frac{1}{2(M+1)} \|v\| \|z\|,$$

by which $g(z)$ meets inequalities (11.11) with $g^i(x, y^i) = g(z)$. This implies that, with noticing that h^i is affine, functions g^i with polynomial growth as $\|(x, y^i)\| \to \infty$ can satisfy Assumption 6 with a barrier such as $\hat{g}$ in (11.12).

11.3 Distributed Optimization Protocol

We provide a protocol by which the agents in $\mathcal{A}$ asymptotically achieve a partial consensus at a Pareto optimum of Problem 1. Let $W = (w^{ij}) \in \mathbb{R}^{N \times N}$ be an arbitrary right stochastic matrix, i.e., $\sum_{j=1}^N w^{ij} = 1$, $i = 1, 2, \ldots, N$, that satisfies the compatibility with edge $\mathcal{E}$ in the sense that $w^{ij} > 0$ if $i = j$ or $(j, i) \in \mathcal{E}$ and $w^{ij} = 0$ otherwise. Then, since $\mathcal{G}$ is strongly connected as stated in Assumption 1, matrix W has an eigenvalue 1 with multiplicity one. Then, $\mathbf{1}_N$ is a right eigenvector corresponding to eigenvalue 1, while there exists a left eigenvector $\pi \in \mathbb{R}^N$ corresponding to eigenvalue 1 that belongs to $\mathbb{P}^N$. Below let $\pi \in \mathbb{P}^N$ be such a left eigenvector.

If W is doubly stochastic, then $\pi = \mathbf{1}^\top$ is a left eigenvector that corresponds to eigenvalue 1. It is known that, if graph $\mathcal{G}$ is strongly connected, then there exists a doubly stochastic matrix that is compatible to $\mathcal{G}$ [11]. We also note, however, that sharing a doubly stochastic matrix W needs extra communication between agents.

Let μ^i, $i = 1, 2, \ldots, N$ be positive numbers, and let N_{iter} be a positive integer.

Assumption 7 Let $\{a_t\}$ and $\{b_t\}$ be sequences of positive numbers that satisfy the following conditions:

$$\lim_{t\to\infty} a_t = 0, \quad \lim_{t\to\infty} b_t = 0, \quad \lim_{t\to\infty} \frac{a_t}{b_t} = 0, \quad \lim_{t\to\infty} \frac{b_t^2}{a_t} = 0,$$

$$\sum_{t=1}^{\infty} a_t = \infty, \quad \sum_{t=1}^{\infty} b_t^2 < \infty.$$

Assumption 7 is satisfied with $a_t = a_1/t^\alpha$ and $b_t = b_1/t^\beta$ with $a_1, b_1 > 0$ and $0.5 < \beta < \alpha \le 1$, for example.

Now, we are ready to present our protocol.

Protocol 1 *Each agent i executes the following with Step 2 done synchronously:*

1. *Initialize $(x_1^i, y_1^i) \in \mathbb{R}^{n+l^i}$, $i = 1, 2, \ldots, N$ arbitrarily and let $t := 1$.*
2. *Receive x_t^j, $j \in \mathcal{J}^i$.*
3. *Compute $\xi_t^i = \sum_{j=1}^{N} w^{ij} x_t^j$.*
4. *Take arbitrary $u_t^i \in \partial f^i(\xi_t^i, y_t^i)$ and $v_t^i \in \partial G^i(\xi_t^i, y_t^i)$.*
5. *Update x_t^i and y_t^i by*

$$(x_{t+1}^i, y_{t+1}^i) := (\xi_t^i, y_t^i) - a_t u_t^i - b_t s_t^i v_t^i, \tag{11.13}$$

where

$$s_t^i := \begin{cases} \mu^i / \|v_t^i\|, & \text{if } \|v_t^i\| > \mu^i, \\ 1, & \text{otherwise.} \end{cases} \tag{11.14}$$

6. *If $t < N_{\text{iter}}$, then let $t := t + 1$ and go to 2.*

The convergence of (x_t^i, y_t^i) in Protocol 1 is stated as follows.

Theorem 1 *Suppose that Assumptions 1–7 are satisfied. Then, for arbitrary initial value $(x_1^i, y_1^i) \in \mathbb{R}^{n+l^i}$, $i = 1, 2, \ldots, N$, there exists a Pareto optimal solution $(x_\infty^i, y_\infty^i) \in \mathbb{R}^{n+l^i}$, $i = 1, 2, \ldots, N$ to Problem 1 with*

$$x_\infty = x_\infty^1 = x_\infty^2 = \cdots = x_\infty^N$$

such that

$$\lim_{t\to\infty} x_t^i = x_\infty, \quad \lim_{t\to\infty} y_t^i = y_\infty^i, \quad i = 1, 2, \ldots, N. \tag{11.15}$$

Proof The proof is provided in the next section.   □

As is seen in Protocol 1, the optimization is executed without disclosing the objective and constraint functions of an agent to others. The consensus and update law at each step are quite easy to compute. One can see that a penalty method is embedded into this protocol as follows. Update law (11.13) can be written as

$$
(x_{t+1}^i, y_{t+1}^i) = (\xi_t^i, y_t^i) - a_t \left\{ u_t^i - \frac{b_t}{a_t} s_t^i v_t^i \right\}
$$

$$
\in (\xi_t^i, y_t^i) - a_t \partial \left[f^i(\xi_t^i, y_t^i) + \frac{s_t^i b_t}{a_t} G^i(\xi_t^i, y_t^i) \right],
$$

where a_t is square summable and not summable. Hence, update law (11.13) can be regarded as a subgradient algorithm applied to the penalized objective function with penalty $s_t^i b_t / a_t$, plugged in the standard linear consensus protocol [12, 23]. The exactness of Problem 4 then yields the convergence of Protocol 1 from this structure of the update law.

The scaling of the length of subgradient v_t^i in (11.13) and (11.14) is not needed if $g^i(x, y^i)$ is globally Lipschitz continuous, whereas v_t^i is bounded. This can be seen from the proof of the consensus and the boundedness shown in Sects. 11.4.2 and 11.4.3, respectively. In this case, the second inequality in (11.11) is not required to assume, as mentioned above in Sect. 11.2.3.

11.4   Proofs

11.4.1   Preliminaries

Define $\overline{\mu} = \max_{1 \le i \le N} \{\mu^i\}$ and $\underline{\mu} = \min_{1 \le i \le N} \{\mu^i\}$. We have

$$
0 < s_t^i \le 1, \quad \|s_t^i v_t^i\| \le \mu^i \le \overline{\mu}, \quad i = 1, 2, \dots, N, \quad t = 1, 2, \dots. \tag{11.16}
$$

From Assumption 3, there exists a constant $L > 0$ such that

$$
|f^i(x^i, y^i) - f^i(\tilde{x}^i, \tilde{y}^i)| \le L \|(x^i, y^i) - (\tilde{x}^i, \tilde{y}^i)\| \tag{11.17}
$$

holds for all $(x^i, y^i), (\tilde{x}^i, \tilde{y}^i) \in \mathbb{R}^{n+l^i}, i = 1, 2, \dots, N$, whereas

$$
\|u_t^i\| \le L, \quad i = 1, 2, \dots, N, \quad t = 1, 2, \dots. \tag{11.18}
$$

Write $u_t^i = (u_t^{1i}, u_t^{2i}), v_t^i = (v_t^{1i}, v_t^{2i})$ with $u_t^{1i}, v_t^{1i} \in \mathbb{R}^n$ and define

$$
\bar{x}_t := \sum_{i=1}^N \pi^i x_t^i, \qquad \bar{u}_t := \sum_{i=1}^N \pi^i u_t^{1i}, \qquad \bar{v}_t := \sum_{i=1}^N \pi^i s_t^i v_t^{1i}.
$$

It holds that

$$
\|\bar{u}_t\| \le \sum_{i=1}^N \pi^i \|u_t^{1i}\| \le L, \qquad \|\bar{v}_t\| \le \sum_{i=1}^N \pi^i \|s_t^i v_t^{1i}\| \le \overline{\mu} \tag{11.19}
$$

from (11.16). Notice that $\sum_{i=1}^{N} \pi^i \xi_t^i = \bar{x}_t$ and $\bar{x}_{t+1} = \bar{x}_t - a_t \bar{u}_t - b_t \bar{v}_t$. Lastly, Assumption 7 implies that there exist positive numbers $r_1, \hat{a}, \hat{b}$ and an integer $\underline{t}_0$ such that

$$a_t \leq r_1 b_t \quad \forall t \geq \underline{t}_0; \quad a_t \leq \hat{a}, \quad b_t \leq \hat{b} \quad \forall t \geq 1. \tag{11.20}$$

Note also that a_t is also square summable from the first inequality in (11.20).

11.4.2 Proof of the Consensus

First, let us consider the consensus of Protocol 1. Define

$$x_t^\delta := \begin{bmatrix} x_t^1 - x_t^2 \\ \vdots \\ x_t^{N-1} - x_t^N \end{bmatrix}.$$

It is easy to see

$$\|x_t^i - \bar{x}_t\| \leq d \|x_t^\delta\|, \qquad \|\xi_t^i - \bar{x}_t\| \leq d \|x_t^\delta\| \tag{11.21}$$

hold for some constant $d > 0$. Hence, the following lemma implies the asymptotic consensus of x_t^i, namely

$$\lim_{t \to \infty} (x_t^i - x_t^j) = 0, \quad i, j = 1, 2, \ldots, N. \tag{11.22}$$

Lemma 1 *Sequence $\{x_t^\delta\}$ is square summable.*

Proof Define

$$Z = \begin{bmatrix} 1 & -1 & & 0 \\ & \ddots & \ddots & \\ 0 & & 1 & -1 \end{bmatrix} \in \mathbb{R}^{N \times (N-1)}, \qquad T = \begin{bmatrix} Z \\ \Pi^\top \end{bmatrix}$$

and $\Pi = \mathrm{diag}\{\pi^1, \pi^2, \ldots, \pi^N\}$. Then, it is easy to see that

$$T^{-1} = \begin{bmatrix} \Pi^{-1} Z^\top (Z \Pi^{-1} Z^\top)^{-1} & \mathbf{1}_N \end{bmatrix}$$

and

$$TWT^{-1} = \begin{bmatrix} W_0 & 0 \\ 0 & 1 \end{bmatrix}, \qquad W_0 = ZW\Pi^{-1}Z^\top (Z\Pi^{-1}Z^\top)^{-1},$$

where the magnitude of the eigenvalues of W_0 is less than 1. Moreover, x_t^δ satisfies

$$x_{t+1}^\delta = (W_0 \otimes I_n)x_t^\delta - a_t u_t^\delta - b_t v_t^\delta, \tag{11.23}$$

where

$$u_t^\delta := \begin{bmatrix} u_t^{11} - u_t^{12} \\ \vdots \\ u_t^{1(N-1)} - u_t^{1N} \end{bmatrix}, \quad v_t^\delta := \begin{bmatrix} s_t^1 v_t^{11} - s_t^2 v_t^{12} \\ \vdots \\ s_t^{N-1} v_t^{1(N-1)} - s_t^N v_t^{1N} \end{bmatrix}.$$

Since a_t and b_t are square summable and u_t^δ and v_t^δ are bounded, (11.23) is a stable linear discrete-time system with l_2 inputs, which implies that x_t^δ is square summable. $\qquad\square$

11.4.3 Proof of the Boundedness

Let $(x_*, y_*^1, y_*^2, \ldots, y_*^N) \in X_{*\pi}$ be arbitrary and define

$$D_t = \sum_{i=1}^N \pi^i \left\| \begin{bmatrix} x_t^i - x_* \\ y_t^i - y_*^i \end{bmatrix} \right\|^2,$$

which plays a role similar to that of a Lyapunov function. From update law (11.13), the first inequality in (11.20), and the convexity of $\|\cdot\|^2$, f^i, G^i, the following inequality holds for all $t \ge t_0$:

$$\begin{aligned}
D_{t+1} &= \sum_{i=1}^N \pi^i \left\| \begin{bmatrix} x_{t+1}^i - x_* \\ y_{t+1}^i - y_*^i \end{bmatrix} \right\|^2 \\
&= \sum_{i=1}^N \pi^i \left\| \begin{bmatrix} \xi_t^i - x_* \\ y_t^i - y_*^i \end{bmatrix} - a_t u_t^i - b_t s_t^i v_t^i \right\|^2 \\
&= \sum_{i=1}^N \pi^i \left\{ \left\| \begin{bmatrix} \xi_t^i - x_* \\ y_t^i - y_*^i \end{bmatrix} \right\|^2 - 2a_t \left\langle u_t^i, \begin{bmatrix} \xi_t^i - x_* \\ y_t^i - y_*^i \end{bmatrix} \right\rangle \right. \\
&\qquad \left. - 2b_t s_t^i \left\langle v_t^i, \begin{bmatrix} \xi_t^i - x_* \\ y_t^i - y_*^i \end{bmatrix} \right\rangle + \|a_t u_t^i + b_t s_t^i v_t^i\|^2 \right\} \\
&\le \sum_{i=1}^N \pi^i \left\{ \left\| \begin{bmatrix} x_t^i - x_* \\ y_t^i - y_*^i \end{bmatrix} \right\|^2 - 2a_t (f^i(\xi_t^i, y_t^i) - f^i(x_*, y_*^i)) \right. \\
&\qquad \left. - 2b_t s_t^i (G^i(\xi_t^i, y_t^i) - G^i(x_*, y_*^i)) \right\} + cb_t^2
\end{aligned}$$

$$\leq D_t - 2a_t \sum_{i=1}^{N} \pi^i (f^i(\xi_t^i, y_t^i) - f^i(x_*, y_*^i))$$

$$- 2b_t \sum_{i=1}^{N} \pi^i s_t^i G^i(\xi_t^i, y_t^i) + cb_t^2, \tag{11.24}$$

where $c := 2(r_1^2 L^2 + \overline{\mu}^2)$. Next, define

$$\overline{\rho} = \max_{1 \leq i \leq N} \rho^i, \quad \underline{\lambda} = \min_{1 \leq i \leq N} \min\{\lambda^i, \sigma^i\},$$

which are strictly positive numbers. Suppose $\|(\xi_t^i, y_t^i)\| > \overline{\rho}$. From Assumption 6 and (11.14), we have $s_t^i G^i(\xi_t^i, y_t^i) = G^i(\xi_t^i, y_t^i) \geq \lambda^i \|(\xi_t^i, y_t^i)\|$ if $\|v_t^i\| \leq \mu^i$, and otherwise $s_t^i G^i(\xi_t^i, y_t^i) = (\mu^i / \|v_t^i\|) G^i(\xi_t^i, y_t^i) \geq \sigma^i \|(\xi_t^i, y_t^i)\|$. Thus, it holds that

$$s_t^i G^i(\xi_t^i, y_t^i) \geq \underline{\lambda} \|(\xi_t^i, y_t^i)\| \tag{11.25}$$

whenever $\|(\xi_t^i, y_t^i)\| > \overline{\rho}$.

Lemma 2 *Sequences $\{x_t^i\}$, $\{y_t^i\}$, $\{\xi_t^i\}$, and $\{\bar{x}_t\}$ are bounded.*

Proof (i) Consider integer $t \geq \underline{t_0}$ for which

$$D_t + cb_t^2 < D_{t+1} \tag{11.26}$$

holds. Denote by $\mathcal{A}_t$ the set of indexes $i \in \mathcal{A}$ for which $\|(\xi_t^i, y_t^i)\| > \overline{\rho}$ holds at t. First, assume $\mathcal{A}_t \neq \emptyset$. From (11.24) and (11.26),

$$0 > \frac{a_t}{b_t} \sum_{i=1}^{N} \pi^i (f^i(\xi_t^i, y_t^i) - f^i(x_*, y_*^i)) + \sum_{i=1}^{N} \pi^i s_t^i G^i(\xi_t^i, y_t^i)$$

$$\geq -\frac{a_t}{b_t} \sum_{i=1}^{N} \pi^i L \left\| \begin{bmatrix} \xi_t^i - x_* \\ y_t^i - y_*^i \end{bmatrix} \right\| + \sum_{i \in \mathcal{A}_t} \pi^i s_t^i G^i(\xi_t^i, y_t^i)$$

$$\geq -\frac{a_t}{b_t} \sum_{i=1}^{N} \pi^i L \left(\left\| \begin{bmatrix} \xi_t^i \\ y_t^i \end{bmatrix} \right\| + \left\| \begin{bmatrix} x_* \\ y_*^i \end{bmatrix} \right\| \right) + \sum_{i \in \mathcal{A}_t} \pi^i \underline{\lambda} \left\| \begin{bmatrix} \xi_t^i \\ y_t^i \end{bmatrix} \right\|,$$

where $G^i(\xi_t^i, y_t^i) \geq 0$ and (11.25) are exploited. Therefore

$$\sum_{i \in \mathcal{A}_t} \pi^i \left(\underline{\lambda} - L \frac{a_t}{b_t} \right) \left\| \begin{bmatrix} \xi_t^i \\ y_t^i \end{bmatrix} \right\| < L \frac{a_t}{b_t} \sum_{i \in \mathcal{A}_t^c} \pi^i \left\| \begin{bmatrix} \xi_t^i \\ y_t^i \end{bmatrix} \right\| + L \frac{a_t}{b_t} \sum_{i=1}^{N} \pi^i \left\| \begin{bmatrix} x_* \\ y_*^i \end{bmatrix} \right\|.$$

Since $a_t/b_t \to 0$ as $t \to \infty$, there exists a $\underline{t_1} \geq \underline{t_0}$ such that $a_t/b_t \leq \underline{\rho}/(2L)$ for all $t \geq \underline{t_1}$. This implies

$$\frac{\lambda}{2} \sum_{i \in \mathcal{A}_t} \pi^i \left\| \begin{bmatrix} \xi_t^i \\ y_t^i \end{bmatrix} \right\| < \frac{\lambda}{2} \sum_{i \in \mathcal{A}_t^c} \pi^i \left\| \begin{bmatrix} \xi_t^i \\ y_t^i \end{bmatrix} \right\| + \frac{\lambda}{2} \sum_{i=1}^{N} \pi^i \left\| \begin{bmatrix} x_* \\ y_*^i \end{bmatrix} \right\| \qquad (11.27)$$

holds for all $t \geq \underline{t}_1$. On the other hand, since $\|(\xi_t^i, y_t^i)\| \leq \overline{\rho}$ for $i \in \mathcal{A}_t^c$, we have

$$\sum_{i \in \mathcal{A}_t^c} \pi^i \left\| \begin{bmatrix} \xi_t^i \\ y_t^i \end{bmatrix} \right\| \leq \sum_{i \in \mathcal{A}_t^c} \pi^i \overline{\rho} \leq \overline{\rho}. \qquad (11.28)$$

Combining (11.27) and (11.28), we obtain

$$\sum_{i=1}^{N} \pi^i \left\| \begin{bmatrix} \xi_t^i \\ y_t^i \end{bmatrix} \right\| < \overline{\rho} + \sum_{i \in \mathcal{A}_t^c} \pi^i \left\| \begin{bmatrix} \xi_t^i \\ y_t^i \end{bmatrix} \right\| + \sum_{i=1}^{N} \pi^i \left\| \begin{bmatrix} x_* \\ y_*^i \end{bmatrix} \right\|$$

$$\leq 2\overline{\rho} + \sup_{(x_*, y_*) \in X_{*\pi}} \sum_{i=1}^{N} \pi^i \left\| \begin{bmatrix} x_* \\ y_*^i \end{bmatrix} \right\| =: R_1.$$

Thus, we obtained an upper bound for $\mathcal{A}_t \neq \emptyset$. If $\mathcal{A}_t = \emptyset$, obviously $\sum_{i=1}^{N} \pi^i \|(\xi_t^i, y_t^i)\| \leq \overline{\rho} < R_1$ and hence we conclude that $\sum_{i=1}^{N} \pi^i \|(\xi_t^i, y_t^i)\| < R_1$ holds whenever (11.26) is true at $t \geq \underline{t}_1$. Moreover, since $\pi^i \|(\xi_t^i, y_t^i)\| < R_1$ for each i,

$$\sum_{i=1}^{N} \pi^i \|(x_{t+1}^i, y_{t+1}^i)\|^2 = \sum_{i=1}^{N} \pi^i \left\| \begin{bmatrix} \xi_t^i \\ y_t^i \end{bmatrix} - a_t u_t^i - b_t s_t^i v_t^i \right\|^2$$

$$\leq \sum_{i=1}^{N} \pi^i \left\{ \left\| \begin{bmatrix} \xi_t^i \\ y_t^i \end{bmatrix} \right\| + \|a_t u_t^i + b_t s_t^i v_t^i\| \right\}^2$$

$$< \sum_{i=1}^{N} \pi^i \left\{ \frac{R_1}{\pi^i} + (\hat{a}L + \hat{b}\overline{\mu}) \right\}^2 =: R_2.$$

(ii) Suppose that (11.26) does *not* hold for $t = t_a, t_a + 1, \ldots, t_b$, that is,

$$D_{t+1} \leq D_t + cb_t, \qquad t = t_a, t_a + 1, t_a + 2, \ldots, t_b, \qquad (11.29)$$

where $t_a \geq \underline{t}_1$. Without loss of generality, we assume (ii-a) $t_a = \underline{t}_1$ or (ii-b) $t_a > \underline{t}_1$ and (11.26) holds at $t = t_a - 1$. From (11.29), it holds for $t \geq \underline{t}_1$ that

$$D_{t+1} \leq D_{t_a} + \sum_{\tau=t_a}^{t} cb_\tau^2, \qquad t = t_a, t_a + 1, t_a + 2, \ldots, t_b.$$

We see that $D_{t_a} = D_{\underline{t}_1}$ for (ii-a) and $D_{t_a} < R_2$ for (ii-b), where the result of (i) provides the latter inequality. Since $\{b_\tau\}$ is square summable,

$$D_{t+1} \le \max\{D_{L_1}, R_2\} + \sum_{\tau=1}^{\infty} cb_\tau^2 < \infty. \tag{11.30}$$

Thus, we have seen that $\{x_t^i\}$ and $\{y_t^i\}$, $i = 1, 2, \ldots, N$ are bounded. The boundedness of $\{\xi_t^i\}$ and $\{\bar{x}_t\}$ are obvious. $\qquad\square$

11.4.4 Proof of the Convergence

Since $\{\bar{x}_t\}$ and $\{y_t^i\}$ are bounded and $X_{*\pi}$ is assumed to be compact, there exists a compact set $K \subset \mathbb{R}^{n+l}$ that satisfies (11.10) and also includes $\{(\bar{x}_t, y_t) : t = 1, 2, \ldots\}$. From Assumption 3 and the definition of G^i, there exists an $L_K > 0$ such that

$$|G^i(x, y^i) - G^i(\tilde{x}, \tilde{y}^i)| \le L_K \|(x, y^i) - (\tilde{x}, \tilde{y}^i)\|, \quad i = 1, 2, \ldots, N \tag{11.31}$$

holds for all $(x, y), (\tilde{x}, \tilde{y}) \in \mathbb{R}^{n+l}$. From this, $\{v_t^i\}$, $i = 1, 2, \ldots, N$ are bounded; for some $M_v > 0$, $\|v_t^i\| \le M_v$ for all t and i. Hence, $s_t^i \ge \underline{s} := \min\{\underline{\mu}/M_v, 1\} > 0$. Then, we can derive the following from (11.24):

$$
\begin{aligned}
D_{t+1} - D_t &\le -2a_t \sum_{i=1}^{N} \pi^i (f^i(\xi_t^i, y_t^i) - f^i(x_*, y_*^i)) - 2b_t \underline{s} \sum_{i=1}^{N} \pi^i G^i(\xi_t^i, y_t^i) + cb_t^2 \\
&\le -2a_t \sum_{i=1}^{N} \pi^i (f^i(\bar{x}_t, y_t^i) - f^i(x_*, y_*^i)) - 2b_t \underline{s} \sum_{i=1}^{N} \pi^i G^i(\bar{x}_t, y_t^i) \\
&\quad + 2 \sum_{i=1}^{N} \pi^i \left\{ a_t L \|\xi_t^i - \bar{x}_t\| + b_t \underline{s} L_K \|\xi_t^i - \bar{x}_t\| \right\} + cb_t^2 \\
&\le -2a_t \left\{ \sum_{i=1}^{N} \pi^i f^i(\bar{x}_t, y_t^i) + \frac{s b_t}{a_t} \sum_{i=1}^{N} \pi^i G^i(\bar{x}_t, y_t^i) - f_* \right\} \\
&\quad + 2(r_1 L + \underline{s} L_K) d \|x_t^\delta\| b_t + cb_t^2 \\
&= -2a_t (F(\bar{x}_t, y_t; p_t) - f_*) + e_t, \tag{11.32}
\end{aligned}
$$

where $y_t = (y_t^1, y_t^2, \ldots, y_t^N)$ and

$$p_t = \frac{s b_t}{a_t}, \qquad e_t = 2(r_1 L + \underline{s} L_K) d \|x_t^\delta\| b_t + cb_t^2.$$

Obviously, $\{e_t\}$ is a summable positive sequence.

To prove (11.15), let us assume the following which leads to a contradiction:

$$\exists \varepsilon > 0 \quad \exists \underline{t}_2 \geq \underline{t}_1 \quad \forall t \geq \underline{t}_2 \quad F(\bar{x}_t, y_t; p_t) - f_* > \varepsilon. \tag{11.33}$$

Then, it holds from (11.32) that $D_{t+1} \leq D_t - 2a_t\varepsilon + e_t$ and hence

$$D_{t+1} < D_{\underline{t}_2} - 2\varepsilon \sum_{\tau=\underline{t}_2}^{t} a_\tau + \sum_{\tau=\underline{t}_2}^{t} e_\tau$$

is true for all $t \geq \underline{t}_2$. From this, it holds for all $t \geq \underline{t}_2$ that

$$\varepsilon < \frac{D_{\underline{t}_2} + \sum_{\tau=\underline{t}_2}^{t} e_\tau}{2 \sum_{\tau=\underline{t}_2}^{t} a_\tau} \tag{11.34}$$

since $\{e_t\}$ is square summable and $\{a_t\}$ is not summable. Therefore, the RHS of (11.34) converges to zero as $t \to \infty$ and contradicts $\varepsilon > 0$. Thus (11.33) is found false, by which we see

$$\forall \varepsilon > 0 \quad \forall j \geq \underline{t}_1 \quad \exists S_{\varepsilon,j} \geq j \quad F(\bar{x}_t, y_t; p_t)\big|_{t=S_{\varepsilon,j}} \leq f_* + \varepsilon.$$

For each positive integer m, let sequence $T_{m,1}, T_{m,2}, \ldots$ be a strictly monotonically increasing subsequence of $S_{(1/m),j}$. Then, it holds that

$$F(\bar{x}_t, y_t; p_t)\big|_{t=T_{m,k}} \leq f_* + \frac{1}{m}, \quad m = 1, 2, \ldots, \quad k = 1, 2, \ldots.$$

From Assumption 5, for K defined at the beginning of this subsection, there exists a positive number $p_{*\pi}^K$ such that $X_{*\pi} = X_{*\pi}^K(p)$ holds for all $p \geq p_{*\pi}^K$. Since $p_t = sb_t/a_t \to \infty$ as $t \to \infty$ from Assumption 7, there exists an integer N_m such that $p_{T_{m,k}} \geq p_{*\pi}^K$ holds for all $k \geq N_m$. Therefore, using also (11.9), we see

$$f_* = \min_{(x,y)\in K} F(x, y; p_{*\pi}^K) \leq F(\bar{x}_t, y_t; p_{*\pi}^K)\big|_{t=T_{m,N_m}}$$

$$\leq F(\bar{x}_t, y_t; p_t)\big|_{t=T_{m,N_m}} \leq f_* + \frac{1}{m}, \quad m = 1, 2, \ldots. \tag{11.35}$$

From the boundedness of $\{\bar{x}_t\}$, there exists a subsequence m_l such that $(\bar{x}_{t_l}, y_{t_l})$ with $t_l = T_{m_l,N_{m_l}}$ converges to some $(x_\infty, y_\infty) \in \mathbb{R}^{n+l}$ as $l \to \infty$. This means that

$$\lim_{l \to \infty} \bar{x}_{t_l} = x_\infty, \quad \lim_{l \to \infty} y^i_{t_l} = y^i_\infty, \quad i = 1, 2, \ldots, N,$$

$$\lim_{l \to \infty} F(\bar{x}_{t_l}, y_{t_l}; p^K_{*\pi}) = \lim_{l \to \infty} F(\bar{x}_{t_l}, y_{t_l}; p_{t_l}) = f_*$$

from (11.35). Moreover, from the continuity of F, we see

$$F(x_\infty, y_\infty; p^K_{*\pi}) = f_*,$$

which implies that x_∞ attains the minimum of $F(x, y; p^K_{*\pi})$. Assumption 5 then guarantees that (x_∞, y_∞) is an optimal solution to Problem 2 and therefore a Pareto optimal solution to Problem 1. Now, let us recall inequality (11.32). Since $(x_*, y_*) \in X_{*\pi}$ is arbitrary in (11.32) and $(x_\infty, y_\infty) \in X_{*\pi}$, (11.32) holds for $(x_*, y_*) = (x_\infty, y_\infty)$. From Assumption 7, there exists $t_3 \geq t_1$ such that $p_t \geq p^K_{*\pi}$ for all $t \geq t_3$. Then, for all $t \geq t_3$, it holds that

$$F(\bar{x}_t, y_t; p_t) \geq \min_{(x,y) \in \mathbb{R}^{n+l}} F(x, y; p_t) = \min_{(x,y) \in \mathbb{R}^{n+l}} F(x, y; p^K_{*\pi}) = f_*.$$

Therefore

$$D_{t+1} \leq D_t + e_t \quad \forall t \geq t_3$$

holds, where

$$D_t = \sum_{i=1}^{N} \pi^i \| (x^i_t - x_\infty, \; y^i_t - y^i_\infty) \|^2.$$

We see

$$\sum_{i=1}^{N} \pi^i \left\| \begin{bmatrix} x^i_{t+1} - x_\infty \\ y^i_{t+1} - y_\infty \end{bmatrix} \right\|^2 \leq \sum_{i=1}^{N} \pi^i \left\| \begin{bmatrix} x^i_{t_l} - x_\infty \\ y^i_{t_l} - x_\infty \end{bmatrix} \right\|^2 + \sum_{\tau=t_l}^{t} e_\tau$$

$$\leq \sum_{i=1}^{N} \pi^i \left\| \begin{bmatrix} x^i_{t_l} - x_\infty \\ y^i_{t_l} - y^i_\infty \end{bmatrix} \right\|^2 + \sum_{\tau=t_l}^{\infty} e_\tau$$

if $t > t_l \geq t_3$ [21]. Hence,

$$\sup_{t \geq t_l} \sum_{i=1}^{N} \pi^i \left\| \begin{bmatrix} x^i_{t+1} - x_\infty \\ y^i_{t+1} - y^i_\infty \end{bmatrix} \right\|^2 \leq \sum_{i=1}^{N} \pi^i \left\| \begin{bmatrix} x^i_{t_l} - x_\infty \\ y^i_{t_l} - y^i_\infty \end{bmatrix} \right\|^2 + \sum_{\tau=t_l}^{\infty} e_\tau.$$

Letting $l \to \infty$ in the above inequality leads to

$$\limsup_{t \to \infty} \sum_{i=1}^{N} \pi^i \left\| \begin{bmatrix} x^i_{t+1} - x_\infty \\ y^i_{t+1} - y^i_\infty \end{bmatrix} \right\|^2 \leq 0,$$

since $\lim_{l \to \infty}(x_{t_l}, y_{t_l}) = (x_\infty, y_\infty)$ and $\{e_t\}$ is positive and summable. Therefore,

$$\lim_{t \to \infty} x_t^i = x_\infty, \quad \lim_{t \to \infty} y_t^i = y_\infty^i, \quad i = 1, 2, \ldots, N.$$

This completes the proof of the convergence (11.15) with the consensus (11.22), which has been proved in Sect. 11.4.2.

Remark 1 Some extensions are available for Protocol 1. (i) In order to facilitate the convergence, we can utilize projection of variable y^i to a closed convex set. Note that those sets are not needed to be common. (ii) The action of consensus in Steps 2 and 3 of Protocol 1 can be intermittent, e.g., one can execute Steps 2 and 3 only once in some iterations while in the other iterations ξ_t^i is set as $\xi_t^i = x_t^i$. More general freedom with W being time-varying is discussed in [21, 22].

11.5 Numerical Examples

In this section, we show two numerical examples of applications of Protocol 1, to an academic example of a constrained minimax optimization problem and a DC-OPF problem with thirty buses.

11.5.1 Minimax Optimization over Unbalanced Network

Let us consider the following distributed minimax optimization problem:

$$\min_{\tilde{x}^i \in \mathbb{R}^5} \max_{i=1,2,\ldots,5} \tilde{f}_i(\tilde{x}^i)$$

$$\text{s.t. } \tilde{g}^i(\tilde{x}^i) \leq 0, \quad i = 1, 2, \ldots, 5,$$
$$\tilde{h}^1(\tilde{x}^1) = 0,$$
$$\tilde{x}^1 = \tilde{x}^2 = \cdots = \tilde{x}^5. \tag{11.36}$$

Objective functions $\tilde{f}^i, i = 1, 2, \ldots, 5$ are given as

$$\tilde{f}^1(x) = x_1 - 3x_2 - 3x_4 + 4x_5 + x_5^2,$$
$$\tilde{f}^2(x) = x_2 + 3x_4 + x_5 + 0.5x_2^2 + 4,$$
$$\tilde{f}^3(x) = 2x_1 + 2x_3 - 7x_4 + 2x_4^2 + 5,$$
$$\tilde{f}^4(x) = -2x_1 - 2x_2 + 4x_3 + x_4 + x_3^2 + 6,$$
$$\tilde{f}^5(x) = -2x_1 + x_2 - x_3 + 2x_5 + 0.5x_1^2 + 12.5,$$

where we represent the elements of $x \in \mathbb{R}^5$ as $x = \begin{bmatrix} x_1 & x_2 & x_3 & x_4 & x_5 \end{bmatrix}^\top$. The inequality constraints $\tilde{g}^i$, $i = 1, 2, \ldots, 5$ and an equality constraint $\tilde{h}^1$ are defined by

$$\tilde{g}^1(x) = \begin{bmatrix} 2x_1 + 3x_3 + 7x_4 + 9x_5 - 8 \\ 3x_1 - 4x_3 + x_4 - 5x_5 - 15 \\ x_4 + 2x_5 - 18 \\ 0.2x_1^2 - 1.2x_1 - 23.2 \end{bmatrix},$$

$$\tilde{g}^2(x) = \begin{bmatrix} -x_1 + 5x_2 + x_3 + x_4 - 3x_5 - 10 \\ -2x_1 - 2x_2 + x_3 + x_4 - 5x_5 - 13 \\ x_1 + 2x_2 - x_3 - 8x_4 + 9x_5 - 14 \\ 0.4x_4^2 + 4x_4 - 90 \end{bmatrix},$$

$$\tilde{g}^3(x) = \begin{bmatrix} 8x_1 - 3x_2 + 4x_3 + 2x_5 - 14 \\ -2x_1 + 5x_2 + x_3 + 4x_4 + 6x_5 - 10 \\ -x_1 + 3x_3 + 2x_4 + x_5 - 10 \\ 0.5x_2^2 - 36 \end{bmatrix},$$

$$\tilde{g}^4(x) = \begin{bmatrix} -x_2 - x_3 - x_4 + 3x_5 - 11 \\ -x_1 + x_2 + 2x_3 + 2x_4 + 5x_5 - 18 \\ x_1 - x_2 - 3x_4 - 13 \\ 0.2x_5^2 + 0.8x_5 - 143.2 \end{bmatrix},$$

$$\tilde{g}^5(x) = \begin{bmatrix} x_1 + 2x_2 + 2x_5 - 13 \\ -x_2 + 2x_4 + x_5 - 17 \\ 2x_1 + x_2 + x_3 - x_4 + 2x_5 - 8 \\ 0.1x_3^2 - 0.6x_3 - 63.1 \end{bmatrix},$$

and $\tilde{h}^1(x) = 2x_1 + x_2 - 2x_4 - 2x_5 + 1$, respectively. An optimal solution is given as

$$\tilde{x}_* = \begin{bmatrix} 0.801079 & -2.48728 & -1.20176 & 1.34615 & -1.28872 \end{bmatrix}^\top,$$
$$\gamma_* = f_* = 7.3557.$$

As we stated in Sect. 11.2, we introduce additional common decision variables $\gamma^i \in \mathbb{R}$, $i = 1, 2, \ldots, 5$ and derive a problem equivalent to (11.36) as

$$\min_{(\tilde{x}^i, \gamma^i) \in \mathbb{R}^5 \times \mathbb{R}} \frac{1}{5} \sum_{i=1}^{5} \gamma^i \quad \text{s.t.} \quad \tilde{f}(\tilde{x}^i) \leq \gamma^i,$$

$$\tilde{g}^i(\tilde{x}^i) \leq 0, \quad i = 1, 2, \ldots, 5,$$
$$\tilde{h}^1(\tilde{x}^1) = 0,$$
$$\tilde{x}^1 = \tilde{x}^2 = \cdots = \tilde{x}^5,$$
$$\gamma^1 = \gamma^2 = \cdots = \gamma^5.$$

Next, let $\mathcal{A} = \{1, 2, \ldots, 5\}$ and

Fig. 11.1 Graph of the network

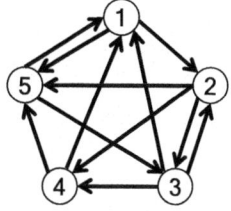

$$\mathcal{J}^1 = \{3, 4, 5\}, \quad \mathcal{J}^2 = \{1, 3\}, \quad \mathcal{J}^3 = \{2, 5\}, \quad \mathcal{J}^4 = \{2, 3\}, \quad \mathcal{J}^5 = \{1, 2, 4\},$$

for which graph $\mathcal{G}$, shown in Fig. 11.1, is strongly connected.

We set

$$W = \begin{bmatrix} 0.4 & 0 & 0.2 & 0.2 & 0.2 \\ 0.2 & 0.6 & 0.2 & 0 & 0 \\ 0 & 0.2 & 0.6 & 0 & 0.2 \\ 0 & 0.2 & 0.2 & 0.6 & 0 \\ 0.2 & 0.2 & 0 & 0.2 & 0.4 \end{bmatrix}.$$

This matrix is only right stochastic and not doubly stochastic, with which the network is unbalanced.

For the following initial solution candidate,

$$x_1^1 = \begin{bmatrix} 0 \\ 0 \\ -2 \\ 0 \\ -1 \end{bmatrix}, \quad x_1^2 = \begin{bmatrix} 0 \\ 2 \\ 0 \\ -1 \\ 0 \end{bmatrix}, \quad x_1^3 = \begin{bmatrix} 1 \\ 0 \\ 0 \\ 0 \\ -1 \end{bmatrix}, \quad x_1^4 = \begin{bmatrix} -1 \\ 0 \\ 0 \\ -2 \\ -1 \end{bmatrix}, \quad x_1^5 = \begin{bmatrix} 0 \\ 0 \\ 1 \\ -1 \\ 0 \end{bmatrix},$$

$$\gamma_1^i = 0, \quad i = 1, 2, \dots, 5,$$

we have executed Protocol 1 with setting

$$a_t = \frac{10}{t}, \quad b_t = \frac{10}{t^{0.7}}, \quad \mu^1 = 5, \quad \mu^2 = 5, \quad \mu^3 = 10, \quad \mu^4 = 5, \quad \mu^5 = 10,$$

and $q = \infty$ in (11.7). After $N_{\text{iter}} = 10^5$ iterations, we have obtained

$$\begin{bmatrix} \tilde{x}_{N_{\text{iter}}}^1 \\ \gamma_{N_{\text{iter}}}^1 \end{bmatrix} = \begin{bmatrix} 0.8082 \\ -2.4873 \\ -1.2002 \\ 1.3349 \\ -1.2903 \\ 7.3572 \end{bmatrix}, \quad \begin{bmatrix} \tilde{x}_{N_{\text{iter}}}^2 \\ \gamma_{N_{\text{iter}}}^2 \end{bmatrix} = \begin{bmatrix} 0.8020 \\ -2.4909 \\ -1.1995 \\ 1.3407 \\ -1.2851 \\ 7.3568 \end{bmatrix}, \quad \begin{bmatrix} \tilde{x}_{N_{\text{iter}}}^3 \\ \gamma_{N_{\text{iter}}}^3 \end{bmatrix} = \begin{bmatrix} 0.8027 \\ -2.4905 \\ -1.1995 \\ 1.3399 \\ -1.2859 \\ 7.3568 \end{bmatrix},$$

$$\begin{bmatrix} \tilde{x}^4_{N_{\text{iter}}} \\ \gamma^4_{N_{\text{iter}}} \end{bmatrix} = \begin{bmatrix} 0.8037 \\ -2.4895 \\ -1.2003 \\ 1.3394 \\ -1.2858 \\ 7.3572 \end{bmatrix}, \quad \begin{bmatrix} \tilde{x}^5_{N_{\text{iter}}} \\ \gamma^5_{N_{\text{iter}}} \end{bmatrix} = \begin{bmatrix} 0.8030 \\ -2.4899 \\ -1.2002 \\ 1.3401 \\ -1.2852 \\ 7.3572 \end{bmatrix}.$$

The plots in Figs. 11.2, 11.3, and 11.4 show the sequences $\{x^i_t\}$, $\{f^i(\xi^i_t)\}$, and relative error $\{\|\tilde{x}_t - x_*\|/\|x_*\|\}$ generated by Protocol 1, respectively, where $\bar{x}_t = \sum_{i=1}^5 x^i_t/5$. One can see the consensus of x^i_t and its convergence to x_* in Fig. 11.2. Figure 11.3 shows that the $\max_{i \in \mathcal{A}} f^i(\xi^i_t)$ converges to f_*. The relative error in Fig. 11.4 is less

Fig. 11.2 Decision variables $\{x^i_t\}$

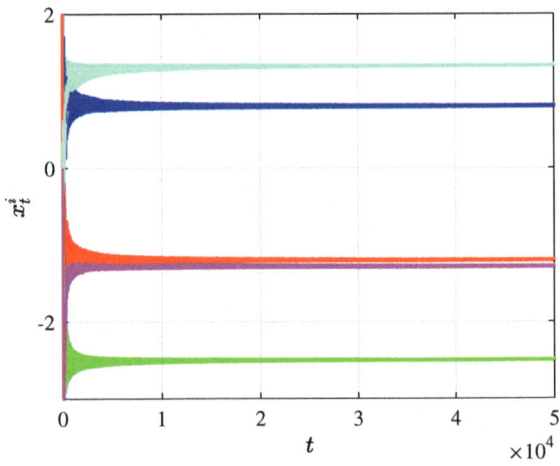

Fig. 11.3 Objective functions $\{f^i(\xi^i_t)\}$

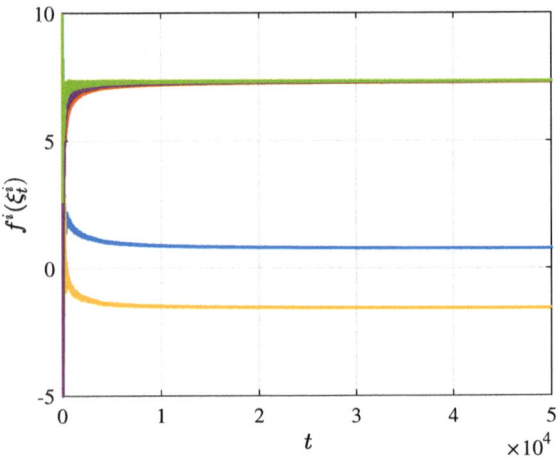

Fig. 11.4 Relative error
$\|\bar{x}_t - x_*\| / \|x_*\|$

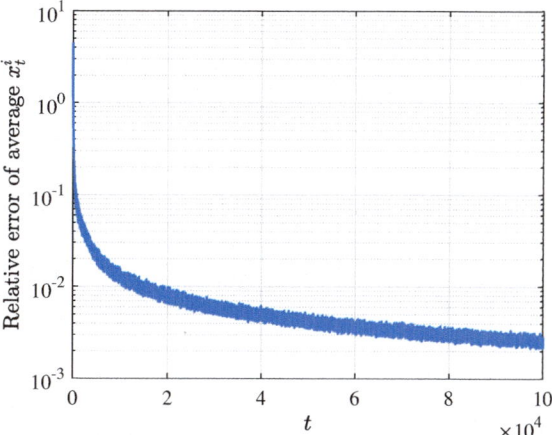

than 1% after 3×10^4 iterations. From these figures, we can verify the convergence of the sequences generated by Protocol 1 to a minimax optimal consensus point.

11.5.2   Application to Energy Management Systems

In this subsection, we consider a DC-OPF problem with thirty buses, six generators, and forty-one transmission lines, where we borrow the parameters of the problem from a case data of MATPOWER [31], whose original data is based on [2]. The problem is described as follows:

$$\min_{\theta \in \mathbb{R}^{30}, P^i, i \in \mathcal{A}_G} \sum_{i \in \mathcal{A}_G} F^i(P^i) \quad \text{s.t.} \quad B\theta = P - D, \quad \theta^1 = 0,$$

$$P^i_{\min} \leq P^i \leq P^i_{\max}, \quad i \in \mathcal{A}_G,$$

$$LF^{ij}_{\min} \leq \frac{\theta^i - \theta^j}{X_{ij}} \leq LF^{ij}_{\max}, \quad \text{if } X_{ij}(i, j) \neq 0,$$

where $\mathcal{A}_G = \{1, 2, 13, 22, 23, 27\}$ is the subset of buses in $\mathcal{A} = \{1, 2, \ldots, 30\}$ that are equipped with a generator. For $i \in \mathcal{A}_G$, P^i [p.u] is the active power of the generator connected to the ith bus and its generation cost is given by F^i [10^5\$/hr], which is a convex quadratic function of P^i. Let P^i and F^i be zero for $i \notin \mathcal{A}_G$. Variable θ [rad] is a vector whose ith element is the phase angle at the ith bus and B represents the susceptances of the transmission lines. More specifically, the (i, j)th element B^{ij} is given by

$$
B^{ij} = \begin{cases} \sum_{i=1}^{30} \dfrac{1}{X_{ij}}, & \text{if } i = j, \\[2ex] -\dfrac{1}{X_{ij}}, & \text{if } i \neq j, \end{cases}
$$

where X_{ij} is the reactance of line (i, j). The third constraint is the limitation on the active power of the generator at bus i, where two parameters $P_{\min}^i$ and $P_{\max}^i$ represent the minimum and maximum of P^i, respectively. The last constraint is the line flow limit, where $LF_{\min}^{ij}$ and $LF_{\max}^{ij}$ are the minimum and maximum of the signed active power of the transmission line (i, j), respectively. The optimal solution of the problem is:

$$
x_* = \begin{bmatrix} -0.0139 \\ -0.0410 \\ -0.0487 \\ -0.0445 \\ -0.0569 \\ -0.0629 \\ -0.0666 \\ -0.0761 \\ -0.0862 \\ -0.0761 \\ -0.0793 \\ -0.0572 \\ -0.0913 \\ -0.0881 \\ -0.0893 \\ -0.0922 \\ -0.1029 \\ -0.1075 \\ -0.1034 \\ -0.0842 \\ -0.0801 \\ -0.0685 \\ -0.0761 \\ -0.0492 \\ -0.0625 \\ -0.0248 \\ -0.0556 \\ -0.0502 \\ -0.0666 \end{bmatrix}, \quad \begin{bmatrix} P_*^1 \\ P_*^2 \\ P_*^{13} \\ P_*^{22} \\ P_*^{23} \\ P_*^{27} \end{bmatrix} = \begin{bmatrix} 0.4473 \\ 0.5826 \\ 0.1578 \\ 0.2231 \\ 0.1578 \\ 0.3232 \end{bmatrix}, \quad f_* = 0.5652.
$$

We separate set $\mathcal{A} = \{1, 2, \ldots, 30\}$ of buses into three groups, according to the area values 1 to 3 of the MATPOWER data, and consider three agents corresponding to each of the areas:

Fig. 11.5 Information
network

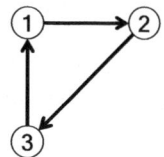

$$\mathcal{A}^1 = \{1, 2, 3, 4, 5, 6, 7, 8, 9, 11, 28\},$$
$$\mathcal{A}^2 = \{12, 13, 14, 15, 16, 17, 18, 19, 20, 23\},$$
$$\mathcal{A}^3 = \{10, 21, 22, 24, 25, 26, 27, 29, 30\},$$
$$\mathcal{A}_G^1 = \{1, 2\}, \quad \mathcal{A}_G^2 = \{13, 23\}, \quad \mathcal{A}_G^3 = \{22, 27\}.$$

Define variables x, y as

$$x = \begin{bmatrix} \theta^1 & \theta^2 & \cdots & \theta^{30} \end{bmatrix}^{\top}, \qquad y^i = \{P^j : j \in \mathcal{A}_G^i\}, \quad i \in \mathcal{A}.$$

According to Remark 1, we treat the constraints on P^i, $i \in \mathcal{A}_G$ by the projection to bounded interval $[P_{\min}^i, P_{\max}^i]$, while the equality constraints of the supply–demand balance and the inequality constraints of transmission power limits are formulated as in Protocol 1. Since F^i is given as a function of P^i and does not explicitly depend on the common variable θ^i, the boundedness of the gradient of the objective functions is guaranteed. A strongly connected graph $G = (\mathcal{A}, \mathcal{E})$ is chosen as Fig. 11.5 with $\mathcal{A} = \{1, 2, 3\}$ and $\mathcal{E} = \{(2, 1), (3, 2), (1, 3)\}$.

Matrix W is selected as

$$W = \begin{bmatrix} 0.5 & 0.5 & 0 \\ 0 & 0.5 & 0.5 \\ 0.5 & 0 & 0.5 \end{bmatrix}.$$

With the following parameters

$$a_t = \frac{100}{t + 5 \times 10^7}, \quad b_t = \frac{1}{(t + 5 \times 10^7)^{0.75}}, \quad \mu^i = 10^5, \quad i = 1, 2, 3, \quad q = 1$$

and initial values

$$x_1^1 = x_1^2 = x_1^3 = 0, \quad P_1^1 = P_1^2 = P_1^{13} = P_1^{22} = P_1^{23} = P_1^{27} = 0,$$

we executed Protocol 1 with $N_{\text{iter}} = 10^8$, where we suppressed the averaging of x_t^i of Steps 2 and 3 of Protocol 1 once in ten iterations, namely Step 2 and 3 are executed at $t = 10k$ with k integer, and in the other iterations, we set $\xi_t^i = x_t^i$. Then, we obtained the following results:

$$
x^1_{N_{\text{iter}}} = \begin{bmatrix}
-0.0138 \\
-0.0407 \\
-0.0483 \\
-0.0442 \\
-0.0563 \\
-0.0624 \\
-0.0662 \\
-0.0753 \\
-0.0852 \\
-0.0752 \\
-0.0791 \\
-0.0573 \\
-0.0910 \\
-0.0878 \\
-0.0887 \\
-0.0913 \\
-0.1023 \\
-0.1068 \\
-0.1025 \\
-0.0832 \\
-0.0791 \\
-0.0686 \\
-0.0758 \\
-0.0502 \\
-0.0635 \\
-0.0265 \\
-0.0553 \\
-0.0519 \\
-0.0682
\end{bmatrix}, \quad
x^2_{N_{\text{iter}}} = \begin{bmatrix}
-0.0138 \\
-0.0407 \\
-0.0483 \\
-0.0442 \\
-0.0563 \\
-0.0624 \\
-0.0662 \\
-0.0752 \\
-0.0852 \\
-0.0753 \\
-0.0790 \\
-0.0573 \\
-0.0910 \\
-0.0878 \\
-0.0887 \\
-0.0913 \\
-0.1023 \\
-0.1068 \\
-0.1025 \\
-0.0832 \\
-0.0791 \\
-0.0686 \\
-0.0758 \\
-0.0502 \\
-0.0635 \\
-0.0265 \\
-0.0553 \\
-0.0519 \\
-0.0682
\end{bmatrix}, \quad
x^3_{N_{\text{iter}}} = \begin{bmatrix}
-0.0138 \\
-0.0407 \\
-0.0483 \\
-0.0442 \\
-0.0563 \\
-0.0624 \\
-0.0662 \\
-0.0752 \\
-0.0853 \\
-0.0753 \\
-0.0791 \\
-0.0573 \\
-0.0910 \\
-0.0878 \\
-0.0887 \\
-0.0912 \\
-0.1023 \\
-0.1068 \\
-0.1025 \\
-0.0831 \\
-0.0792 \\
-0.0685 \\
-0.0759 \\
-0.0502 \\
-0.0635 \\
-0.0265 \\
-0.0553 \\
-0.0519 \\
-0.0682
\end{bmatrix},
$$

$$
P^1_{N_{\text{iter}}} = 0.4438, \quad P^2_{N_{\text{iter}}} = 0.5787, \quad P^{13}_{N_{\text{iter}}} = 0.1552,
$$
$$
P^{22}_{N_{\text{iter}}} = 0.2220, \quad P^{23}_{N_{\text{iter}}} = 0.1551, \quad P^{27}_{N_{\text{iter}}} = 0.3137
$$

with $f_{N_{\text{iter}}} = \sum_{i=1}^{5} \tilde{f}^i(\tilde{x}^i_t, y^i_t)|_{t=N_{\text{iter}}} = 0.5567$. The sequences of variables x_t, y_t, relative error $\|\bar{x}_t - x_*\|/\|x_*\|$, equation error $\|B\theta_t - P_t + D\|$, and relative error of the sum of the objective functions $(f_t - f_*)/f_*$ are depicted in Figs. 11.6, 11.7, 11.8, 11.9, and 11.10, respectively. From these figures, we can see that the obtained solution is close to the optimal solution with relative errors in one percent order.

Fig. 11.6 Common decision variables (angles) $\{x_t^i\}$

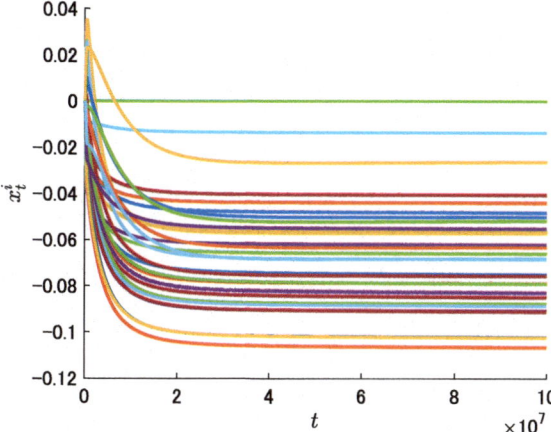

Fig. 11.7 Active power generation $\{P_t^i\}$

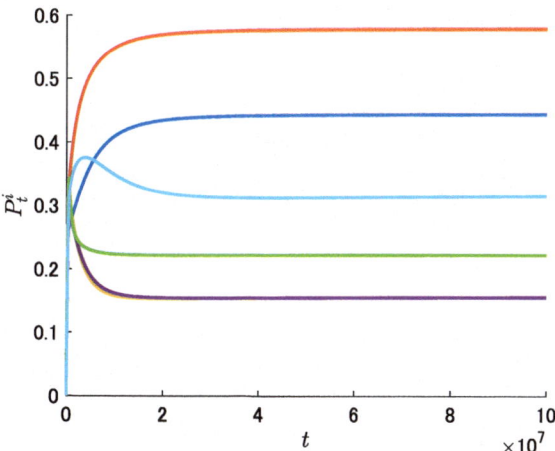

Fig. 11.8 Relative error $\|\bar{x}_t - x_*\|/\|x_*\|$

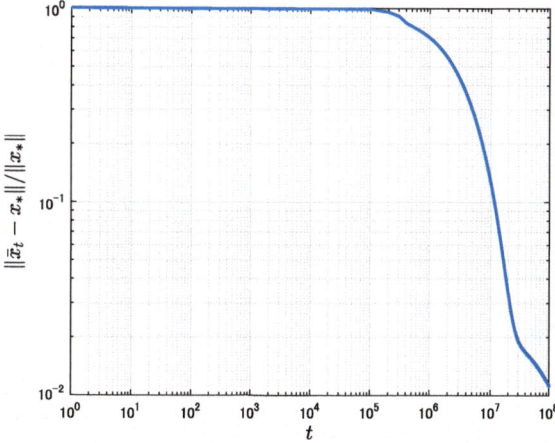

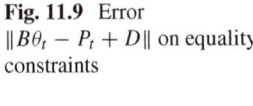

Fig. 11.9 Error $\|B\theta_t - P_t + D\|$ on equality constraints

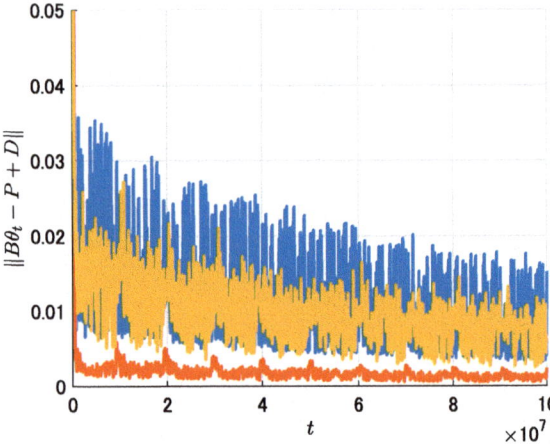

Fig. 11.10 Error $(f_t - f_*)/f_*$ on objective function

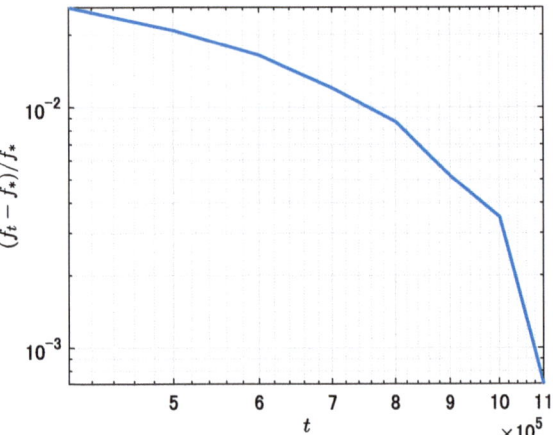

11.6 Conclusion

We have presented a distributed multi-agent optimization protocol that works over possibly unbalanced networks that achieves partial consensus of decision variables of agents at a Pareto optimal solution under inequality and equality conditions. The agents in the network can execute the protocol without disclosing their objective and constraint functions to others, which is an important issue in decision making over complex networks including recent energy management systems. The protocol is based on the linear consensus protocol and exact penalty methods, where the protocol is designed so that the penalized objective function emerges in the analysis, which leads to the concrete proof of the convergence. The protocol works out over networks represented with a directed graph. Even though the network is unbalanced, minimax and min-sum problems can be formulated with including auxiliary variables appropriately. The protocol is verified via numerical examples of a constrained minimax problem and a DC-OPF problem.

References

1. Abboud A, Couillet R, Debbah M, Siguerdidjane H (2014) Asynchronous alternating direction method of multipliers applied to the direct-current optimal power flow problem. In: Proceedings of the 2014 IEEE international conference on acoustics, speech and signal processing (ICASSP), pp 7764–7768
2. Alsac O, Stott B (1974) Optimal load flow with steady-state security. IEEE Trans Power Apparus Syst PAS-93(3):745–751
3. Bertsekas DP (1999) Nonlinear programming. Athena Scientific
4. Borwein J, Lewis A (2006) Convex analysis and nonlinear optimization. CMS Books in Mathematics. Springer, NY
5. Boyd S, Parikh N, Chu E, Peleato B, Eckstein J (2010) Distributed optimization and statistical learning via the alternating direction method of multipliers. Found Trends® Mach Learn 3(1):1–122
6. Chang T-H, Nedić A, Scaglione A (2014) Distributed constrained optimization by consensus-based primal-dual perturbation method. IEEE Trans Autom Control 59(6):1524–1538
7. Di Pillo G (1994) Exact penalty methods. Algorithms for continuous optimization. Springer, Netherlands, Dordrecht, pp 209–253
8. Di Pillo G, Grippo L (1988) On the exactness of a class of nondifferentiable penalty functions. J Optim Theory Appl 57(3):399–410
9. Di Pillo G, Grippo L (1989) Exact penalty functions in constrained optimization. SIAM J Control Optim 27(6):1333–1360
10. Han SP, Mangasarian OL (1979) Exact penalty functions in nonlinear programming. Math Program 17(1):251–269
11. Hartfiel DJ, Spellmann JW (1972) A role for doubly stochastic matrices in graph theory. Proc Am Math Soc 36(2):389–394
12. Jadbabaie A, Lin J, Morse A (2003) Coordination of groups of mobile autonomous agents using nearest neighbor rules. IEEE Trans Autom Control 48(6):988–1001
13. Kim B, BaldickR, (2000) A comparison of distributed optimal power flow algorithms. IEEE Trans Power Syst 15(2):599–604
14. Marler RT, Arora JS (2010) The weighted sum method for multi-objective optimization: new insights. Struct Multidiscip Optim 41(6):853–862
15. Masubuchi I, Wada T, Asai T, Nguyen, THL, Ohta Y, Fujisaki Y (2015) Distributed constrained optimization protocol via an exact penalty method. In: Proceedings of the 2015 European control conference
16. Masubuchi I, Wada T, Asai T, Nguyen THL, Ohta Y, Fujisaki Y (2016) Distributed multi-agent optimization based on an exact penalty method with equality and inequality constraints. SICE J Control, Meas, Syst Integr 9(4):179–186
17. Masubuchi I, Wada T, Fujisaki Y, Dabbene F (2018a) Distributed multi-agent optimization for Pareto optimal problem over unbalanced networks via exact penalty methods with equality and inequality constraints. In: Proceedings of the 23rd international symposium on mathematical theory of networks and systems, pp 447–452
18. Masubuchi I, Wada T, Fujisaki Y, Dabbene F (2018b) Improvement of distributed multi-agent optimization protocol based on exact penalty method. Proc SICE Annu Conf 2018:262–265
19. Masubuchi I,Wada T, FujisakiY, Dabbene F (2019a) Exact-penalty based distributed multi-agent optimization protocol with partial consensus. In: Proceedings of the SICE annual conference 2019, pp 1667–1670

20. Masubuchi I, Wada T, Fujisaki Y, Dabbene F (2019b) A new distributed constrained multi-agent optimization protocol with convergence proof via exactness of penalized objective function. In: Proceedings of the 12th Asian control conference, pp 19–24
21. Nedić A, Ozdaglar A (2009) Distributed subgradient methods for multi-agent optimization. IEEE Trans Autom Control 54(1):48–61
22. Nedić A, Ozdaglar A, Parrilo PA (2010) Constrained consensus and optimization in multi-agent networks. IEEE Trans Autom Control 55(4):922–938
23. Olfati-Saber R, Murray RM (2004) Consensus problems in networks of agents with switching topology and time-delays. IEEE Trans Autom Control 49(9):1520–1533
24. Rockafellar RT (1970) Convex analysis. Princeton University Press
25. Srivastava K, Nedić A, Stipanović D (2013) Distributed Bregman-distance algorithms for min-max optimization. Agent-Based Optimization. Springer, Berlin, Heidelberg, pp 143–174
26. Stott B, Jardim J, Alsaç O (2009) DC power flow revisited. IEEE Trans Power Syst 24(3):1290–1300
27. Wada T, Fujisaki Y (2016) Distributed optimization over directed unbalanced networks. In: Proceedings of the SICE international symposium on control systems 2016, SY0003/16/0000–0870
28. Wood AJ, Wollenberg BF, Sheblé GB (2013) Power generation, operation, and control, 3rd edn. Wiley
29. Xie P, You K, Tempo R, Song S, Wu C (2018) Distributed convex optimization with inequality constraints over time-varying unbalanced digraphs. IEEE Trans Autom Control 63(12):4331–4337
30. Zhu M, Martínez S (2012) On distributed convex optimization under inequality and equality constraints. IEEE Trans Autom Control 57(1):151–164
31. Zimmerman RD, Murillo-Sanchez CE, Thomas RJ (2010) MATPOWER: steadystate operations, planning, and analysis tools for power systems research and education. IEEE Trans Power Syst 26(1):12–19

Chapter 12
A Passivity-Based Design of Cyber-Physical Building HVAC Energy Management Integrating Optimization and Physical Dynamics

Takeshi Hatanaka, Tomohiro Ikawa and Na Li

Abstract This chapter investigates cyber-physical system (CPS) design for enhancing energy efficiency of heating, ventilation, and air-conditioning (HVAC) systems in a building consisting of multiple zones. We first present a thermodynamics model, called resistance–capacitance circuit model, where the inter-zone heat transfer is identified with current flow on an electrical circuit. We then formulate a set point optimization problem to balance the human comfort and energy saving while satisfying several constraints. We then design a CPS which integrates optimization dynamics based on primal-dual gradient dynamics and the physical dynamics with a local controller. The resulting CPS is then shown to be interpreted as an interconnection of passive systems. Accordingly, convergence of the room temperatures to the optimal solution and input–output stability are rigorously proved based on the passivity paradigm. The present framework is further extended to a scenario of co-optimizing energy management in multiple connected buildings. The present CPS is finally demonstrated on a simulator developed by combining a variety of software.

12.1 Introduction

Stimulated by the fact that buildings are one of the sections consuming the largest energy, smart building energy management algorithms have been developed both in industry and academia. In particular, about half of the power consumption in buildings is occupied by heating, ventilation, and air-conditioning (HVAC) systems,

T. Hatanaka (✉) · T. Ikawa
School of Engineering, Tokyo Institute of Technology, 2-12-1 S5-16, Ookayama,
Meguro-ku, Tokyo 152-8550, Japan
e-mail: hatanaka@sc.e.titech.ac.jp

T. Ikawa
e-mail: ikawa@hfg.sc.e.titech.ac.jp

N. Li
School of Engineering and Applied Sciences, Harvard University, 33 Oxford St,
Cambridge, MA 02138, USA
e-mail: nali@seas.harvard.edu

© Springer Nature Singapore Pte Ltd. 2020
T. Hatanaka et al. (eds.), *Economically Enabled Energy Management*,
https://doi.org/10.1007/978-981-15-3576-5_12

and a great deal of advanced algorithms to enhance energy efficiency of the HVAC systems has been reported in the literature [1].

One of the most promising control methodologies for HVAC optimization and control is model predictive control (MPC) [1, 2, 6, 15, 16, 20]. While several successful results by MPC have been reported, it inherently requires state measurements and prediction of disturbances including heat gains associated with the human activities in order to predict the future state evolution. State feedback may be problematic if the state includes temperatures of walls, floors, windows, and ceilings since embedding sensors in such zones would increase the install and maintenance costs of the system. Regarding the heat gain prediction, learning techniques have been commonly employed in the literature [16]. However, collecting a large volume of measurements is required in the learning stage, which may be a barrier against installing the energy management systems. More essentially, any learning techniques more or less presume that a physical quantity follows the same pattern as the past data, which is not always true for some buildings.

In this chapter, we employ another optimization-based approach investigated in [13, 14, 18, 22, 25, 29, 30] that is applicable even in the presence of unmeasurable/unpredictable states and disturbances. The approach first formulates an optimization problem to optimize the steady-state input and/or output, and then designs a dynamics to solve the optimization, where the primal-dual dynamics [5] or its variations are normally employed [13, 14, 18, 22, 25, 29–31]. The dynamical solution to optimization is then integrated with the physical dynamics as an interconnection of dynamical systems. The real-time path from physics to optimization dynamics renders the optimal steady states adaptive to the situation changes, e.g. caused by unmeasurable disturbances.

Shiltz et al. [22] interconnected a dynamic optimization process for smart grid control with locally controlled grid dynamics, and the resulting system is demonstrated through simulation. Tsumura et al. [25] presented an extended version of [22], and stability of the system is rigorously proved. The authors of [18, 30] incorporate the grid dynamics into the optimization dynamics by identifying the physical dynamics with a subprocess in the primal-dual dynamics. A scheme to eliminate the structural constraints required in [18, 30] is presented by Zhang et al. [29] while instead assuming measurements of disturbances. The authors of [13, 14] present more general results that enable order reduction of the controller. The paper [31] employs the approach for the building HVAC optimization, but the control mechanism inherently requires state measurements in order to recover the unmeasurable disturbances, which may be problematic as mentioned above.

In this chapter, we present a novel control architecture for the HVAC optimization based on the above paradigm. We start with presenting the so-called resistance–capacitance (RC) circuit model to describe the thermodynamics of a building with multiple zones, while emphasizing existence of unmeasurable disturbances associated with the human activities. We then formulate a set point optimization problem to balance the human comfort and energy saving while satisfying several constraints including the stationary equation of the RC circuit model.

We next design optimization dynamics to solve the above set point optimization based on the primal-dual gradient algorithm [5], where it is pointed out that running the dynamics requires information of the unmeasurable disturbances. A local PI controller is then designed to control the physical dynamics. The controller produces an estimate of the unmeasurable disturbances and makes the room temperatures track the reference ones. Combining the above optimization and physical dynamics, we then develop a novel cyber-physical system (CPS) architecture. In the system, the reference room temperature needed for the local controller is provided by the optimization dynamics in the cyber world, and the disturbance estimates needed for the optimization dynamics are provided from the local controller.

It is then proved that the present CPS is an interconnection of passive systems [7, 27]. Accordingly, we prove convergence of the room temperatures to the optimal ones that reflect unmeasurable disturbances, and input–output stability as well.

We further present an extension of the present CPS to a scenario of co-optimizing the energy management of multiple buildings.

Finally, the present CPS is demonstrated through simulation of HVAC energy management for a building in Tokyo Institute of Technology. To this end, we build a real-time control and optimization simulator, wherein the building dynamics is simulated by a standard building energy simulator, EnergyPlus [26], the RC circuit model to be embedded in the control algorithm is computed by Building Resistance-Capacitance Modeling (BRCM) toolbox [23], the control algorithm is coded on Simulink, and the EnergyPlus and Simulink programs are interconnected by the so-called MLE+ [3].

The contents of this chapter were in part presented in the authors' conference papers [10, 11]. The major incremental contributions relative to the conference versions are to develop a novel simulator for a real building, to present the design procedure for the realistic model, and to demonstrate the algorithm through simulation on the simulator. Actually, the conference papers [10, 11] conduct simulations only for small-scale mathematical models. The authors also entirely revised the texts in order to better present the contents and modified the procedure to prove the main theoretical result.

12.2   Preliminary

In this section, we introduce some terminologies used in this paper.

We first introduce passivity. Consider a system with a state-space representation

$$\dot{x} = \phi(x, u), \quad y = \varphi(x, u), \tag{12.1}$$

where $x(t) \in \mathbb{R}^N$ is the state, $u(t) \in \mathbb{R}^p$ is the input, and $y(t) \in \mathbb{R}^p$ is the output. Then, passivity is defined as below.

Definition 1 The system (12.1) is said to be passive if there exists a positive semi-definite function $S : \mathbb{R}^N \to \mathbb{R}_+ := [0, \infty)$, called storage function, such that

$$S(x(t)) - S(x(0)) \leq \int_0^t y^\top(\tau)u(\tau)d\tau \tag{12.2}$$

holds for all inputs $u : [0, t] \to \mathbb{R}^p$, all initial states $x(0) \in \mathbb{R}^N$ and all $t \in \mathbb{R}^+$.

In the case of a static system $y = \varphi(u)$, it is passive if $y^\top u = \varphi^\top(u)u \geq 0$ for all $u \in \mathbb{R}^p$, which can be shown by taking $S \equiv 0$ as the storage function. As widely known, if S is differentiable, (12.2) can be replaced by

$$\dot{S}(x(t)) \leq y^\top(t)u(t). \tag{12.3}$$

Passivity is known to be preserved for feedback interconnections of passive systems as follows. Consider two passive systems from u_1 to y_1 and from u_2 to y_2 with storage functions S_1 and S_2, respectively. Then, the feedback interconnections $u_1 = r - y_1$ and $u_2 = y_1 + d$ of these systems with exogenous inputs r and d provide

$$\dot{S}_1 + \dot{S}_2 \leq y_1^\top u_1 + y_2^\top u_2 = y_1^\top r + y_2^\top d, \tag{12.4}$$

which means passivity from $[r^\top \ d^\top]^\top$ to $[y_1^\top \ y_2^\top]^\top$. In the absence of r and d ($r = d = 0$), the energy dissipation $\dot{S}_1 + \dot{S}_2 \leq 0$ is moreover proved, and accordingly closed-loop stability is ensured under additional assumptions on strict energy dissipation and observability. Please refer to [7, 27] or other seminal books cited therein for more details on passivity.

We next introduce another notion closely related to passivity, namely incremental passivity. In the context of this paper, it is sufficient to define incremental passivity for a static system.

Definition 2 A static system $y = \varphi(u)$ is said to be incrementally passive if the function φ satisfies

$$(\varphi(u_1) - \varphi(u_2))^\top(u_1 - u_2) \geq 0$$

for all $u_1 \in \mathbb{R}^p$ and $u_2 \in \mathbb{R}^p$.

We also use the following fundamental tool in convex optimization.

Definition 3 A function $f : \mathbb{R}^N \to \mathbb{R}$ is said to be convex if the following inequality holds for any $x, y \in \mathbb{R}^N$.

$$(\nabla f(x))^\top(y - x) \leq f(y) - f(x) \tag{12.5}$$

The function is said to be strongly convex if there exists $\delta > 0$ such that

$$(\nabla f(x))^\top (y - x) \le f(y) - f(x) - \delta \|x - y\|^2 \tag{12.6}$$

The following well-known result links the convex functions and passivity theory.

Lemma 1 *[4] Consider a convex function $f : \mathbb{R}^N \to \mathbb{R}$. Then, its gradient $\nabla f : \mathbb{R}^N \to \mathbb{R}^N$ is incrementally passive; i.e., the following inequality holds for any $x, y \in \mathbb{R}^N$.*

$$(\nabla f(x) - \nabla f(y))^\top (x - y) \ge 0 \tag{12.7}$$

The notation $[g]_\lambda^+$ for real vectors g and λ with the same dimension provides a vector whose lth element, denoted by $([g]_\lambda^+)_l$, is given by

$$([g]_\lambda^+)_l = \begin{cases} 0, & \text{if } \lambda_l = 0 \text{ and } g_l < 0 \\ g_l, & \text{otherwise} \end{cases}, \tag{12.8}$$

where g_l and λ_l are the lth elements of g and λ, respectively. The symbol $\circ$ represents the Hadamard product, namely $g \circ \lambda$ is a real vector with the same dimension as g and λ whose lth element is $g_l \lambda_l$.

12.3   Physical Dynamics and Set Point Optimization

In this chapter, we consider a building with multiple zones $i = 1, 2, \ldots, n$ that are divided into two categories, namely active zones and inactive zones. Active zones with labels $i = 1, 2, \ldots, n_1$ $(n_1 \le n)$ correspond to rooms equipped with active variable air volume HVAC systems, while inactive zones $i = n_1 + 1, \ldots, n$ $(n_2 := n - n_1)$ include walls, windows, corridors, ceilings, floors, roofs, and rooms not in use.

12.3.1   Physical Dynamics

We first present a model of the building thermal dynamics. Throughout this chapter, we consider only the scene of cooling the rooms but heating is also treated in the same way. The notations to describe physical quantities are summarized in Table 12.1.

Suppose that the thermal dynamics in building b is modeled by the so-called RC circuit model. The model for active zone i is formulated as

$$C_i \dot{T}_i = \frac{T^a - T_i}{R_i} + \sum_{j \in \mathcal{N}_i} \frac{T_j - T_i}{R_{ij}} + c(S_i - T_i)m_i + q_i, \tag{12.9}$$

and that for the inactive zone i is

Table 12.1 Notations

$T_i[°C]$	Temperature of zone i
$m_i[kg/s]$	Mass flow rate of the HVAC system in zone i
$q_i[W]$	Heat gain from external sources in zone i
$s_j[W]$	Solar radiation from jth exterior surface
$S_i[°C]$	Temperature of the air supplied to zone i
$C_i[J/°C]$	Thermal capacitance of zone i
$R_i[°C/W]$	Thermal resistance of zone i corresponding to walls/windows
$R_{ij}[°C/W]$	Thermal resistance between zone i and j
a_i	Absorption coefficient of zone i corresponding to walls/windows
$T^a[°C]$	Ambient temperature
$c[J/kg·°C]$	Specific heat of the air

$$C_i \dot{T}_i = \frac{T^a - T_i}{R_i} + \sum_{j \in \mathcal{N}_i} \frac{T_j - T_i}{R_{ij}} + \sum_{j=1}^{n_s} e_{ij} s_j, \qquad (12.10)$$

where $\mathcal{N}_i$ is the set of zones neighboring to zone i, n_s is the number of exterior surfaces of the building, and $e_{ij} \neq 0$ only if zone i is a part of the jth exterior surface. In the sequel, S_i $(i = 1, \ldots, n_1)$ and c are assumed to be constant.

Let us now denote, by $T \in \mathbb{R}^n$, $q \in \mathbb{R}^{n_1}$, $m \in \mathbb{R}^{n_1}$, and $s \in \mathbb{R}^{n_s}$, collections of T_i $(i = 1, 2, \ldots, n)$, q_i $(i = 1, 2, \ldots, n_1)$, m_i $(i = 1, 2, \ldots, n_1)$, and s_j $(j = 1, 2, \ldots, n_s)$, respectively. The collective dynamics of (12.9) for all $i = 1, \ldots, n_1$ and (12.10) for all $i = n_1 + 1, \ldots, n$ is then formulated as

$$C\dot{T} = RT^a \mathbf{1} - (R + L)T + B_1 G(T)m + B_1 q + B_2 Es, \qquad (12.11)$$

where the matrices C and R are diagonal matrices with diagonal elements C_i $(i = 1, 2, \ldots, n)$, $\frac{1}{R_i}$ $(i = 1, 2, \ldots, n)$, and E is a matrix whose (i, j) element is e_{ij}, L describes the weighted graph Laplacian whose (i, j) elements $(i \neq j)$ are $\frac{1}{R_{ij}}$, $G(T) \in \mathbb{R}^{n_1 \times n_1}$ is a block diagonal matrix with diagonal elements equal to $c(S_i - T_i)$ $(i = 1, 2, \ldots, n_1)$. The matrices $B_1 \in \mathbb{R}^{n \times n_1}$ and $B_2 \in \mathbb{R}^{n \times n_2}$ are defined as $B_1 = [I_{n_1} \ 0]^\top$ and $B_2 = [0 \ I_{n_2}]^\top$. The symbol $\mathbf{1}$ is the all-ones vector with an appropriate dimension. Note that the model parameters in (12.11) are automatically identified using the toolbox in [23] and assumed to be available for control.

We next linearize the model (12.11) around an equilibrium $\bar{T}$ corresponding to some equilibrium inputs $T^a \equiv \bar{T}^a, m \equiv \bar{m}, q \equiv \bar{q}$, and $s \equiv \bar{s}$ and transform the variables as

$$x := C^{1/2} \delta T, \quad u := B_1^\top C^{-1/2} B_1 G(\bar{T}) \delta m,$$
$$w_a := C^{-1/2} R \delta T^a \mathbf{1} + B_2^\top C^{-1/2} B_2 E \delta s, \qquad (12.12)$$
$$w_q := B_1^\top C^{-1/2} B_1 \delta q,$$

where δT, δm, δT^a, δq, and δs describe the errors between T, m, T^a, q, and s and the corresponding equilibrium states and inputs, respectively. The linearized model is then given as

$$\dot{x} = -Ax + B_1 u + B_1 w_q + w_a, \quad x := \begin{bmatrix} x_1 \\ x_2 \end{bmatrix}, \tag{12.13}$$

where $x \in \mathbb{R}^n$, $u \in \mathbb{R}^{n_1}$, $w_q \in \mathbb{R}^{n_1}$, $w_a \in \mathbb{R}^n$, $x_1 \in \mathbb{R}^{n_1}$, $x_2 \in \mathbb{R}^{n_2}$. The matrix A is defined as $A := C^{-1/2}(R + L)C^{-1/2} + \bar{U}$, where $\bar{U}$ is a diagonal matrix whose diagonal elements are $\bar{m}_1, \ldots, \bar{m}_{n_1}, 0, \ldots, 0$, and $\bar{m}_i$ is the ith element of the equilibrium input $\bar{m}$. Remark that A is symmetric and positive definite [12] as long as $\bar{m}_i \geq 0 \; \forall i$.

In the sequel, we assume that the temperatures of active zones T_i ($i = 1, 2, \ldots, n_1$), the ambient temperature T^a and the solar radiation s, namely x_1 and w_a, are measurable. On the other hand, the temperatures of inactive zones x_2 and heat gain w_q are assumed to be unmeasurable in order to avoid additional costs for installing the energy management systems. Some readers may wonder if measuring the solar radiations is demanding for users, but they can be estimated using current technology like [24] whose data can be downloaded in real time.

12.3.2   Set Point Optimization

In this subsection, we formulate an optimization problem to determine the reference set points of active zones $i = 1, \ldots, n_1$. The decision variables corresponding to the zone temperature x and mass flow rate u are denoted by $z_x \in \mathbb{R}^n$ and $z_u \in \mathbb{R}^{n_1}$, respectively, and the pair of z_x and z_u is denoted by $z \in \mathbb{R}^{n+n_1}$. The fundamental issue to be addressed in the HVAC energy management lies in managing the tradeoff between the human comfort and energy saving. In order to reflect the tradeoff, we define the cost function to be minimized by the sum of the two terms as

$$f(z) := \frac{1}{2} \|z_{x1} - h\|^2 + P(z_u), \tag{12.14}$$

where $z_{x1} \in \mathbb{R}^{n_1}$ is a collection of elements in z_x corresponding to the active zone temperatures x_1. The symbol h denotes the collection of the most comfortable temperatures h_i for active zones $i = 1, \ldots, n_1$, which may be selected by the occupants in the zones similar to the existing systems or be computed using some human comfort metric like predicted mean vote. Namely, the first term evaluates the human comfort. The function P in (12.14) describes the power required for the HVAC systems to supply the air with temperature S_i at the rates in z_u. It may be required to put weights on each element of $z_{x1} - h$ to give priority to each zone, but this can be done by appropriately scaling the power consumption of each room in the function P. We thus take the present simple formulation.

In order to describe the hardware constraints associated with the mass flow rate, we also introduce a constraint function $g : \mathbb{R}^{n_1} \to \mathbb{R}^{n_c}$ to constrain z_u as

$$g(z_u) \leq 0. \tag{12.15}$$

The variables z_x and z_u are also required to meet the equality constraint

$$-Az_x + B_1 z_u + B_1 d_q + d_a = 0, \tag{12.16}$$

which corresponds to the stationary equation of the dynamical model (12.13). The symbols $d_q \in \mathbb{R}^{n_1}$ and $d_a \in \mathbb{R}^n$ are DC components of the disturbances w_q and w_a, respectively, which are needed to describe the stationary equation.

In summary, the optimization problem to be addressed in this chapter is formulated as below.

$$\min_{z \in \mathbb{R}^{n+n_1}} f(z) \text{ subject to (15) and (16).} \tag{12.17}$$

In the sequel, we employ the following assumption.

Assumption 1 The function P and each element of g are convex, continuously differentiable, and their gradients are locally Lipschitz. There exists z_u such that $g(z_u) < 0$.

The papers [2, 19, 21, 28] simply take a linear or quadratic function of control efforts as a cost function and then Assumption 1 is trivially satisfied. Also, as mentioned in [16, 31], the power consumption of supply fans is approximated by the cube of the sum of the mass flow rates, which also satisfies Assumption 1. A simple model of consumption at the cooling coil is given by the product of the mass flow rate and the error between the ambient temperature and the zone temperature [17], which also belongs to the intended class.

For the subsequent discussions, we eliminate the variable z_x from (12.17) by substituting (12.16) into the first term of (12.14) as

$$\min_{z_u \in \mathbb{R}^{n_1}} \phi(z_u) + P(z_u) \text{ subject to } g(z_u) \leq 0, \tag{12.18}$$

where

$$\phi(z_u) := \frac{1}{2} \| M(z_u + d_q) + B_1^\top A^{-1}(d_a - \bar{h}) \|^2$$

with $M := B_1^\top A^{-1} B_1$ and $\bar{h} := AB_1 h$. Since A is positive definite, M is also positive definite. This means that the function ϕ is strongly convex in z_u, and there must exist a unique solution to (12.18), denoted by z_u^*, under Assumption 1. The solution z_u^* must satisfy the KKT condition [4]; namely, there exists $\lambda^* \in \mathbb{R}^{n_c}$ such that

$$M^2(z_u^* + d_q) + N(d_a - \bar{h}) + \nabla P(z_u^*) + \nabla g(z_u^*)\lambda^* = 0, \tag{12.19a}$$

$$g(z_u^*) \leq 0, \ \lambda^* \geq 0, \ \lambda^* \circ g(z_u^*) = 0, \tag{12.19b}$$

where $N := MB_1^\top A^{-1}$.

Since (12.18) is essentially equivalent to the original problem (12.17), if we define

$$z_x^* := A^{-1}(B_1 z_u^* + B_1 d_q + d_a) \in \mathbb{R}^n,$$

the pair of z_u^* and z_x^* must be a solution to (12.17). Now, divide z_x^* as $z_x^* := [(z_{x1}^*)^\top \ (z_{x2}^*)^\top]^\top$ ($z_{x1}^* \in \mathbb{R}^{n_1}$, $z_{x2}^* \in \mathbb{R}^{n_2}$). We then have

$$z_{x1}^* = M(z_u^* + d_q) + B_1^\top A^{-1} d_a \in \mathbb{R}^{n_1}. \tag{12.20}$$

The control goal of this chapter is to design a controller so as to ensure convergence of the room temperature x_1 to the optimal one z_{x1}^*, the solution to (12.30), without direct measurements of w_q.

12.4 CPS Design

In this section, we design a CPS interconnecting the optimization dynamics in the cyber world and physical dynamics in the physical world.

12.4.1 Optimization Dynamics

Let us first present a dynamical solution to the optimization problem (12.18). Once the parameters d_q and d_a are fixed, the problem (12.18) is just a convex optimization problem. A dynamic solution to the problem is known to be given by the primal-dual gradient algorithm [5] formulated as:

$$\dot{z}_u = -\alpha\{M^2(z_u + d_q) + K(d_a - \bar{h}) + \nabla P(z_u) + \mu\}, \ \alpha > 0, \tag{12.21a}$$

$$\dot{\lambda} = [g(z_u)]_\lambda^+, \ \mu = \nabla g(z_u)\lambda, \ \lambda(0) \geq 0, \tag{12.21b}$$

where $z_u \in \mathbb{R}^{n_1}$ and $\lambda \in \mathbb{R}^{n_c}$ are estimates of z_u^* and λ^*, respectively.

Given d_q and d_a, the variable z_u generated by the algorithm (12.24) is known to converge to the optimal solution z_u^* [5]. However, in practice, d_q and d_a are required to be updated in real time according to the changes of disturbances. In terms of d_a, it is sufficient just to replace d_a by the real-time measurement of w_a. However, the same approach is not applied to d_q since w_q is not directly measurable. We thus need to estimate d_q from the measurements of physical quantities, which motivates us to interconnect the physical dynamics with the optimization dynamics (12.21).

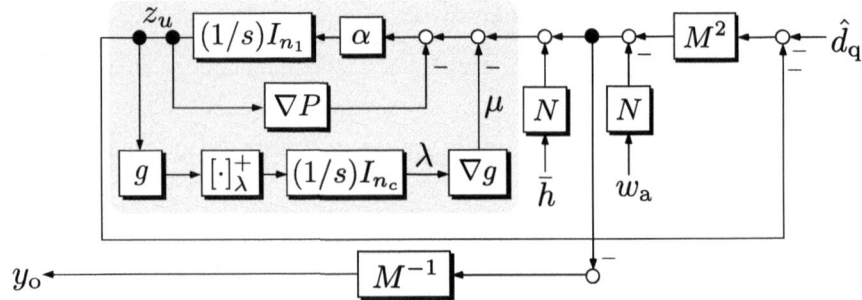

Fig. 12.1 Block diagram of optimization dynamics (12.22)

Let us now denote the estimate of d_q by $\hat{d}_q$. We then present the following dynamics by just replacing d_q and d_a by $\hat{d}_q$ and w_a, respectively.

$$\dot{z}_u = -\alpha\{M^2(z_u + \hat{d}_q) + N(w_a - \bar{h}) + \nabla P(z_u) + \mu\}, \qquad (12.22a)$$
$$\dot{\lambda} = [g(z_u)]_\lambda^+, \quad \mu = \nabla g(z_u)\lambda, \qquad (12.22b)$$

Note that the disturbance estimate $\hat{d}_q$ is regarded as an external input to the system (12.22).

In view of the definition of the optimal room temperature (12.20), we define an estimate of the optimal room temperatures z_{x1}^* as

$$y_o = M(z_u + \hat{d}_q) + B_1^\top A^{-1} w_a. \qquad (12.23)$$

We then let the signal y_o be sent to the physical dynamics as a reference signal to be tracked since y_o is estimated to be optimal. The block diagram of the system is illustrated in Fig. 12.1.

In the same way as [5], we can prove the following lemma by using (12.19) together with (12.20).

Lemma 2 *Consider the system (12.22) with constant inputs $w_a \equiv d_a$ and $\hat{d}_q \equiv d_q$. Then, the pair of $z_u = z_u^*$ and $\lambda = \lambda^*$ is a steady state of the system (12.22). Then, the output y_o at the steady state is equal to z_{x1}^*.*

12.4.2 Physical Dynamics with Local HVAC Control

In this subsection, we design a local HVAC controller to let the output x_1 of the physical dynamics (12.13) track a reference signal r. Inspired by the fact that proportional-integral (PI) controllers have been employed for local control in many existing systems, we design the following PI controller while adding reference and disturbance feedforward control u_{ff}.

$$\dot{\xi} = k_{\text{I}}(r - x_1), \tag{12.24a}$$

$$u = k_{\text{P}}(r - x_1) + \xi + u_{\text{ff}}, \tag{12.24b}$$

$$u_{\text{ff}} = \kappa r + F w_{\text{a}}, \tag{12.24c}$$

where k_{P}, k_{I}, $\kappa > 0$ and $F := [-I_{n_1} \ A_2^\top A_3^{-1}]$ with A_2 and A_3 defined as

$$A = \begin{bmatrix} A_1 & A_2^\top \\ A_2 & A_3 \end{bmatrix}, \quad A_1 \in \mathbb{R}^{n_1 \times n_1}, \quad A_3 \in \mathbb{R}^{n_2 \times n_2}.$$

Substituting (12.24) in (12.13) yields

$$\dot{x} = -Ax + k_{\text{P}} B_1 (r - x_1) + B_1 \xi + \kappa B_1 r + B_1 w_{\text{q}} + (B_1 F + I_n) w_{\text{a}}, \tag{12.25a}$$

$$\dot{\xi} = k_{\text{I}}(r - x_1). \tag{12.25b}$$

By calculation, we immediately have the following result.

Lemma 3 *Consider the system (12.25) with constant inputs $r \equiv r^*$, $w_{\text{a}} \equiv d_{\text{a}}$ and $w_{\text{q}} \equiv d_{\text{q}}$. Then, the steady states of the system, denoted by x^* and ξ^*, are given as follows.*

$$x^* = \begin{bmatrix} r^* \\ A_3^{-1}(-A_2 r^* + B_2 d_{\text{a}}) \end{bmatrix} \tag{12.26a}$$

$$\xi^* = (M^{-1} - \kappa I_{n_2}) r^* - d_{\text{q}}. \tag{12.26b}$$

Let us now define the output signal of (12.25) as

$$y_{\text{p}} := (\kappa I_{n_1} - M^{-1}) r + \bar{k}_{\text{P}}(r - x_1) + \xi \tag{12.27}$$

with $\bar{k}_{\text{P}} := k_{\text{P}} + \kappa$. We then immediately see from Lemma 3 that the output y_{p}^* at the steady state for $r \equiv r^*$, $w_{\text{a}} \equiv d_{\text{a}}$ and $w_{\text{q}} \equiv d_{\text{q}}$ is given as

$$y_{\text{p}}^* = (\kappa I_{n_1} - M^{-1}) r^* + \xi^* = -d_{\text{q}}, \tag{12.28}$$

Hence, the signal (12.27) is regarded as an estimate of $-d_{\text{q}}$. The block diagram of the physical dynamics (12.25) is shown in Fig. 12.2.

Remark 1 The reference feedforward κr is introduced not to improve the reference tracking performance but to ensure the subsequent theoretical results. In practice, it is sufficient to assign a sufficiently small value to κ.

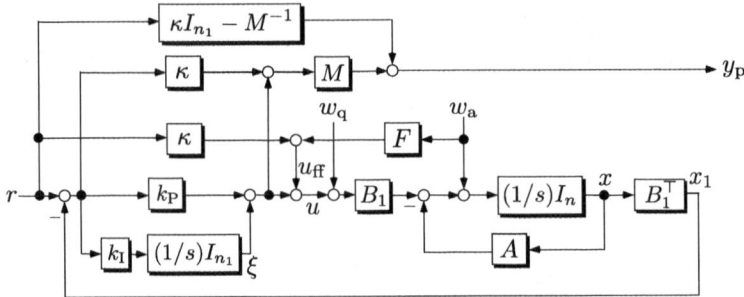

Fig. 12.2 Block diagram of physical dynamics with local controller (12.25)

12.4.3 CPS Design

Let us now interconnect (12.22) and (12.25) via negative feedback

$$\hat{d}_{\mathrm{q}} = -y_{\mathrm{p}}, \ r = y_{\mathrm{o}}. \tag{12.29}$$

The overall system is then illustrated in Fig. 12.3.

Although another implementation would be viable, we basically consider the system architecture in Fig. 12.3, where the control algorithm consists of zone-level local controllers and central building energy management system (BEMS). For each zone $i = 1, 2, \ldots, n_1$, the local controller measures the room temperature T_i, receives ith elements of $r = y_{\mathrm{o}}$ and u_{ff} from the central BEMS, computes the control input u_i from these information, and communicates u_i to the BEMS, where u_i is the ith element of u. The BEMS measures the disturbance w^{a}, receives u from the local controllers, and computes the disturbance estimate $\hat{d}_{\mathrm{q}}$ in addition to the feedforward control u_{ff} through (12.24c). It then updates z_u according to (12.22) based on the disturbance estimate $\hat{d}_{\mathrm{q}}$, generates the estimated optimal room temperatures y_{o} in (12.23), and communicates $r = y_{\mathrm{o}}$ and u_{ff} to the local controllers.

The major advantage of the above architecture is that the local controllers do not need to have any information on the building model. For example, when the building structure changes, we have only to change parameters in BEMS and do not need to re-program embedded controllers in each room, which would drastically reduce the workload of the engineers.

Let us now focus on the closed-loop system with the physical dynamics and the local controller (12.24a) and (12.24b), namely the light gray system. The transfer function from w_{q} to the estimate $\hat{d}_{\mathrm{q}}$ then takes the form of the complementary sensitivity function and hence only the low frequency components of w_{q} are estimated by the present algorithm. This is why $\hat{d}_{\mathrm{q}}$ is regarded as an estimate of the DC components of w_{q}. Although the high frequency components of the disturbance are filtered out, it does not always bring any drawback at least qualitatively. Indeed, the complementary sensitivity function from r to x_1 takes a similar form as above since the diagonal

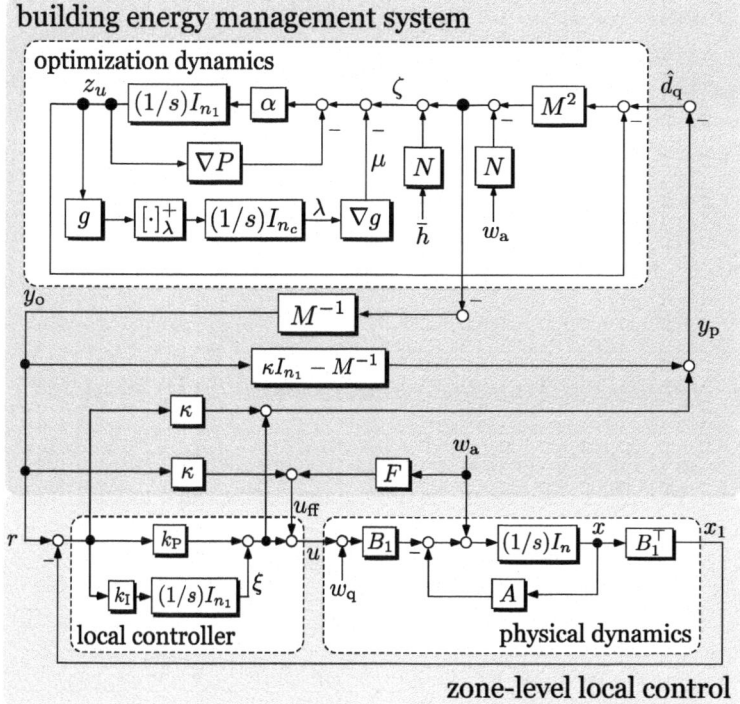

Fig. 12.3 Architecture of the present CPS

elements tend to be dominant in A and the local controller blocks are diagonal. This means that even if optimal solutions reflecting much faster disturbance variations are provided from BEMS, the physical states, say x_1, cannot track such fast variations exceeding the bandwidth.

In the present CPS, the above control and optimization algorithms in the cyber world bidirectionally interact with the physical dynamics (12.13). In other words, it is a feedback system, and hence, we have a concern on closed-loop stability even if both of the cyber- and physical systems are stable as open-loop systems. This is the issue to be addressed in the next section.

Remark 2 In Fig. 12.3, both BEMS and the local controller are biproper and hence a problem of algebraic loops can occur. This, however, does not matter in practice since the information transmissions between them usually suffer from possibly small delays. Although BEMS itself contains an algebraic loop, it is easily confirmed that the loop can be solved by direct calculations of the algebraic constraints.

12.5 Passivity, Optimality, and Stability

In this section, we analyze passivity of the components in the present CPS, and asymptotic optimality and stability of the system. Before that, we present the following lemma, which is immediately proved by just combining Lemmas 2 and 3.

Lemma 4 *Consider the system (12.22), (12.25), and (12.29) with constant inputs $w_a \equiv d_a$ and $w_q \equiv d_q$. Then, the collection of $z_u = z_u^*$, $\lambda = \lambda^*$, x^*, and ξ^* in (12.26) with $r^* = z_{x1}^*$ is a steady state of the system.*

12.5.1 Equivalent Transformation of Building Dynamics

For the sake of the subsequent discussions, we equivalently transform Fig. 12.3 to the form of Fig. 12.4 consisting of the three subsystems shaded by dark, middle, and light gray.

The system colored by dark gray is formulated as

$$\dot{z}_u = \alpha\{-\nabla P(z_u) - \mu + v + N\bar{h}\}, \tag{12.30a}$$

$$\dot{\lambda} = [g(z_u)]_\lambda^+, \quad \mu = \nabla g(z_u)\lambda, \quad \lambda(0) \geq 0, \tag{12.30b}$$

whose input is v and output is z_u. The middle gray system is the collection of static equations which are simplified to

$$\begin{bmatrix} v \\ y_o \end{bmatrix} = -\frac{1}{\kappa} \begin{bmatrix} M & -I_{n_1} \\ -I_{n_1} & M^{-1} \end{bmatrix} \begin{bmatrix} z_u \\ \eta \end{bmatrix} - \frac{1}{\kappa} \begin{bmatrix} M^{-1} \\ M^{-2} \end{bmatrix} N w_a. \tag{12.31}$$

The light gray is represented as

$$\dot{x} = -Ax + k_P B_1 (y_o - x_1) + \kappa B_1 y_o + B_1 \xi + B_1 w_q + (B_1 F + I_n) w_a$$

$$= -\bar{A}x + \bar{k}_P B_1 (y_o - x_1) + B_1 \xi + B_1 w_q + \bar{F} w_a, \tag{12.32a}$$

$$\dot{\xi} = k_I (y_o - x_1), \tag{12.32b}$$

$$\eta = \bar{k}_P M (y_o - x_1) + M\xi, \tag{12.32c}$$

whose input is y_o and output is η. The symbols $\bar{A}, \bar{k}_P$, and $\bar{F}$ denote $\bar{A} := A - \kappa B_1 B_1^\top$, $\bar{k}_P := k_P + \kappa$, and $\bar{F} := B_1 F + I_n$, respectively. In the sequel, we assume that the feedforward gain κ is chosen so that $\bar{A}$ is positive definite.

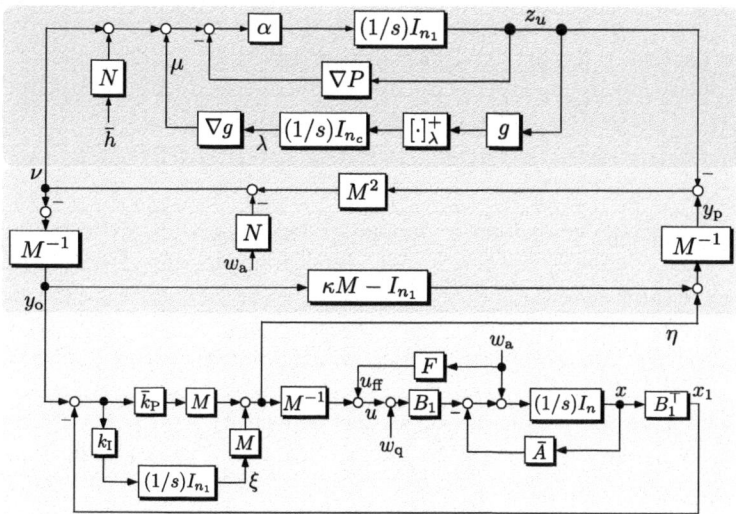

Fig. 12.4 Block diagram of physical dynamics with local controller (12.25). The system shaded by the dark gray is passive from $\tilde{v} := v - v^*$ to $\tilde{z}_u := z_u - z_u^*$ with $v^* := -Mz_{x1}^*$ (Lemma 5), that by middle gray is passive from $-[\tilde{z}_u^\top \ \tilde{\eta}^\top]^\top$ to $[\tilde{v}^\top \ \tilde{y}_o^\top]^\top$ with $\tilde{\eta} := \eta - \eta^*$, $\tilde{y}_o := y_o - y_o^*$, $\eta^* := -Md_q + (I_{n_1} - \kappa M)z_{x1}^*$ and $y_o^* := z_{x1}^*$ (Lemma 6), and the system by light gray is passive from $\tilde{y}_o$ to $\tilde{\eta}$ (Lemma 8)

12.5.2 Passivity in Optimization and Physical Dynamics

In this subsection, we prove passivity of the subsystems (12.30), (12.31), and (12.32).

Let us first focus on the system (12.30). Defining

$$\mu^* := \nabla g(z_u^*)\lambda^*, \quad v^* := -M^2(z_u^* + d_q) - Nd_a, \tag{12.33}$$

we have the following lemma.

Lemma 5 *Consider the system (12.30). Then, under Assumption 1, the system is passive from $\tilde{v} := v - v^*$ to $\tilde{z}_u := z_u - z_u^*$.*

Proof From KKT condition (12.19a) and the definition of v^* and μ^* in (12.33), we immediately see

$$v^* + N\bar{h} - \nabla P(z_u^*) - \mu^* = 0.$$

Subtracting this from the right-hand side of (12.30a) yields

$$\dot{\tilde{z}}_u = -\alpha(\nabla P(z_u) - \nabla P(z_u^*)) - \alpha\tilde{\mu} + \alpha\tilde{v}, \tag{12.34}$$

where $\tilde{\mu} := \mu - \mu^*$. Define $V_{\mathrm{f}} := \frac{1}{2\alpha}\|\tilde{z}_u\|^2 = \frac{1}{2\alpha}\|z_u - z_u^*\|^2$. The time derivative of V_{f} along the trajectories of (12.34) is then given as

$$\begin{aligned}
\dot{V}_{\mathrm{f}} &= -(z_u - z_u^*)^\top (\nabla P(z_u) - \nabla P(z_u^*)) + \tilde{z}_u^\top (-\tilde{\mu} + \tilde{v}) \\
&\leq \tilde{z}_u^\top (-\tilde{\mu} + \tilde{v}),
\end{aligned} \tag{12.35}$$

where the inequality holds from Assumption 1 and Lemma 1.

Let us next define $V_{\mathrm{c}} := \frac{1}{2}\|\lambda - \lambda^*\|^2$. Then, following the same procedure as Lemma 9 in [8], it follows

$$D^+ V_{\mathrm{c}} \leq \tilde{\mu}^\top \tilde{z}_u, \tag{12.36}$$

where the symbol D^+ denotes the upper Dini derivative.

Define $S_{\mathrm{o}} := V_{\mathrm{f}} + V_{\mathrm{c}}$. Then, combining (12.35) and (12.36), it follows

$$D^+ S_{\mathrm{o}} \leq \tilde{z}_u^\top \tilde{v}. \tag{12.37}$$

Integrating this in time proves this lemma.

Let us next consider the static system (12.31). Define

$$v^* := -M z_{x1}^*, \quad y_{\mathrm{o}}^* := z_{x1}^*, \quad \eta^* := -M d_q + (I_{n_1} - \kappa M) z_{x1}^*. \tag{12.38}$$

We then have the following lemma.

Lemma 6 *Consider the system (12.31) with a constant $w_{\mathrm{a}} \equiv d_{\mathrm{a}}$. Then, the system is passive from $-[\tilde{z}_u^\top \ \tilde{\eta}^\top]^\top$ to $[\tilde{v}^\top \ \tilde{y}_{\mathrm{o}}^\top]^\top$ with $\tilde{\eta} := \eta - \eta^*$ and $\tilde{y}_{\mathrm{o}} := y_{\mathrm{o}} - y_{\mathrm{o}}^*$.*

Proof By calculation, we obtain

$$\begin{aligned}
-M z_u^* + \eta^* - M^{-1} N d_{\mathrm{a}} &= -M(z_u^* + d_q) + (I_{n_1} - \kappa M) z_{x1}^* - B_1^\top A^{-1} d_{\mathrm{a}} \\
&= B_1^\top A^{-1} d_{\mathrm{a}} - z_{x1}^* + (I_{n_1} - \kappa M) z_{x1}^* - B_1^\top A^{-1} d_{\mathrm{a}} \quad (\because (20)) \\
&= -\kappa M z_{x1}^* = \kappa v^*
\end{aligned} \tag{12.39}$$

and $y_{\mathrm{o}}^* = -M^{-1} v^*$. It thus follows

$$\begin{bmatrix} v^* \\ y_{\mathrm{o}}^* \end{bmatrix} = -\frac{1}{\kappa} \begin{bmatrix} M & -I_{n_1} \\ -I_{n_1} & M^{-1} \end{bmatrix} \begin{bmatrix} z_u^* \\ \eta^* \end{bmatrix} - \frac{1}{\kappa} \begin{bmatrix} M^{-1} \\ M^{-2} \end{bmatrix} N d_{\mathrm{a}}. \tag{12.40}$$

Subtracting (12.40) from (12.31) with $w_{\mathrm{a}} \equiv d_{\mathrm{a}}$ yields

$$\begin{bmatrix} \tilde{v} \\ \tilde{y}_{\mathrm{o}} \end{bmatrix} = -\frac{1}{\kappa} \begin{bmatrix} M & -I_{n_1} \\ -I_{n_1} & M^{-1} \end{bmatrix} \begin{bmatrix} \tilde{z}_u \\ \tilde{\eta} \end{bmatrix}. \tag{12.41}$$

Now, the following equations hold.

$$- \begin{bmatrix} \tilde{\nu} \\ \tilde{y}_o \end{bmatrix}^{\top} \begin{bmatrix} \tilde{z}_u \\ \tilde{\eta} \end{bmatrix} = \frac{1}{\kappa} \begin{bmatrix} \tilde{z}_u \\ \tilde{\eta} \end{bmatrix}^{\top} \begin{bmatrix} M & -I_{n_1} \\ -I_{n_1} & M^{-1} \end{bmatrix} \begin{bmatrix} \tilde{z}_u \\ \tilde{\eta} \end{bmatrix} = \frac{1}{\kappa} (\tilde{z}_u - M^{-1}\tilde{\eta})^{\top} M (\tilde{z}_u - M^{-1}\tilde{\eta})$$

$$\geq \frac{\sigma}{\kappa} \|\tilde{z}_u - M^{-1}\tilde{\eta}\|^2 = \kappa \sigma \|\tilde{y}_o\|^2 \geq 0, \tag{12.42}$$

where $\sigma > 0$ is the minimal eigenvalue of M. This completes the proof.

We finally prove passivity of (12.32). To this end, we employ the following additional assumption.

Assumption 2 The matrix $M A_1 + A_1 M - 2\kappa M$ is positive definite.

Once κ is selected to be sufficiently small, the assumption is almost equivalent to $M A_1 + A_1 M > 0$. The matrix inequality does not always hold even if the matrices A_1 and M are positive definite, but it is expected to be true in many practical cases since the diagonal elements tend to be dominant both for A_1 and $M = (A_1 - A_2^{\top} A_3^{-1} A_2)^{-1}$ in this application [31].

Under Assumption 2, the following lemma is shown to be true.

Lemma 7 *Under Assumption 2, the system $(\tilde{A}_3, A_2 X^{-1/2})$ is stabilizable and the system $(X^{-1/2} M A_2^{\top}, \tilde{A}_3)$ is detectable for any positive definite matrix $X > 0$, where $\tilde{A}_3 := -A_3 + A_2 M X^{-1} A_2^{\top}$.*

Proof Define $\Phi_s := X^{1/2} M X^{-1} A_2^{\top}$ and $\Phi_d := A_2 M X^{-1} M^{-1} X^{1/2}$. Then,

$$\tilde{A}_3 - A_2 X^{-1/2} \Phi_s = -A_3, \quad \tilde{A}_3 - \Phi_d X^{-1/2} M A_2^{\top} = -A_3$$

hold and $-A_3$ is stable since A_3 is a diagonal block of $A > 0$ and hence positive definite. This completes the proof.

Using Lemma 7, the system (12.32) is shown to be passive as below.

Lemma 8 *Consider the system (12.32) with constant inputs $w_a \equiv d_a$ and $w_q \equiv d_q$. Then, under Assumption 2, the system is passive from $\tilde{y}_o$ to $\tilde{\eta}$.*

Proof Subtracting the stationary equation of (12.25) in Lemma 4 from (12.32a) and (12.32b) yields

$$\dot{\tilde{x}} = -\bar{A}\tilde{x} + \bar{k}_P B_1 (\tilde{y}_o - \tilde{x}_1) + B_1 \tilde{\xi} \tag{12.43a}$$

$$\dot{\tilde{\xi}} = k_1 (\tilde{y}_o - \tilde{x}_1), \tag{12.43b}$$

where $\tilde{x} := x - x^*$, $\tilde{x}_1 := x_1 - z_{x1}^*$ and $\tilde{\xi} := \xi - \xi^*$. We also have

$$\tilde{\eta} = \bar{k}_P M (\tilde{y}_o - \tilde{x}_1) + M \tilde{\xi} \tag{12.44}$$

from (12.32c), (12.38) and (12.26) with $r^* = y_o^*$.

Take a positive semi-definite matrix $\Psi \in \mathbb{R}^{n_2 \times n_2}$ and define $\bar{\Psi} := \begin{bmatrix} M & 0 \\ 0 & \Psi \end{bmatrix} \in$ $\mathbb{R}^{n \times n}$. Then, by calculation, we have

$$\bar{\Psi}\bar{A} + \bar{A}\bar{\Psi} = \begin{bmatrix} Y & MA_2^\top + A_2^\top \Psi \\ \Psi A_2 + A_2 M & \Psi A_3 + A_3 \Psi \end{bmatrix}$$

with $Y := MA_1 + A_1 M - 2\kappa M$ which is positive definite under Assumption 2. Using Schur complement, $\bar{\Psi}\bar{A} + \bar{A}\bar{\Psi} > 0$ is equivalent to the following Riccati inequality.

$$-\tilde{A}_3 \Psi - \Psi \tilde{A}_3 + \Psi A_2 P^{-1} A_2^\top \Psi + A_2 M P^{-1} M A_2^\top < 0 \tag{12.45}$$

A positive semi-definite solution Ψ to (12.45) is shown to exist from Lemma 6.

Let us now define the energy function

$$S_{\mathrm{p}} := \frac{1}{2}\tilde{x}^\top \bar{\Psi}\tilde{x} + \frac{1}{2k_1}\tilde{\xi}^\top M\tilde{\xi}, \tag{12.46}$$

which is positive semi-definite because of $M > 0$. Then, the time derivative along with the trajectories of (12.43) is given by

$$\begin{aligned}
\dot{S}_{\mathrm{p}} &= -\frac{1}{2}\tilde{x}^\top(\bar{\Psi}\bar{A} + \bar{A}\bar{\Psi})\tilde{x} + \bar{k}_{\mathrm{P}}\tilde{x}^\top \bar{\Psi} B_1(\tilde{y}_{\mathrm{o}} - \tilde{x}_1) + \tilde{x}^\top \bar{\Psi} B_1\tilde{\xi} + \tilde{\xi}^\top M(\tilde{y}_{\mathrm{o}} - \tilde{x}_1), \\
&\leq \bar{k}_{\mathrm{P}}\tilde{x}_1^\top M(\tilde{y}_{\mathrm{o}} - \tilde{x}_1) + \tilde{x}_1^\top M\tilde{\xi} + \tilde{\xi}^\top M(\tilde{y}_{\mathrm{o}} - \tilde{x}_1), \\
&= \bar{k}_{\mathrm{P}}\tilde{x}_1^\top M(\tilde{y}_{\mathrm{o}} - \tilde{x}_1) + \tilde{\xi}^\top M\tilde{y}_{\mathrm{o}}, \\
&= -\bar{k}_{\mathrm{P}}\tilde{x}_1^\top M\tilde{x}_1 - \bar{k}_{\mathrm{P}}\tilde{y}_{\mathrm{o}}^\top M\tilde{y}_{\mathrm{o}} + 2\bar{k}_{\mathrm{P}}\tilde{x}_1^\top M\tilde{y}_{\mathrm{o}} + \{\bar{k}_{\mathrm{P}}(\tilde{y}_{\mathrm{o}} - \tilde{x}_1)^\top M + \tilde{\xi}^\top M\}\tilde{y}_{\mathrm{o}}, \\
&= -\bar{k}_{\mathrm{P}}(\tilde{y}_{\mathrm{o}} - \tilde{x}_1)^\top M(\tilde{y}_{\mathrm{o}} - \tilde{x}_1) + \tilde{\eta}^\top \tilde{y}_{\mathrm{o}} \\
&\leq -\bar{k}_{\mathrm{P}}\sigma \|y_{\mathrm{o}} - x_1\|^2 + \tilde{\eta}^\top \tilde{y}_{\mathrm{o}} \leq \tilde{\eta}^\top \tilde{y}_{\mathrm{o}} \tag{12.47}
\end{aligned}$$

where we use $\tilde{x}^\top \bar{\Psi} B_1 = \tilde{x}_1^\top M$ in the second equation and (12.44) in the fifth equation. This completes the proof.

In summary, we conclude that Fig. 12.4 is an interconnection of passive systems (12.30), (12.31), and (12.32).

12.5.3 Asymptotic Optimality and Stability

We are now ready to prove the main result of this section.

Theorem 1 *Consider the system consisting of (12.30)–(12.32) with $w_{\mathrm{a}} \equiv d_{\mathrm{a}}$ and $w_{\mathrm{q}} \equiv d_{\mathrm{q}}$. Then, under Assumptions 1 and 2, the state trajectories of the system are bounded and satisfy $x_1 \rightarrow z_{x1}^*$.*

Proof Define $S := S_o + S_p$. Combining (12.37) and (12.47), we immediately have

$$D^+ S \le \tilde{z}_u^\top \tilde{v} + \tilde{\eta}^\top \tilde{y}_o - \bar{k}_P \sigma \| y_o - x_1 \|^2. \tag{12.48}$$

Substituting (12.42) into (12.48) yields

$$D^+ S \le -\kappa \sigma \| \tilde{y}_o \|^2 - \bar{k}_P \sigma \| y_o - x_1 \|^2$$
$$= -\kappa \sigma \| y_o - z_{x1}^* \|^2 - \bar{k}_P \sigma \| y_o - x_1 \|^2. \tag{12.49}$$

From the definition of S and (12.49), all the state variables except for x_2 must be bounded. Extracting the dynamics for x_2 from (12.13) with $w_a \equiv d_a$, we have

$$\dot{x}_2 = -A_3 x_2 - A_2 x_1 + B_2^\top d_a. \tag{12.50}$$

The matrix A_3 is a diagonal block of the positive definite matrix A and hence positive definite. Thus, the system (12.50) is a stable linear time-invariant system with bounded input $-A_2 x_1 + B_2^\top d_a$, which means that x_2 is also bounded. This completes the proof of the boundedness of the state trajectories.

Integrating (12.49) in time also mean that $y_o - z_{x1}^* \in \mathcal{L}_2$ and $y_o - x_1 \in \mathcal{L}_2$. In the case of $w_a \equiv d_a$ and $w_q \equiv d_q$, $\dot{y}_o = \dot{z}_u$ holds from (12.23). Because of the boundedness of the states, the right-hand side of (12.22a) is bounded, and hence, the time derivative of $y_o - z_{x1}^{b*}$ is concluded to be bounded. Since the right-hand side of (12.13) is bounded, the time derivative of $y_o - x_1$ is also bounded. Invoking Barbalat's lemma, we have

$$\lim_{t \to \infty} y_o = z_{x1}^*, \ \lim_{t \to \infty} (y_o - x_1) = 0. \tag{12.51}$$

Combining them completes the proof.

The above results are obtained assuming that both of w_q and w_a are constant. Since the ambient temperature is in general slowly varying, analysis under a constant w_a would be approximately applicable to the actual situations. However, the heat gain w_q may contain high frequency components. To address the issue, we decompose the signal w_q into the DC components d_q and others $\tilde{w}_q$ as $w_q = d_q + \tilde{w}_q$. Then, a trivial extension of the well-known passivity theorem [7, 27] proves the following result.

Corollary 1 *Consider the system in Theorem 1 with $w_a \equiv d_a$ and $w_q = d_q + \tilde{w}_q$. Then, if Assumptions 1 and 2 hold, the system has a finite $\mathcal{L}_2$ gain from $\tilde{w}_q$ to $x_1 - z_{x1}^*$.*

Proof In the present case, the term $B_1 \tilde{w}_q$ appears in the right-hand side of (12.43a), and the term $\tilde{x}_1^\top M \tilde{w}_q$ is added to the right-hand side of (12.47) and hence to that of (12.49). Accordingly, (12.49) is reformulated as

$$D^+ S \leq \tilde{x}_1^\top M \tilde{w}_q - \sigma(\kappa \|y_o - z_{x1}^*\|^2 + \bar{k}_P \|y_o - x_1\|^2)$$

$$\leq \tilde{x}_1^\top M \tilde{w}_q - \frac{\sigma \kappa k_P}{\kappa + k_P} \|x_1 - z_{x1}^*\|^2$$

$$= \tilde{x}_1^\top M \tilde{w}_q - \beta \|\tilde{x}_1\|^2, \quad \beta := \frac{\sigma \kappa k_P}{\kappa + k_P} > 0. \tag{12.52}$$

Inequality (12.52) is further rewritten as

$$D^+ S = -\frac{\sqrt{\beta}}{\sqrt{2}} \|\tilde{x}_1 - (1/\beta) M \tilde{w}_q\|^2 - \frac{\beta}{2} \|\tilde{x}_1\|^2 + \frac{1}{2\beta} \|M \tilde{w}_q\|^2$$

$$\leq \frac{1}{2} \left(-\beta \|\tilde{x}_1\|^2 + \gamma \|\tilde{w}_q\|^2 \right), \tag{12.53}$$

where $\gamma := \frac{\bar{\sigma}^2}{\beta}$ and $\bar{\sigma}$ is the maximal eigenvalue of M. Integrating (12.53) in time completes the proof.

12.6 Extension to Co-Optimization of Multiple Buildings

In this section, we show that the present framework is extendable to the scenario of co-optimizing multiple buildings' energy management. The IDs of the buildings are denoted as $b = 1, 2, \ldots, \bar{b}$. In the sequel, every variable, function, and parameter associated with building b is denoted by adding superscript b to the same symbol as the previous sections.

Suppose that the dynamics of building b corresponding to (12.13) is described as

$$\dot{x}^b = -A^b x^b + B_1^b u^b + B_1^b w_q^b + w_a^b, \quad x^b := \begin{bmatrix} x_1^b \\ x_2^b \end{bmatrix}. \tag{12.54}$$

The collective dynamics for all buildings is then given in the form of (12.13) if we define

$$x := \begin{bmatrix} x^1 \\ \vdots \\ x^{\bar{b}} \end{bmatrix}, w_q := \begin{bmatrix} w_q^1 \\ \vdots \\ w_q^{\bar{b}} \end{bmatrix}, w_a := \begin{bmatrix} w_a^1 \\ \vdots \\ w_a^{\bar{b}} \end{bmatrix}, A := \begin{bmatrix} A^1 & & O \\ & \ddots & \\ O & & A^{\bar{b}} \end{bmatrix}, B_1 := \begin{bmatrix} B_1^1 & & O \\ & \ddots & \\ O & & B_1^{\bar{b}} \end{bmatrix}. \tag{12.55}$$

We assume that every building b has a local cost function and constraints in the form of (12.14) and (12.15), which are, respectively, described as below.

$$f^b(z^b) := \frac{1}{2}\|z^b_{x1} - h^b\|^2 + P^b(z^b_u), \ b = 1, 2, \ldots, \bar{b}, \quad (12.56)$$

$$g^b(z^b_u) \le 0, \ b = 1, 2, \ldots, \bar{b}. \quad (12.57)$$

Let us now denote the actual power consumption of building b without any scaling by $\tilde{P}^b(z^b_u)$. Then, in order to constrain the total power consumption of all buildings, we employ a global constraint formulated as

$$g^0(z_u) := \sum_{b=1}^{\bar{b}} \tilde{P}^b(z^b_u) - \bar{P} \le 0 \quad (12.58)$$

for some $\bar{P} > 0$, where $z_u := [(z^1_u)^\top \cdots (z^{\bar{b}}_u)^\top]^\top$. In the sequel, we also use the notation $z := [(z^1)^\top \cdots (z^{\bar{b}})^\top]^\top$.

Define

$$f(z) := \sum_{b=1}^{\bar{b}} f^b(z^b), \ g(z_u) := \begin{bmatrix} g^0(z_u) \\ g^1(z^1_u) \\ \vdots \\ g^{\bar{b}}(z^{\bar{b}}_u) \end{bmatrix}. \quad (12.59)$$

Then, adding the equality constraint (12.16) for the matrices A and B_1 defined in (12.55), we obtain the optimization problem (12.17) and hence (12.18). Accordingly, the same solution as Sect. 12.4 is applicable and all of the theoretical results in Sect. 12.5 are preserved as long as the optimization problem and dynamics meet Assumptions 1 and 2, respectively.

Due to the separability of the cost and constraint functions and independence of the individual building dynamics, the dynamics (12.22), (12.25), and (12.29) are decomposed as follows. The dynamics (12.22) with (12.29) is decomposed into

$$\dot{z}^b_u = -\alpha\{(M^b)^2(z^b_u - y^b_p) + N^b(w^b_a - \bar{h}^b) + \nabla P^b(z^b_u) + \mu^b + \nabla \tilde{P}^b(z^b_u)\lambda^0\} \quad (12.60a)$$

$$\dot{\lambda}^b = [g^b(z^b_u)]^+_{\lambda^b}, \ \mu^b = \nabla g^b(z^b_u)\lambda^b, \ \lambda^b(0) \ge 0 \quad (12.60b)$$

$$y^b_o = M^b(z^b_u - y^b_p) + (B^b_1)^\top(A^b)^{-1}w^b_a \quad (12.60c)$$

$$u^b_{ff} = \kappa^b r^b + F^b w^b_a \quad (12.60d)$$

$$p^b = \tilde{P}^b(z^b_u) \quad (12.60e)$$

for $b = 1, 2, \ldots, \bar{b}$ and

$$\dot{\lambda}^0 = \left[\sum_{b=1}^{\bar{b}} p^b\right]^+_{\lambda^0}, \ \lambda^0(0) \ge 0, \quad (12.61)$$

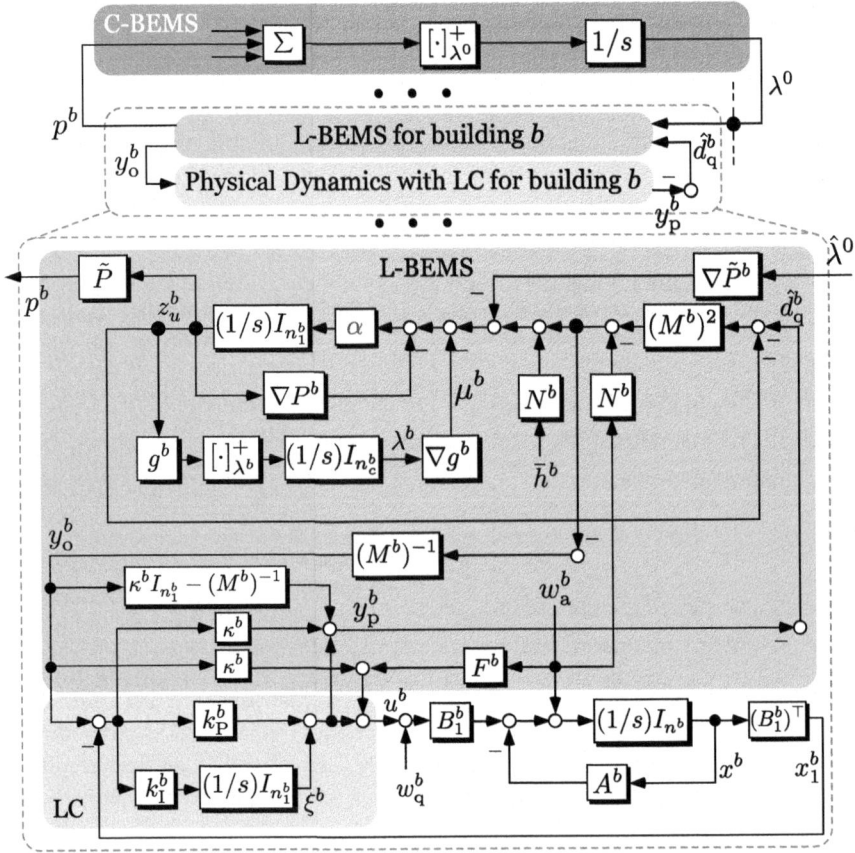

Fig. 12.5 Hierarchical/distributed optimization and control algorithm

where λ^b and λ^0 are the Lagrange multipliers corresponding to (12.57) and (12.58), respectively. The dynamics (12.25) with (12.29) is also decomposed as

$$\dot{\xi}^b = k_I^b(y_o^b - x_1^b) \tag{12.62a}$$

$$u^b = k_P^b(y_o^b - x_1^b) + \xi^b + u_{\text{ff}}^b \tag{12.62b}$$

$$y_P^b = (\kappa^b I_{n_1^b} - (M^b)^{-1})y_o^b + \bar{k}_P^b(y_o^b - x_1^b) + \xi^b, \tag{12.62c}$$

which is independent of the buildings other than b. The overall system is then illustrated in Fig. 12.5.

The top block colored by dark gray, namely (12.61), is assumed to be implemented on a comprehensive building management system (C-BEMS). It collects the (estimated) power consumption $p^b = \tilde{P}^b(z_u^b)$ from all buildings $b = 1, 2, \ldots, \bar{b}$ and updates λ^0 by (12.61). Hence, C-BEMS does not need any local information on the building structures, heat gains, the desirable set points h, and whether each HVAC system is active or not. C-BEMS sends λ^0, which is regarded as a price of the power [4], to every building b.

The middle block is implemented on the local building energy management system (L-BEMS) installed in each building. L-BEMS measures w_a^b and receives λ^0 from C-BEMS and u^b from low-level controllers (LC) that will be mentioned below. It then computes y_p^b from u^b and updates z_u^b according to (12.60a) based on the power price λ^0 and y_p^b. L-BEMS also computes the estimated optimal room temperatures y_o^b in (12.60c), the feedforward control u_{ff}^b in (12.60d), and the estimated power consumption p^b in (12.60e). Then, the outputs y_o^b and u_{ff}^b are sent to the zone-level local controllers (LC), and p^b is sent back to C-BEMS. The L-BEMS does not need any local variables, costs, and constraints of the other buildings, namely the algorithm is distributed in the level of buildings.

The bottom block colored by light gray is implemented by the zone-level LC. LC measures room temperature x_1^b and determines the mass flow rate u^b by (12.62) from y_o^b and u_{ff}^b received from L-BEMS. The process (12.62) is decentralized in the level of zones $i = 1, 2, \ldots, n_1^b$, which allow one to implement them by the local controller (LC) in each room similar to the existing systems. It then sends back y_p^b to L-BEMS.

12.7 Simulation

This section demonstrates the present CPS through simulation.

12.7.1 Development of Real-Time Building Control Simulator

In this subsection, we present a simulator developed to implement the present algorithm. Here, we take, as the building to be simulated, the Energy, Environment, and Innovation (EEI) Building in Tokyo Institute of Technology shown in the left of Fig. 12.6. The building has 7 floors and 98 rooms. A 3-D building structure model built on software called Sketchup (Trimble Navigation Limited) from the dimensional drawing is shown in the right of Fig. 12.6.

While the RC circuit model (12.11) is suitable for the algorithm design, it is too simple to use for demonstrations. We thus take a standard building energy simulator, EnergyPlus [26], to simulate the building thermal dynamics. It accepts more detailed external parameters like relative humidity, atmospheric station pressure, wind direction/speed and accordingly is expected to simulate the temperature evolutions more accurately than the RC circuit model. The EnergyPlus model of the building is immediately obtained by directly loading the structure model in the right of Fig. 12.6.

EnergyPlus installs a variety of existing HVAC systems with local controllers to track externally added reference set points. However, the controllers are black boxed,

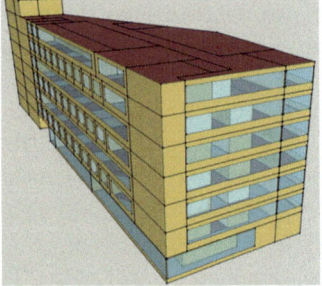

Fig. 12.6 Energy, Environment, and Innovation (EEI) Building in Tokyo Institute of Technology (left: real building. right: building structure model on SketchUp)

Fig. 12.7 Development environment of MLE+

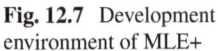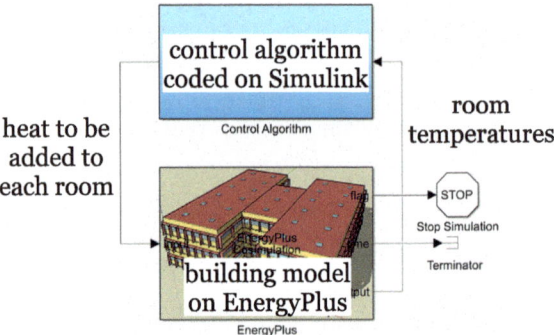

heat to be added to each room

room temperatures

and it is hard to customize the controller and to extract internal signals like the mass flow rate m_i. We thus add an input port, termed OtherEquipment, to directly input the heat to each room, and then implement the mathematical model of the actuator in (12.11) at the cost of accurate modeling of the actuator. The parameters of the actuator model in (12.11) are set to $S_i = 12.8°C$ and $c = 1005J/kg·°C$.

We then interconnect the EnergyPlus model with MALTAB/Simulink (Mathworks Inc.) for coding control algorithms by using the toolbox MLE+ [3] (Fig. 12.7) in order to run the real-time simulation. Note that the sampling period of the EnergyPlus block is set to 60s, which is the smallest value accepted by EnergyPlus. The sampling period of the Simulink block is set to 1s, which is much longer than real computation, just to save time to run the simulation.

The control algorithm requires the parameters C_i, R_{ij}, B_i, and D_i in the RC circuit model. These parameters are automatically identified from the structure model in Fig. 12.6 by BRCM toolbox [23]. In the case of the building in Fig. 12.6, the number of the zones n, which coincides with the dimension of the state T, is 1460. The zone division of the model is illustrated in Fig. 12.8.

We compare the RC circuit model with the EnergyPlus model, where we use the ambient temperature data in Fig. 12.9 from July to September at 36.18 degrees north latitude and 140.42th meridian east without adding solar radiation. The other

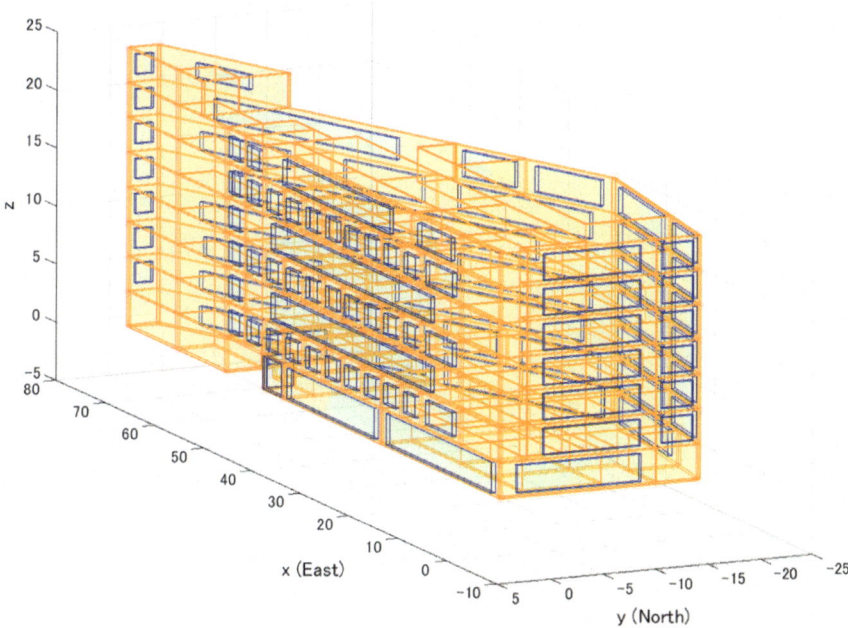

Fig. 12.8 BRCM model of EEI Building in Fig. 12.6

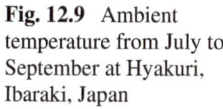

Fig. 12.9 Ambient temperature from July to September at Hyakuri, Ibaraki, Japan

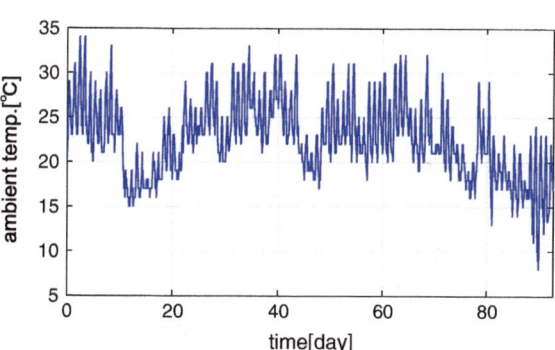

parameters of the EnergyPlus model are set to default values. The temperature evolutions at randomly chosen four rooms are then illustrated in Fig. 12.10. We see from the figures that the RC circuit model simulates the temperature evolution almost as accurately as EnergyPlus.

The resulting RC circuit model is finally linearized around the equilibrium corresponding to $\bar{T}^{\mathrm{a}} = 30$, $\bar{q} = 0$ and $\bar{m} = 0$. The sigma plot of the resulting linear time-invariant system from u to x_1 is illustrated by red curves in Fig. 12.11.

Remark 3 In the sequel, we run the simulation without adding the solar radiation. It is pointed out in [9] that the RC circuit model produced by BRCM toolbox tends to

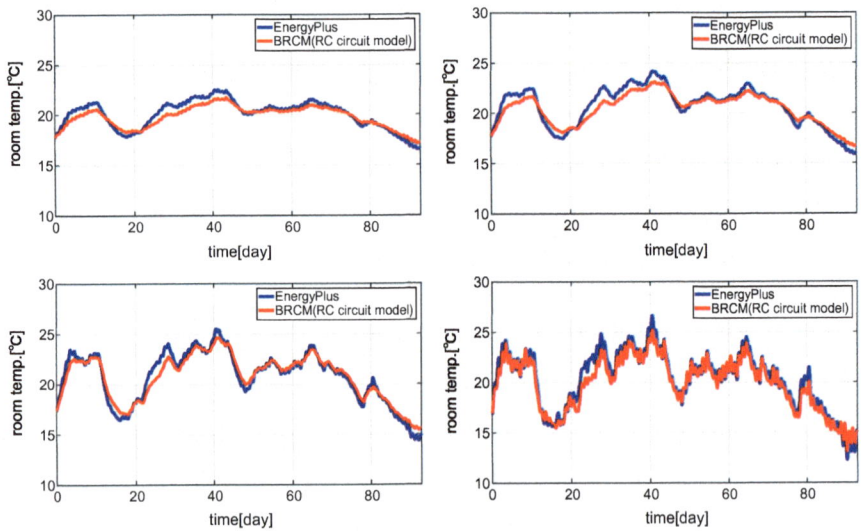

Fig. 12.10 Temperature evolutions generated by the RC circuit model and EnergyPlus, where the blue lines show the evolutions produced by EnergyPlus and the red lines show those by the RC circuit model. The top left figure corresponds to a room in the first floor, the top right to a room in the second floor, the bottom left to a room in sixth floor and the bottom right to a room in the seventh floor

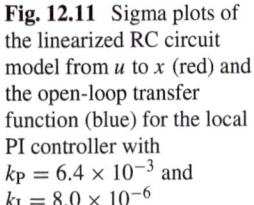

Fig. 12.11 Sigma plots of the linearized RC circuit model from u to x (red) and the open-loop transfer function (blue) for the local PI controller with $k_P = 6.4 \times 10^{-3}$ and $k_I = 8.0 \times 10^{-6}$

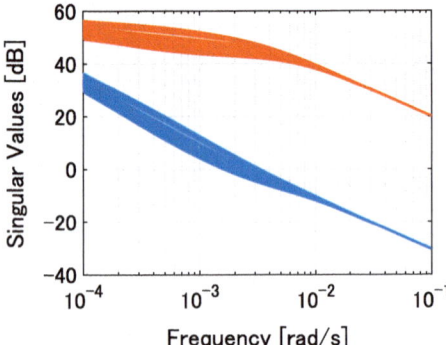

overestimate the effect of the radiation and does not always fit the data of EnergyPlus. We present a data-based re-modeling scheme based on a regularization technique in [9]. It is expected that combining this allows one to treat the case with radiation, but it is left as future work.

12.7.2 Algorithm Design

We first design the local PI controllers. The gains k_P and k_I are now determined according to the following policy.

- The DC gain of the controller is first determined so that the gain crossovers lie around $2\pi/(30 \times 60) \approx 3.5 \times 10^{-3}$ rad/s expecting that the settling time is almost the same as the standard systems, 30 min or a little earlier than it.
- The zeros of the controller are then determined to lead the phase at around the gain crossover to gain the stability margin.

Accordingly, the gains are fixed to $k_P = 6.4 \times 10^{-3}$ and $k_I = 8.0 \times 10^{-6}$. The sigma plot of the open-loop transfer function matrix with the PI controller is illustrated by blue curves in Fig. 12.11. The feedforward gain κ is set to $\kappa = 5.0 \times 10^{-3}$. Then, the sigma plot of the closed-loop transfer function from the reference set points r to the room temperatures x_1 is illustrated in Fig. 12.12. It is confirmed that the bandwidth is around 2.0×10^{-3} rad/s.

We next formulate the set point optimization problem (12.17). The power consumption of the HVAC systems in the building is selected based on [15] as

$$P_{\mathrm{m}}(m) = \sum_{i=1}^{n} \left(266.7 + 14.7 m_i + 63 m_i^2 + \frac{c(T^{\mathrm{a}} - T^{\mathrm{s}}) m_i}{4} \right),$$

where the first three terms describe the consumption at the fans and the last one is that needed to cool the air at the cooling coil. The function P in (12.18) is set to

$$P(z_u) = 3000 P_{\mathrm{m}}(G^{-1}(\bar{T})(B_1^{\top} C^{-1/2} B_1)^{-1} z_u)$$

by adding the weight 3000 to balance the energy saving with the human comfort quantified by the first term of (12.18). In order to bound the total power consumption, we also employ the constraint function g as

Fig. 12.12 Sigma plots of the closed-loop system from reference temperatures r to room temperatures x_1

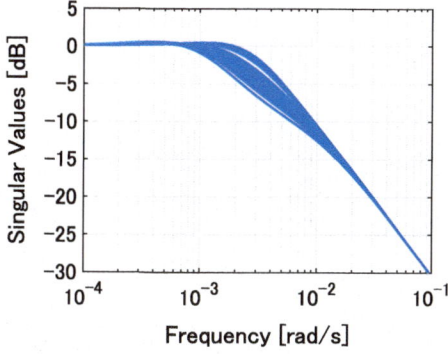

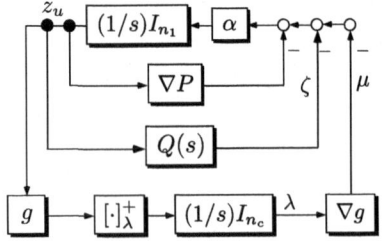

Fig. 12.13 Simplification of Fig. 12.3

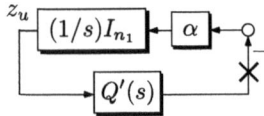

Fig. 12.14 Transformation of Fig. 12.13 with the outermost loop eliminated into a standard form of the closed-loop systems

$$g(z_u) = \tilde{P}(G^{-1}(\bar{T})(B_1^\top C^{-1/2} B_1)^{-1} z_u) - \bar{P}, \quad \bar{P} := 130 \times 10^3.$$

The parameter h is selected so that the reference set point coincides with 23°C.

Let us next design the optimization dynamics. The aforementioned theorems do not require any gain condition, but the simulator contains discretization not only in physical dynamics but also in the optimization dynamics. Accordingly, choosing too large gains may cause instability. Let us now focus on Fig. 12.3. Unifying the bottom blocks including the physical dynamics to a single block $Q(s)$ from z_u to η, then Fig. 12.3 is simplified to Fig. 12.13, where the blocks are linear time-invariant due to the quadratic cost function except for the outermost loop. Eliminating the loop with nonlinearity and combining the other loops to a single transfer function matrix $Q'(s)$, we have a standard closed-loop system in Fig. 12.14. The gain α is then selected to $\alpha = 4 \times 10^{-7}$ so that the gain crossovers for the open-loop transfer function with the loop cut at the mark $\times$ in Fig. 12.14 do not exceed one-tenth of the Nyquist frequency π for the sampling period 1s (Fig. 12.15). In order to avoid that the signals on the outer feedback path in Fig. 12.13 get too large to cause instability, the constraint function g is scaled as

$$g(z_u) = 0.0275 \tilde{P}(G^{-1}(\bar{T})(B_1^\top C^{-1/2} B_1)^{-1} z_u) - 0.0275 P_{\text{max}}$$

through trial and error.

The sigma plot for the overall CPS designed as above from w_q to $\hat{d}_q$ is shown in Fig. 12.16. As expected in Sect. 12.4.3, we see from the figure that the system works as a low pass filter for the unmeasurable disturbance w_q and the low frequency components are fed back to the optimization dynamics as $\hat{d}_q$. It is also confirmed that the cutoff frequency almost coincides with the bandwidth in Fig. 12.12.

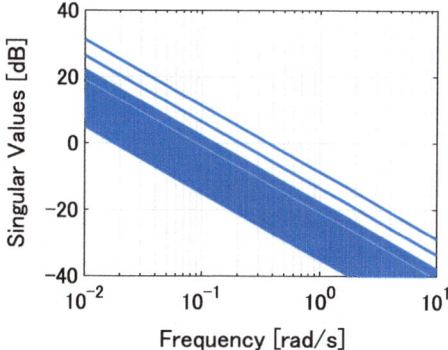

Fig. 12.15 Sigma plots of the open-loop transfer function in Fig. 12.14

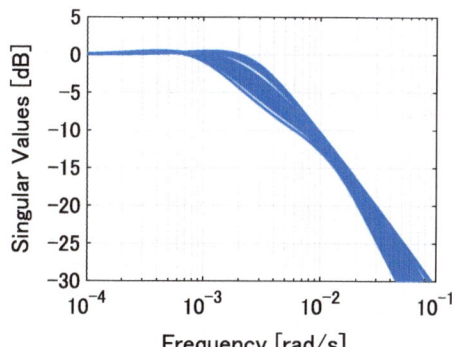

Fig. 12.16 Sigma plots of the overall CPS from w_q to $\hat{d}_q$

12.7.3 Demonstration

Let us finally run the simulation of the designed CPS over 4 days. In the simulation, we take the ambient temperature data in Fig. 12.9. The heat gain q at each room is randomly changed at every 3 h only during the daytime from 8:00am to 20:00pm.

Let us first confirm the accuracy of the disturbance estimation. Figure 12.17 shows the actual heat gain (blue) and its estimate (red) in the same rooms as Fig. 12.10. In view of the fact that a human generates 100 W on average, we conclude from the figures that the present CPS almost correctly estimates the unmeasurable disturbance.

The evolutions of the total power consumption $\tilde{P}$ and the room temperatures T_i are shown in Fig. 12.18 and Fig. 12.19, respectively. We see from the figures that, due to the correct disturbance estimation in Fig. 12.17, the CPS increases the power consumption during the time with large heat gains in order to ensure the human comfort. It is also confirmed that the CPS tries to avoid the constraint violation at the cost of the human comfort when $\tilde{P}$ touches the upper bound, differently from the dotted line without taking account of the upper limit of the power consumption. During the other periods, it balances the human comfort and energy saving since the function P is added to the cost function, where the balance is determined by the weight on the function P. How to prioritize each factor strongly depends on

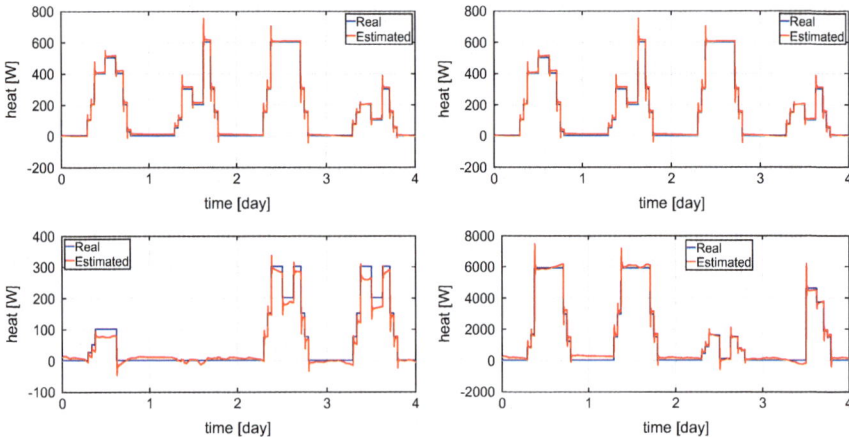

Fig. 12.17 Evolution of the actual heat gains w_q (blue) and its estimate $\hat{d}_q$ in the same rooms as Fig. 12.10

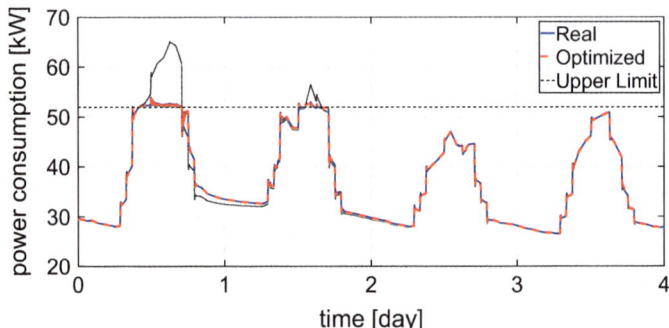

Fig. 12.18 Evolution of the total power consumption, where the red curve illustrates the estimated power consumption in the optimization dynamics, the blue the actual power consumption in the physical dynamics coded on EnergyPlus, the dotted curve the trajectory of the actual consumed power in the absence of the constraint $g(z_u) \leq 0$, and the dashed line the upper bound $\bar{P} = 130 \times 10^3$

the policy of the building, and it is out of the scope of this article. We also see unevenness in the temperatures of the rooms and that large rooms tend to record high temperatures. This is because we do not scale the power consumption in each room in the cost function. By appropriately scaling the consumption of each room, the problem would be eliminated, and moreover, we would be able to prioritize the comfort in some specific rooms.

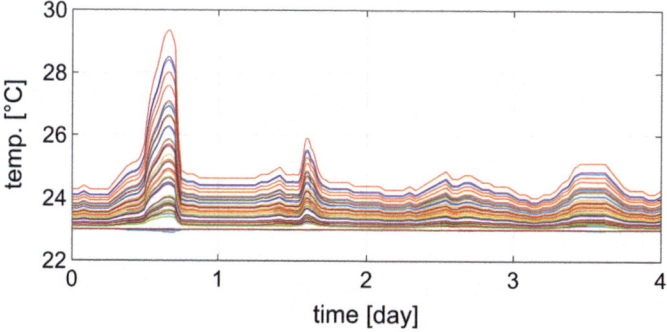

Fig. 12.19 Evolution of the all room temperatures generated by EnergyPlus

12.8   Conclusion

In this chapter, we addressed CPS design for enhancing energy efficiency of HVAC systems in a building. The building thermodynamics was modeled by the RC circuit model. We then formulated a set point optimization to balance the human comfort and energy saving involving the stationary equation of the dynamic model as a constraint. Primal-dual dynamics was then applied to solve the optimization. We then designed CPS by integrating the optimization dynamics and physical dynamics with local PI controllers. The resulting CPS was then shown to be interpreted as an interconnection of passive systems. Accordingly, convergence of the room temperatures to the optimal solution and input–output stability were rigorously proved based on the passivity paradigm. The present framework was further extended to a scenario of co-optimizing energy management in multiple connected buildings, wherein we revealed that a hierarchical/distributed control architecture is naturally provided by the present scheme. The present CPS was finally demonstrated on a simulator developed by combining a variety of softwares.

The authors would like to thank Mr. S. Yamashita and Mr. R. Okumura for their efforts to complete the simulation.

References

1. Afram A, Janabi-Sharif F (2014) Theory and applications of HVAC control systems—a review of model predictive control (MPC). Build Environ 72:343–355
2. Aswani A, Master N, Taneja J, Culler D, Tomlin C (2012) Reducing transient and steady state electricity consumption in HVAC using learning-based model-predictive control. Proc IEEE 100(1):240–253
3. Bernal W, Behl M, Nghiem T, Mangharam R (2012) MLE+: a tool for integrated design and deployment of energy efficient building controls. In: Real-time systems symposiuml work in progress
4. Boyd S, Vandenberghe L (2004) Convex optimization. Cambridge University Press

5. Cherukuri A, Mallada E, Cortés J (2016) Asymptotic convergence of constrained primal-dual dynamics. Syst Control Lett 87:10–15
6. Goyal S, Ingley H, Barooah P (2012) Zone-level control algorithms based on occupancy information for energy efficient buildings. In: Proceedings of 2012 American control conference, pp 3063–3068
7. Hatanaka T, Chopra N, Fujita M, Spong MW (2015) Passivity-based control and estimation in networked robotics. Springer-Verlag
8. Hatanaka T, Chopra N, Ishizaki T, Li N (2018) Passivity-based distributed optimization with communication delays using PI consensus algorithm. IEEE Trans Autom Control 63(12):4421–4428
9. Hatanaka T, Ikawa T and Okamoto D (2019) Remodeling of RC circuit building thermodynamics model with solar radiation based on a regularization-like technique. In: Proceedings of Asian control conference, pp 7–12
10. Hatanaka T, Zhang X, Shi W, Zhu M, Li N (2017) An integrated design of optimization and physical dynamics for energy efficient buildings: a passivity approach In: Proceedings of 1st IEEE conference on control technology and applications, pp 1050–1057
11. Hatanaka T, Zhang X, Shi W, Zhu M, Li N (2017) Physics-integrated hierarchical/distributed HVAC optimization for multiple buildings with robustness against time delays In: Proceedings of 56th IEEE conference on decision and control, pp 6573–6579
12. Hong Y, Hu J, Gao L (2006) Tracking control for multi-agent consensus with an active leader and variable topology. Automatica 42(7):1177–1182
13. Lawrence LSP, Nelson ZE, Mallada E, Simpson-Porco JW (2018) Optimal steady-state control for linear time-invariant systems. arXiv preprint arXiv:1810.03724
14. Lawrence LSP, Simpson-Porco JW, Mallada E (2018) The optimal steady-state control problem. arXiv preprint arXiv:1810.12892
15. Ma Y, Kelman A, Daly A, Borrelli F (2012) Predictive control for energy efficient buildings with thermal storage: modeling, stimulation, and experiments. IEEE Control Syst 32(1):44–64
16. Ma Y, Matusko J, Borrelli F (2015) Stochastic model predictive control for building HVAC systems: Complexity and conservatism. IEEE Trans Control Syst Technol 23(1):101–116
17. Maasoumya M, Razmara M, Shahbakhti M, Vincentelli AS (2014) Handling model uncertainty in model predictive control for energy efficient buildings. Energy Build 77:377–392
18. Mallada E, Zhao C, Low S (2014) Optimal load-side control for frequency regulation in smart grids. In: Proceedings of 52nd annual allerton conference on communication, control, and computing, pp 731–738
19. Morosan P-D, Bourdais R, Dumur D, Buisson J (2010) Building temperature regulation using a distributed model predictive control. Energy Build 42:1445–1452
20. Oldewurtel F, Parisio A, Jones CN, Gwerder Gyalistras DM, Stauch V, Lehmann B, Morari M (2012) Use of model predictive control and weather forecasts for energy efficient building climate control. Energy Build 45:15–27
21. Privara S, Siroky J, Ferkl L, Cigler J (2011) Model predictive control of a building heating system: the first experience. Energy Build 43:564–572
22. Shiltz DJ, Cvetkovic M, Annaswamy AM (2016) An integrated dynamic market mechanism for real-time markets and frequency regulation. IEEE Trans Sustain Energy 7(2):875–885
23. Sturzenegger D, Gyalistras D, Semeraro V, Morari M, Smith RS (2014) BRCM Matlab toolbox: Model generation for model predictive building control. In: Proceedings of 2014 American control conference, pp 1063–1069
24. Takenaka H, Nakajima TY, Higurashi A, Higuchi A, Takamura T, Pinker RT, Nakajima T (2011) Estimation of solar radiation using a neural network based on radiative transfer. J Geophys Res 116:D08215
25. Tsumura K, Baros S, Okano K, Annaswamy AM (2018) Design and stability of optimal frequency control in power networks: a passivity-based approach. In: Proceedings of 2018 European control conference, pp 2581–2586
26. US Dept. of Energy's: EnergyPlus, https://energyplus.net/

27. van der Schaft AJ (2000) L2-gain and passivity techniques in nonlinear control. 2nd edn. Springer-Verlag
28. Yuan S, Perez R (2006) Multiple-zone ventilation and temperature control of a single-duct VAV system using model predictive strategy. Energy Build 38:1248–1261
29. Zhang X, Papachristodoulou A, Li N (2015) Distributed optimal steady-state control using reverse- and forward-engineering. In: Proceedings of 54th IEEE conference on decision and control, pp 5257–5264
30. Zhao C, Topcu U, Li N, Low S (2014) Design and stability of load-side primary frequency control in power systems. IEEE Trans Autom Control 59(5):1177–1189
31. Zhang X, Shi W, Li X, Yan B, Malkawi A, Li N (2017) Decentralized and distributed temperature control via HVAC systems in energy efficient buildings. arXiv preprint arXiv:1702.03308

Printed by Printforce, the Netherlands